WORLD SCIENTIFIC SERIES ON NONLINEAR SCIENCE

Editor: Leon O. Chua
University of California, Berkeley

*To view the complete list of the published volumes in the series, please visit:
http://www.worldscibooks.com/series/wssnsa_series.shtml

WORLD SCIENTIFIC SERIES ON
NONLINEAR SCIENCE

Series A   Vol. 77

Series Editor: Leon O. Chua

MATHEMATICAL MECHANICS
From Particle to Muscle

Ellis D. Cooper

World Scientific

NEW JERSEY • LONDON • SINGAPORE • BEIJING • SHANGHAI • HONG KONG • TAIPEI • CHENNAI

Published by

World Scientific Publishing Co. Pte. Ltd.

5 Toh Tuck Link, Singapore 596224

USA office: 27 Warren Street, Suite 401-402, Hackensack, NJ 07601

UK office: 57 Shelton Street, Covent Garden, London WC2H 9HE

British Library Cataloguing-in-Publication Data
A catalogue record for this book is available from the British Library.

World Scientific Series on Nonlinear Science Series A — Vol. 77
MATHEMATICAL MECHANICS
From Particle to Muscle

ISBN-13 978-981-4289-70-2
ISBN-10 981-4289-70-1

Printed in Singapore.

To My Mathematics Teachers

Contents

Acknowledgments

I thank my friend Florian Lengyel for countless fun-filled mathematical conversations and my brother Steve Cooper for his critical and material support. I thank my wife Carolyn for her boundless confidence in me, and for our long laughs together.

Many thanks to Benoit Mandelbrot for ceding to me his copy of the collected works of Josiah Willard Gibbs.

PART 1
Introduction

Chapter 1

Introduction

If an individual is to gain a
thorough understanding of a
hard-won concept he must
personally experience some of
the painful intellectual struggle
that led to it. A lecture or a
text can do no more than point
the way.

[Lin and Segel (1988)] p. 504

"No one studies willingly, the
hard, slow lesson of Sophocles
and Shakespeare – that one
grows by suffering."

Clara Park, in [Manin (2007)]

Aside 1.0.1. When I was in college I wrote a term paper for a philosophy course and received a grade of B. The professor wrote, "The trouble with your paper is apparently that you exhausted yourself in devising a style that combines the worst elements of [Rudolf] Carnap and Earl Wilson and have not as yet come up with something for which this style might serve as a vehicle." Then he added, "I retract this characterization of your style which I do like although I think it gets somewhat out of hand in places."

I cannot resist the impulse to employ two different styles, one to somehow reflect my everyday man-on-the-street personality, the other to exhibit my aspiration to think abstractly, clearly, and impersonally.

Some readers may not care for the mathematical abstractions and will prefer the copiously captioned illustrations flowing with the ideas in this book. These asides are written for that sort of reader. Only here do I permit myself use of the first person form of address, so that my personality can appear – as if by magic – in the otherwise impersonal landscape of my abstract so-called *magnum opus*.

Then again, there may be a reader who would prefer to avoid distractions in the midst of abstractions. Fine, such a one may merrily skip across the asides.

1.1   Why Would I Have Valued This Book in High School?

Aside 1.1.1. Here is a personal story that antedates my high school years, but actually sets the stage for all my subsequent interests.

When I was kid growing up in Flatbush, that's in Brooklyn, New York City, I had a little group of friends interested in science. "Dickie" Matthews was a bit older, and I remember his huge workroom, strung with wires and ham radio equipment and glass tubes for chemistry experiments. "Bobbie" Reasenberg was more my age, about 13, and his father was an engineer, and they had a really nice house. One day in 1955 he showed me a book with a yellow paper jacket called "Giant Brains," by Edmund C. Berkeley[Berkeley (1949)]. It fired my imagination big time. When my mother saw this flame she did something amazing. She found out that Berkeley had an office in Greenwich Village on West 11 Street, which was about a 45 minute subway ride from our house. One Saturday we went to visit. At a very nice brownstone on West 11 Street we went up to the second floor and there were these men surrounding a terrific looking machine on a tabletop, densely packed with relays and wire bundles. They were excited about adding a magnetic memory drum to it. That machine was "Simon," considered by some historians to be the world's first personal computer. Mr. Berkeley taught me Boolean algebra and the habit of putting the date on my notes. So began my interests in the brain, mind, machines and consciousness.

In a sense I am writing this book to myself when I was in high school. I would have enjoyed the illustrations and diagrams even if I could not follow all the mathematics. But even the mathematics would have been gripping, because I would have seen that the author is not holding back details, nor condescending to tell me that something patently opaque to my untrained young eyes is "obvious." I would have found value in references to articles and books where I could search further for understanding. I would have enjoyed the quotations from principal researchers, not only for the pleasure of their distinctive writing styles, but also the specific insights only they could convey. And I would have really appreciated how the book brought between two covers a very interesting range of scientific subjects but all to one goal, the understanding of one important topic. The stories and asides would have served my desire to know the author and make with him a personal connection. It would have been fine if they are sometimes entertaining. More importantly, it would have been encouraging to know it is possible to gain understanding if one sets a high standard and relentlessly persists.

1.2 Who Else Would Value This Book?

Aside 1.2.1. I think this book would be valuable for someone who has already studied thermodynamics, but has always felt unsatisfied – even if they can remember the equations and solve the problems. That dissatisfaction seethes beneath the sense of disconnect between the intuitive physics and the formal mathematics, between the calculus as practiced by authors of thermodynamics textbooks, and the calculus as taught in mathematics courses.

If you are a high school student of mathematics or physics you will like the streamlined, no frills, brisk presentation of basic calculus and linear algebra – with all the easy proofs.

High school science teachers should definitely be up to speed on modern mathematical technology in general, and know more about their subject than they are responsible for teaching. Hence, this book provides auxiliary material for teachers of mathematics, physics, chemistry, computer programming and biology.

Likewise, researchers might find new material here, if only from a different viewpoint.

The book also provides a new model-building technology called "timing machinery" implemented in MATLAB that is potentially useful for simulating alternative muscle contractions models. Hence, this book is also for scientists interested in how global behavior emerges from local rules, for that is what happens when myosin molecules – coupled by feedback from their effects – *cooperate* to generate the relatively huge forces that are responsible for animal behavior [Jülicher *et al.* (1997)][Duke (2000)][Lan and Sun (2005)].

1.3   Physics & Biology

> Since antiquity, motion has been looked upon as the index of life. The organ of motion is muscle.
>
> ————————————
>
> [Szent-Györgyi (2004)]

Aside 1.3.1. The discovery of DNA and the subsequent explosion in molecular biology research was instigated by a handful of people, not the least of whom was Francis Crick – a physicist. Obviously, the scientists studying muscle contraction must know their physics well, since muscle contraction is the primary cause of all motion in animals and physics is the scientific study of motion.

The range of physics tools used for studying muscle contraction is prodigious. All the way from basic classical components like polarizing filters, dichroic mirrors, centrifuges and viscometers, to classical instruments like microscopes, to electronic instruments such as the oscilloscope, the photomultiplier tube and three-dimensional electron tomography, to machinery that depends on X-rays such as X-ray crystallography and X-ray diffraction probes and sub-millisecond time-sliced synchrotron X-ray sources, to modern equipment that could not be understood and deployed without quantum mechanics, such as photodiodes, lasers, and optical tweezers for manipulation of individual muscle contraction molecules. (The story of "X-ray diffraction work on muscle structure and the contraction mechanism" is reviewed in [Huxley (2004)].)

When I was in high school and strongly interested in physics I had an overwhelming feeling that there was no way I could truly understand it without personally repeating all of the basic experiments performed throughout its history [Strong (1936)][Shamos (1959)]. I got over that, of course, by realizing that it is necessary to think of modern tools and instruments as "black boxes" which have a history and a structure, sure, but for the experimental purpose at hand must be considered as opaque relations between inputs and outputs. Unless something weird happens, in which case a jump into the box is necessary.

1.4 Motivation

Aside 1.4.1. Masters of scientific research are finally investigating human consciousness. There have been some amazing insights [Blackmore (2004)] [Metzinger (2009)] but still, there is no scientific consensus on how to define consciousness, let alone a technology based on a science of consciousness for engineering useful consciousness structures.

But if I had a scientific theory of consciousness I would certainly want everyone to know about it. How would I do that? I would speak and write about it. These communication activities – indeed all intentional human activities – are implemented by conscious control of muscles, whether in the throat or at the fingertips, and so on. So, even though there is no scientific consensus about consciousness, it is at least reasonable to ask, is there a scientific consensus about muscle contraction? Fortunately, there is on this topic a great wealth of scientific understanding – if no consensus. I write this book to provide in one place a story of muscle contraction starting from first principles in mathematics and physics.

Since antiquity, motion has been looked upon as the index of life. The organ of motion is muscle. Our present understanding of the mechanism of contraction is based on three fundamental discoveries, all arising from studies on striated muscle. The modern era began with the demonstration that contraction is the result of the interaction of two proteins, actin and myosin with ATP, and that contraction can be reproduced in vitro with purified proteins. The second fundamental advance was the sliding filament theory, which established that shortening and power production are the result of interactions between actin and myosin filaments, each containing several hundreds of molecules and that this interaction proceeds by sliding without any change in filament lengths. Third, the atomic structures arising from the crystallization of actin and myosin now allow one to search for the changes in molecular structure that account for force production [Szent-Györgyi (1974)].

The fun of it is that such a theory, expressed in conformity with modern standards of streamlined mathematical rigor, must refer to basic concepts in several branches of science normally taught in separate courses of separate departments of educational institutions. A standard textbook in mathematics, or particle mechanics, or thermodynamics, or chemistry, or physiology of muscle, in general cannot – and perhaps should not – be responsible for joining into a seamless whole some fundamentals of the science of muscle contraction. Even though an outsider, I set myself the challenge of obtaining a clear sense of that multidisciplinary science, and this book is the result.

The other half of the story, how conscious human beings choose – or sometimes forced – to perform particular activities with their muscles, remains to be told.

Sharply distinguishing thought from action divides the labor of understanding. Understanding their relationship is a goal for a different book. Indeed,

> [O]n the sensorimotor theory the primary function of the central representational system is the planning and control of voluntary action. Hence all representations should be viewed as available for playing the functional role of action plans, which can lead to the development of motor programs that, when activated, trigger motor behavior. This means that representations would normally involve higher-level action planning in the frontal cortex. It also means that language production, a species of motor output, shares the action-planning representational system with nonverbal behavior systems [Newton (1996)].

The word "arm" may be associated with armed force, army, armada, armistice, and armor. Not to mention Armageddon. Focus on the human arm, which has many muscles. For example, the biceps brachii muscle. This is the muscle in the front of the upper arm, the one that bulges in arm-wrestling, or when someone demands, "Let's see your muscle."

> Human muscles provide the mechanical energy necessary to set the body in motion. Some muscles, such as those in the lower limbs, provide large

forces required to walk or run while others, like
those around the wrist, have the dexterity needed
to perform complex tasks. Understanding how
these muscles function is an integral component of
comprehending skeletal motion. Models and simulations of muscles are used not only to analyze
human locomotion, but also to design robotic de-
vices or to treat orthopaedic abnormalities [Aigner
and Heegaard (1999)].

An arm muscle consists of muscle fiber cells within which are many
parts that together produce force. How does that work? You could dismiss
the question by saying that a supernatural force directs everything in the
natural world, and it is not the job of human beings to question the supernatural. That line of thought is not scientifically interesting, so move on to
speculate on the *mechanism* of muscle contraction.

Perhaps along with gravitational force, electrical force, and magnetic
force, there is a new kind, muscle force. Perhaps there is a rack and pinion
inside a muscle, in other words, a gear that has teeth rotating while meshed
with a straight series of matching teeth. Or maybe there is a system of
rubber-like bands or springs that are controlled by the brain. Or maybe
there is a system of balloons that expand but are linked in a way to produce
contraction. Or, assuming there is a more fundamental level at which force
is generated, maybe there is a kind of molecule that shrinks when it receives
a special signal.

It came as a shock when the electron microscope revealed muscle to consist of "thick" myosin
and "thin" actin filaments, which did not shorten
on contraction, but only slid alongside one another.
Initially no connection was seen between the two
[Szent-Györgyi (2004)].

A great nineteenth century self-taught English educator, researcher, and
public intellectual named Thomas Henry Huxley – known as "Darwin's
bulldog" for his support of Charles Darwin's theory of evolution – among
other accomplishments coined the word "agnostic" to denote his personal
view on theology, and originated the idea that birds evolved from dinosaurs.
His grandchildren included Aldous Huxley, author of "Brave New World,"
and Andrew Huxley.

Andrew F. Huxley was awarded with Alan Lloyd Hodgkin the *1963 Nobel Prize in Physiology or Medicine* for work in mathematical biology on nerve conduction. They provided a system of differential equations that model the action potentials – a.k.a. "spikes" – of electro-chemical activity that zoom among neurons in the brain and along nerves in the body. In particular, spikes trigger muscles to contract.

Andrew Huxley's answer to the question, what is the mechanism for muscle contraction, is *not* that there is a kind of shrinking molecule, but that there are two kinds of molecular filaments bridged by tiny little molecular "arms" between them which – like rowers in a racing boat – collaborate to propel one filament relative to the other. In other words, the answer to how an arm works is that inside of it there are many, many tiny, tiny arms.

Aside 1.4.2. This answer reminds me of the idea that "causality is circular," that "all explanations deriving events from something completely other than themselves become explanations because somewhere along the way they introduce the outcome itself and thus turn the account into one in which the outcome is already contained in the ground." [Rosch (1994)]

1.5   The Principle of Least Thought

Aside 1.5.1. The name of this principle is supposed to be reminiscent of the "Principle of Least Action." Whereas that principle belongs to particle mechanics – indeed, to quantum mechanics via Feynman's path integral approach – my Principle of Least Thought belongs to the psychology of teaching and learning.

I crave understanding through the smallest number of steps, where the steps are no larger than a certain size. That size is, for me, perhaps smaller than it is for more intuitive people, the people who can make big leaps in thought without much exertion. By "smallest number" I mean there should be no extraneous fluff, no extra symbols, no irrelevant or misleading ideas. For me, understanding is not so much a demand for mathematical rigor; rather, it is an anxiety to grasp intuitive plausibility. Then again, the very effort to achieve rigor has been for me a terrific boost to intuition. Rigor cleans the window through which intuition shines.

Even the slightest increment of understanding requires a leap of intuition, that is, an un-reportable thought process somehow "behind the scenes" of reportable conscious thought. For different individuals and for different topics, there is a largest possible leap, beyond which one may feel not only lack of understanding, but even frustration and discouragement. An exposition readily apprehended by one class of individuals might be impenetrable to another. This exposition is tuned to my own capacity for intuitive leaps, hence may be too pedestrian for some, perhaps over the heads of others.

The mean of two numbers is the point half-way between them. If an increment of understanding is called a step, then one step is a means towards a goal. Think of clearing a new path through a jungle towards a treasure you know is there. *Meaning* is the process of finding means between what is understood and what is not understood. The Principle of Least Thought says, "Find the smallest number of steps, none of which exceeds your stride." Once the path is found there is an urge to return to earlier steps and explore side paths. This enlarges understanding.

1.6   Measurement

To make a measurement in the everyday world of objective procedures – algorithms for doing things with muscles – requires counting. Time measurements end up as counting marks as a physical body passes by (think of seconds-hand sweeping past numerals around circumference of a circular clock). Space measurements also end up as counting marks as a physical body passes by (think of fingertip of hand moving past markers on a ruler set against a rectangular solid body).

A physical body "passing by" implies a physical body (pointer) moving relative to another physical body (mark), even if the pointer is the orientation of the eyeball when attention is on the mark. For measurement, ultimately all that matters is the "counting of proximities" between pointers and marks.

1.7   Conceptual Blending

Aside 1.7.1. There is a large computer science/mathematics research industry that centered around the concept of "finite state machine." This is a mathematical abstraction of the idea that a system consists of states – usually represented by circles or dots in a diagram – connected by arrows representing the possibilities for transitions between states. There are many, many variations on this idea including triggered transitions, nested transitions, conditional transitions, and probabilistic transitions. I developed timing machinery as a conceptual blend [Fauconnier and Turner (2002)] of *time* with *finite state machine*, but I had not familiarized myself with the published literature on that idea.

At first timing machinery was called "symbol train processing." I programmed a simulator in *Mathematica* in 1990 and submitted an article to *The Mathematica Journal* under the title, "An Object-Oriented Interpreter for Symbol Train Processing." The referees were negative. One reviewer said I was unfamiliar with the literature on neural networks, and that that was obvious from my lack of references. Another reviewer wrote, "From the abstract alone, I can tell you that the article is flaky." A reviewer who wrote more lengthy comments said there wasn't much originality present and that there are "controversial and unsupported statements that would concern experts in the field." He said he suspected that the system would be very slow. He went on to emphasize my apparent lack of familiarity with literature on finite state machines and artificial intelligence. The reviewer closed by suggesting changes including a reference to "message passing parallel systems" and wrote that nevertheless he was "giving a tepid recommendation to publish."

The MATLAB timing machine simulator in this book is most definitely a message passing parallel system, and is expected to be very fast.

In this book I also advance a Theory of Substances as a foundation for macroscopic thermodynamics in terms of a conceptual blend of prior research [Fermi (1956)][Tisza (1961)][Falk *et al.* (1983)][Schmid (1984)][Herrmann and Schmid (1984)][Callen (1985)][Fuchs (1996)][Fuchs (1997)][Haddad *et al.* (2005)][Herrmann and Schmid (Year not available)].

Aside 1.7.2. Very early in my life I was gripped by the desire to understand what seemed to me then, and what still seems to me now, the absolutely interesting theories collected under the general heading of "mathematical physics." Specifically, I wanted to understand special relativity and quantum mechanics. I have made some progress on the former, essentially because its intuitive foundations in thoughts about electricity, magnetism, and light are accessible to the visual imagination. All that is not relevant to this book. But quantum mechanics is a different order of difficulty, because my understanding is that historically it was a deep problem in thermodynamics – the problem of black body radiation - that Max Planck solved by inventing quantization [Kuhn (1978)].

Thermodynamics is a very, very hard subject to understand, and not until twenty years ago did I feel sufficiently competent to renew my effort.

Since then I encountered a breakthrough in the little book [Fermi (1956)]. I became convinced – by his proofs unlike I had seen elsewhere – that there is an essentially *algebraic* root at the bottom of thermodynamics.

1.8   Mental Model of Muscle Contraction

There are many trillions of cells in a human body. Cells are thousands of times smaller than bodies, and the molecules out of which cells are made are thousands of times still smaller. Each cell contains over a billion non-water molecules [Goodsell (1998)] and also envelops over a thousand times as many water molecules ([Tester and Modell (2004)] p. 433).

At body temperature liquid water molecules are in-between and ceaselessly banging into all the other molecules at random. Therefore movement of the much larger molecules is impeded – the net effect is like a scuba diver trying to swim in molasses. The faster a molecule moves, the more it is impeded. But even without moving on its own, say, due to changing position or shape because of a chemical reaction , the bombardment by water molecules makes it jiggle around. In mathematical physics the scenario of a collection of large moving spheres being banged around at random by a great many smaller spheres is modeled by a "stochastic differential equation" called the Langevin Equation. This equation has a rich history in physics and – it might seem strangely – in mathematical finance [Lemons and Gythiel (1997)][Reimann (2001)] [Perrin (2005)][Davis and Etheridge (2006)][Lindén (2008)].

The smallest complete unit of muscle contraction is the sarcomere, which is most certainly not a collection of spheres. There are tens of thousands of sarcomeres in a single muscle cell. Actually there is a hierarchy of distinguishable structures from arm to muscle to muscle fiber to myofibril to sarcomere [Nelson and Cox (2005)].

A sarcomere is a complex but spectacularly symmetrical, periodic arrangement of several kinds of molecules along filaments. The crystalline regularity of this intricate nano-structure is what makes it possible by X-ray crystallography to measure precisely the relative positions and angles of the molecules [Reconditi *et al.* (2002)] [Nelson and Cox (2005)].

The sarcomere is not a static structure, of course, since it is the ultimate source of animal motion. Over many decades in the twentieth century, hundreds of scientists from all over the world contributed to views on how chemical reactions are linked to mechanical actions of molecules [Szent-

Györgyi (2004)]. The result is a set of simple and very beautiful ideas with a substantial mathematical and experimental foundation. For example,

> **The ultra-structure of a sarcomere. (a) The thick and thin filaments between the Z-disk and the M-line. (b) Side-view of the sarcomere. myosin motors are arranged all along the thick filament. The thick filament is connected to the Z-disk via the elastic titin molecules. Titin molecules restrain the movement of the Z-disk away from the thick filament, thus they are passive force generators. (c) Geometry of the actin-myosin interaction. (d) The myosin motor consists of three domains. The angle, u, between the motor domain and the light-chain domain (LCD) changes during the power-stroke. The stalk, which consists of a coiled-coil motif, actually continues into the thick filament. A bend is thought to occur in the light-chain, angling it upward to actin [Lan and Sun (2005)].**

myosin is the molecule that is responsible for muscle contraction. There are different ways to graphically represent the structures of molecules. Generally the idea is to capture the essence of shapes sometimes involving thousands of atoms without actually drawing each atom. It turns out that representations in terms of just two main kinds of overall shapes – springs called "α−helices" and ribbons called β−sheets – composed of many linked atoms are extremely useful. In particular the distinctive shape of the myosin molecule has been determined and represented in this way [Reedy (2000)].

The mechanical operation of the myosin molecule that produces force is called the "powerstroke," and is like a tiny arm that bends at its tiny elbow.

One theory of the transduction of energy stored in a chemical called ATP (adenosine triphosphate) into mechanical motion of the tiny arm is often presented in textbooks [Voet and Voet (1979)] and the cooperation of many tiny arms to produce large forces can be simulated on a computer:

> **Basic cycle of the swinging cross-bridge model. The myosin molecule makes stochastic transitions between a detached state D, and two attached states, $A1$ and $A2$, which are structurally distinct.**

In general, the transition rates f, α, g and the corresponding reverse rates depend on the strain of the elastic element. Owing to the free-energy change associated with ATP hydrolysis, the forward rates are predominately faster than the reverse rates and the molecule is driven one way around the cycle: $D \to A1 \to A2 \to D$. One ATP molecule is split during each cycle [Duke (2000)].

This books is directed towards answering the question, what is the mathematical science education necessary for understanding such mechanisms?

1.9   Organization

The parts of this book are as follows:

Chapters 2–4 **Mathematics** is shot through with diagrams of dots connected by arrows, both of which are usually decorated with labels. Such diagrams are used to represent structures (think of airline flight paths), relationships (think family trees), and operations (think flowcharts). These kinds of diagrams also represent interconnected algebraic operations, and are themselves collectively amenable to algebraic operations. Experience with diagrams has condensed into a mathematical industry called **Category Theory**. This book provides just enough category theory to modernize the presentation of thermodynamics: there is even a proposal in modern theoretical physics that physical theories are best understood as diagrams in a certain category. This book implicitly adopts that framework [Baez and Lauda (2009)][Döring and Isham (2008)].
Calculus is the mathematical theory of change. The book adopts a modern formulation of calculus in terms of rigorously defined infinitesimals, and many proofs involving calculus are explicitly justified by algebraic rules of calculation with numbers or infinitesimals, or both.

Chapters 5–10 **Physics defined is the scientific study of motion**. The deterministic theory of particles is simple and beautiful. Newton's Laws of mechanics, Lagrange's Reformulation, Hamilton's Principle, and Hamilton's Equations are presented in detail. The treatment of Legendre Transform is important in this story and even more so in the thermodynamics to follow, and so is explained very carefully to lay bare its algebraic essence and boiled down to an algorithm [Alberty (2001)].

Aristotle discussed "substance" in his *Metaphysics*. That is a rather profound philosophical work. In this book I offer a mathematical **Theory of Substances**.

Physics is shot through with analogies [Muldoon (2006)]. Many seemingly different physical processes are quite analogous. Energy transduction links the energy in processes of different kinds, including thermal, motion, deformation, chemical, and electrical processes Fig. 1.1. In other words, thermodynamics links the dif-

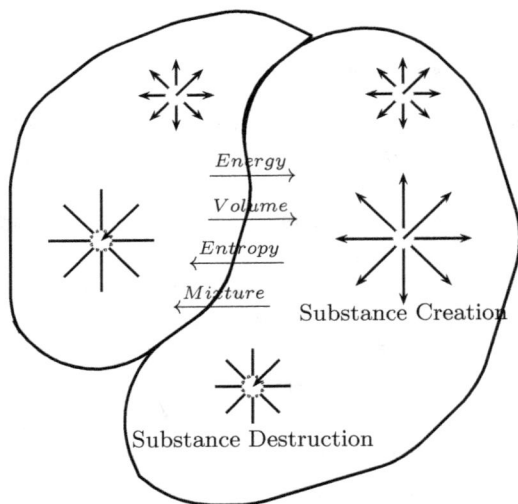

Fig. 1.1 Bodies in contact that are sustaining internal creation and destruction of substances as well as external flows of substances.

ferent ways that energy may be carried around, and hence, in a sense, is the center of physics. There are material substances, such as chemical species, and immaterial substances, such as energy, volume, momentum, probability, and entropy. All substances are tied together by the conserved substance, energy, which in thermal processes is carried by the indestructible immaterial substance, entropy Fig. 1.2.

$$\dot{E}_{\mathcal{A}} = \text{Th}_{\mathcal{A}} + \text{Vo}_{\mathcal{A}} + \text{Ch}_{\mathcal{A}} + \text{Me}_{\mathcal{A}} + \text{El}_{\mathcal{A}}$$

Fig. 1.2 The Power Balance Equation. $\dot{E}_{\mathcal{A}}$ is the rate of change of energy – the power – of system $\mathcal{A}$. Successive terms on the right denote the rate of change of energy in the system due to thermal, volume, chemical, mechanical, and electrical processes, respectively.

If $\mathcal{A}$ is an isolated system then its total energy is constant, so $\dot{E}_{\mathcal{A}} = 0$. In that case, by the Power Balance Equation, changes in system energy by one type of process must be compensated by changes in energy due to one or more other processes. This is the meaning of *energy transduction* and in particular the transduction of energy in chemical processes to mechanical processes underlies muscle contraction.

Chapters 6–7 I invented "timing machinery" in 1989 while working as a computer programmer at The Rockefeller University in a laboratory where the neurophysiology of vision was studied. I wanted to abstract the essence of inter-communicating neurons. The result was the idea of a multitude of states timing out and consequently emitting signals that triggered other states (or even themselves) into commencement of timing out. This idea blossomed in my mind into a personal industry leading over the years to a general, parallel, graphic programming language.

These chapters offer **timing machinery** for computer simulation of systems such as the molecular machinery of muscle contraction. The timing machine interpreter is written in **MATLAB** to take advantage of its inherent parallelism.

Chapter 11 My fascination with **muscle contraction** and the reason for writing this book stems from the facts that
(1) it is crucial for nearly all vertebrate animal function behavior;
(2) it is of significant interest to health professionals and sports physiologists; (3) its mechanism has been intensely studied by top-notch scientists for many decades and there are some very interesting alternative – perhaps even controversial – theories of how it works; (4) some basic mathematics and physics for understanding it can be but never have been published in a single book.

The closing chapter is a decade-by-decade chronology of muscle contraction research from the 19th century up through 2010. My idiosyncratic selections from the literature are biased by my interests in thermodynamics and simulation.

1.10   What is Missing?

Aside 1.10.1. It is a quirk of mine that, upon getting the gist of some theory or general explanation in the literature, I ask myself, "What is missing?" This is partly to counter my natural tendency to become wildly enthusiastic about the theory. But attempting to answer the question helps me put the story into better perspective. So, for example, if I read an article about a new theory of consciousness, I want to know how it relates, say, "attention" to "qualia," or, what consciousness experiments it cites.

This same habit is certainly applicable to my own work. There are numerous primary omissions:

Proofs The book has many proofs of theorems, some rather severely pedantic calculations, some more detailed than those in the literature, and some intrinsically very nice. But there are also theorems whose proofs are omitted because – dare I say it – the proofs are immediately obvious from the antecedent definitions, or, the proofs are exceedingly routine calculations. These omissions may seem to fly in the face of my Principle of Least Thought, but actually they just reveal my own – albeit small – intuitive leaps.

Adjoint Functors Arguably the most beautiful, universal, and natural concept introduced by category theory into mathematics is that of *adjoint functor*. The book does not define this abstract concept. Rather, it suggests with several leading examples that such a definition is inevitable.

Infinitely Slow Processes Classical thermodynamics is shot through with the intuition of an "infinitely slow process." In the Theory of Substances in Chapters 7–10 all of that is absorbed into a single term that may be set equal to zero in the Power Balance Equation. This amounts to ignoring the generation of entropy in the process under study. Indeed, it ignores the Second Law of Thermodynamics which says in part that *all* natural processes generate some entropy. The virtue of this unnatural assumption is that a large swath of classical thermodynamics is presented under the rubric of "equilibrium thermodynamics." This is not necessary for the intuitive and rigorous understanding of thermodynamics.

Heat Engines & Absolute Temperature Although the practice and theory of heat engines have been absolutely crucial in the history of thermodynamics, it turns out they are not essential in this book's

Theory of Substances. The same goes for the concept of "absolute temperature." This is not to say, of course, that heat engines and absolute temperature cannot be introduced within the framework of this book. It is just that those concepts may be left to self-imposed challenges for the reader. Likewise for the "Zeroth Law of Thermodynamics."

Numerical Calculations Not only are there no numerical calculations in this book, there are no standard units of measurement such as "Newton" or "Joule" or for that matter, "meter." Instead, generic placeholders for units of measurement such as **[FRC]** or **[NRG]** or **[DST]** are employed. Any convention may substitute standard units for these placeholders. As for calculations, there are a great many standard textbooks to consult, and some of the best are referenced in this book.

Phase Transformation The important theory of phase transformations is omitted because so far as I know phase transformations – gas to liquid to solid, and so on – play no role in muscle contraction.

Statistical Thermodynamics One of the challenges I set to myself in writing this book was to become convinced that it is possible to offer a rigorous mathematical theory of classical thermodynamics and chemical thermodynamics without mention of the molecular theory and accompanying statistical considerations. Therefore, Particle Mechanics in Chapter 6 and Stochastic Timing Machinery in Chapters 11–12 are entirely separate from the Theory of Substances in Chapters 7–10.

Muscle Contraction Simulation Code This book is not a report on original scientific research on muscle contraction that culminates in a new model for simulation. It is focused on the parts of mathematical science education that this outsider considers relevant to any such simulation, and so stops short of adopting a particular viewpoint among those cited in the last chapter.

Small Systems Biological motors such as the muscle contraction system are not classical chemical-mechanical thermodynamic systems. Discussion of cutting-edge mathematical physics of non-equilibrium thermodynamics of small systems as in, for example [Jarzynski (1997)] and [Astumian and Hänggi (2002)], is beyond the scope of this book.

1.11 What is Original?

Aside 1.11.1. I claim three original contributions to mathematical science education. First, baby steps towards a *Ground* in which to situate a new Foundation of Mathematics. All of the mathematics in this book is built on that foundation. Second, a new graphical programming language – *Stochastic Timing Machinery* – with source code for an interpreter to calculate simulations of concurrent processes such as clouds of particles moving in response to force fields. This is partly motivated by the literature on muscle contraction which includes models based on elaborations of such processes.

Third, my *Theory of Substances* is a new algebraic thermodynamics including an improved version of the "Second Law of Thermodynamics" called the *Entropy Axiom*.

PART 2
Mathematics

This part introduces the algebraic focus of the mathematics in the book. Specifically, the unifying apparatus of basic category theory is revealed, calculus is viewed as an algebraic tool for calculating with infinitesimals, and ordinary lists and tables actually offer a wealth of instruments for calculation.

Chapter 2

Ground & Foundation of Mathematics

2.1 Introduction

Aside 2.1.1. I've had the good fortune of being encouraged by great teachers and Ralph Abraham is among the best and friendliest. At Columbia University I had to drop out for a while due to an illness, and came back in 1963. On a bulletin board at the mathematics department was the announcement of a course on differential geometry and general relativity. I made an appointment to visit with the professor. He wore sneakers and dungarees and a white shirt open at the collar. This was definitely different from the usual attire of a professor. There were about seven other students.

The course was heavily diagram-oriented in the sense of category theory. The greatest fun was an algebraic calculation I did with another student. We filled several blackboards with our calculations to give a coordinate-free proof of a standard curvature identity that is usually demonstrated by flipping indices around. I took the notes for the course, and Ralph encouraged me with $A+$ for my course work.

Fast forward to 1968. Ralph invited me to join his group formed with the brilliant logicianKen McAloon to write a series of college mathematics textbooks. That was *Eagle Mathematics*, a hearty band of professors (the "heads") and students (the "hands") who went off to Berkeley, California one summer-time, to write mathematics textbooks. The "heads" had a book contract and we wrote and published the books. What a great experience. Ralph has published classics in the literature of mathematical mechanics [Abraham and Shaw (1983)][Abraham (1967)][Abraham and Robbin (1967)].

In the Spring of 1969 I went to Paris for Eagle Mathematics to work with Ken on an advanced calculus textbook. It turned out that my part of it was totally inadequate. I had a mental-block trying to explain the connection between the geometry of oriented surfaces and the algebraic representation of orientation in terms of permutations. (Today I would say that the geometric intuition is formalized – made rigorous – by the algebra.)

Nevertheless, three important things happened for me in Paris. One was that Ken delivered – in his enthusiastic fluent French – a course on infinitesimals. I remember sitting in the back row of a large ancient French lecture hall, sneezing my head off due to the dust. Another was a lecture series by Barry Mitchell at L'École Polytechnique, in his impeccable clipped Canadian French, on homological algebra. That's how I met Barry. Third, Ken introduced me to F. William Lawvere, the charismatic creator of hot ideas in category theory. I really can't remember what I might have said that could have impressed this man, but anyhow he invited me to join the category theory research group he was forming in Halifax, Nova Scotia. That is where Barry Mitchell with superhuman patience supervised my graduate work.

Aside 2.1.2. Samuel Eilenberg and another eminent mathematician Saunders Mac Lane invented category theory around 1945. At Columbia University around 1968–9 I was granted permission to audit Professor Eilenberg's course on category theory. What a privilege that was. My friend Nicholas Yus prepared me for that. Let me explain.

I had to drop out of school for a semester due to an illness and when I came back I audited an undergraduate course in axiomatic set theory [Suppes (1960)]. In the back of the classroom was Nick, sitting in on the course even though he was a graduate student. We became friends, and one of his great gifts to me was an introduction to algebraic topology using category theory diagrams. Shortly afterwards I had the great fortune to enroll in a course on advanced calculus given by Ian R. Porteous based on a mimeographed draft of a groundbreaking text by Nickerson, Spencer and Steenrod, vintage 1959. It uses arrows for maps!

> **Category theory starts with the observation that many properties of mathematical systems can be unified and simplified by a presentation with diagrams of arrows. ... Many properties of mathematical constructions may be represented by universal properties of diagrams** [Mac Lane (1971)].

Maybe I always get a big charge out of diagrams because my interest in the mind and consciousness [Cooper (1996)] brought me to encounter Charles S. Sherrington's famous metaphor for what happens in the brain upon waking.

The great topmost sheet of the mass, that where hardly a light had twinkled or moved, becomes now a sparkling field of rhythmic flashing points with trains of traveling sparks hurrying hither and thither. The brain is waking and with it the mind is returning. It is as if the Milky Way entered upon some cosmic dance. Swiftly the head mass becomes an enchanted loom where millions of flashing shuttles weave a dissolving pattern, always a meaningful pattern though never an abiding one; a shifting harmony of subpatterns [Sherrington (2009)].

"Loom" refers to the 19th century machine for weaving fabric into complex patterns. It was a mechanically programmed robot with thousands of shuttling parts. The image Sherrington invokes has stayed with me, and probably underlies not only my interest in category theory with its *dynamic* diagrams of algebraic relationships, but also my invention of "timing machinery," which was stimulated by observations of neurophysiology experiments on animal brains. More on that later.

Sitting in on Professor Eilenberg's course was a super thrill. He invited me to give a presentation on a fundamental theorem called the "adjoint functor theorem." (We were using Barry Mitchell's epochal textbook.) I was stumbling my way through the steps of the proof. Eilenberg interrupted and explained that the construction involved building a very huge space, and then cutting back to the thing we needed to complete the proof. This is a perfect example of my Principle of Least Thought. Here I was, going through these tiny steps in the proof, and Eilenberg leaped across, carrying me on his back.

Socializing is the lubricant of mathematics. I attended some pretty great mathematics parties. One time at a party in New York I gingerly asked Eilenberg if he had read this thing I submitted to him, entitled "Chronicle of the Writer's Involvement in Linguistics." It included the story of how I got interested in category theory, but it was written – pretentiously – in the third person. He grunted, "Building monuments to yourself, eh?"

At another party I told him I wanted to create mathematical metaphysics. He harumphed, "Better to create metaphysical mathematics." Frankly, I think F. William Lawvere is the best at that [Lawvere (1969)][Lawvere (1976)].

I visited Eilenberg in the hospital after I learned of his stroke. It had rendered him mute. He gestured for me to massage his right hand. It was a very beautiful hand. Okay, I just want you to know that I loved Professor Eilenberg. He taught and encouraged me, he told me the truth without hurting my feelings – he was never condescending.

2.2   Ground: Discourse & Surface

Aside 2.2.1. I have to start the technicalities somehow, and it seems natural to assume a human being may be equipped with writing tools and materials. I separate the idea of writing about something and the something written about. So Discourse is supposed to be the text, notations, and diagrams, say, on paper or in a window on a computer screen. The Discourse is about the text, notations, and diagrams on a separate piece of paper, or in a separate window – the Surface.

Some mathematicians and philosophers of mathematics engage in a quest to determine the nature of mathematical existence, proof, and truth. Traditionally there are four basic views: Platonism, Formalism, Intuitionism, and Logicism [Lambek (1994)]. I insist there is something beneath such foundations: the Ground in which they are founded.

In this section of the book the first occurrence of a technical word is rendered in italics. This by no means implies that italicized terms are defined. The primary distinction of the *Ground of Mathematics* is between *Discourse* and *Surface*. Every *expression* occurs in a *region* of the Surface. The Discourse may specify expressions and regions in the Surface.

A *context* is a specified region of the Surface within which smaller regions may or may not contain expressions. In any case, the extent of such a context is clearly marked, for example, by Chapter, Section, Subsection, or Paragraph headings. Thus, contexts may be nested. Sometimes a single expression is considered to be a context. Care must be taken to observe context boundaries.

The choice of a symbol to represent an idea is arbitrary – provided the symbol is used consistently. More precisely, in a specified context the choice of a symbol to represent an idea – including all its copies in the context –

may be replaced by some other symbol in all of its occurrences within the context, provided the replacing symbol occurs nowhere else in the context. In this sense the replaced symbol is called *bound*. For every symbol there is a sufficiently large context in which it is bound. The Ground includes human *cognitive ability* capable of answering the following questions:

(1) What is the specified context of the Surface?
(2) What is the specified region of the Surface?
(3) What is the specified expression?
(4) For a specified region of the Surface is there some expression *occurring* the region?
(5) Is a specified expression occurring in a specified region of the Surface?
(6) Of two specified regions is one *left*, *right*, *above* or *below* the other?
(7) Of two expressions in distinct regions, is one a *copy* of the other?
(8) What is the total count of expressions in a row, column, or other specified region?
(9) Is a *Context* nested within another Context?

The Ground includes human *muscle contraction* capable of performing the following actions:

(1) Introduce an expression specified in Discourse into a specified region of Surface. For example, to introduce a copy of an expression of Discourse in a blank region to the right of a specified region. Repeating this action yields a *list* expression on the Surface.
(2) Copy the expression in a specified region into a distinct specified region.
(3) Mark the start of a *Context*.
(4) Mark the end of a started Context.
(5) Delete the expression – if any – occurring in a specified region.

These capabilities are called the **Ground Rules of Discourse**.

2.2.1 *Symbol & Expression*

Aside 2.2.2. At a conference in Bolzano, Italy some years ago I gave a talk to some philosophers, linguists and mathematicians. I put before them my idea that every word has an ancient literal root. A linguist immediately shouted, "Oh yeah? What about unesco?" Briey cowed, I had to qualify my statement to exclude acronyms. But what I had in mind were examples such as the word "spirit," as in, say, "spirituality." Its ancient literal root is a Latin word meaning "to breathe."

The ancient literal root of "express," – as in, say, "express yourself" – is Latin for "to squeeze out." Mathematicians express themselves in writing by producing mathematical expressions: an expression is an arrangement of symbols. Symbols include the usual alphabetical and numerical symbols, punctuation marks, and numerous specially contrived symbols. Like the different segments of a television show – title, first commercial, show, second commercial, show, credits – a mathematical exposition is divided into clearly marked successive scopes or contexts. Expressions conform to a grammar – formation rules – that are explicitly announced or implicitly understood for each context. Expressions are a sculptural medium, with a difference. Like, say, clay, expressions are a moldable, additive, and subtractive medium, except more supple. A porous plaster mold for clay is used to make copies of a master shape. A highly viscous material called "slip" mixed from clay and water is poured into the mold, and over a period of time not more than hours water seeps out of the slip into the plaster and eventually evaporates, leaving a relatively durable solid clay shape conforming to the interior of the mold. But a mathematician produces a copy of an expression by just looking at it and writing on a clear space on the page. A potter can add a handle to a previously molded piece by shaping a lump of clay and adhering it to the piece using some slip as "glue," but a mathematician need only add a subscript to a symbol. A sculptor may gouge out and discard a lump from a block of clay to begin shaping it into a bust, while a mathematician need merely wield an eraser to delete a part of an expression.

A *symbol* is a small expression usually but not always drawn with a single continuous motion of the writing instrument – dotting an "**i**" or crossing a "**t**" requires lifting the instrument to complete the symbol. Mathematical expressions are composed of symbols drawn successively in nearby regions. The basic expression is a *list* which is by definition a row of symbols – or other expressions – placed successively in a region from left to right. A *diagram* is an expression with arrows connecting dots, and may spread out upon the Surface.

2.2.2 *Substitution & Rearrangement*

The basic operations on expressions are copying and deleting. The supreme combination of these operations is substitution. The inputs for the substitution algorithm are two given expressions and a designated sub-expression of the first one. Substitution of the second expression for the sub-expression in the first one consists of the following sequence of basic operations: copy the entire first expression, delete the sub-expression, and copy the second expression in its place.

This notion of substitution is amenable to deeper analysis. A difficulty with the "definition" arises when the second expression is not exactly the same shape as the part for which it is to be substituted in the first expression. There is less of a problem if the second expression is much smaller than the part, but if it is larger then some copy and deletion operations must be performed in the copy of the first expression to make room, so to speak, for the replacement.

The primary means for declaring that an expression may substitute for another is the equation. An equation defined consists of an equals symbol flanked by expressions. Within a given context, an equation declares the option – but not the obligation – to substitute either flanking expression for the other.

For example, the equation $15 = (3 \times 5)$ says that any occurrence of (3×5) as a sub-expression may be replaced by 15. Or, vice versa. The 15×15 "multiplication table" that all high school students should know is a convenient representation of 225 such equations.

Another combination operation on expressions is rearrangement. A part of an expression is copied, then the remainder is also copied, but in a different position relative to the copy of the first part. "Rules for doing arithmetic" are algorithms based on such tables and on combinations of substitution and rearrangement operations. These rules include summing columns of multiple-digit numbers, multiplication of multiple-digit numbers, long division, square-root approximation, and so on. An arithmetic expression is composed of numbers and arithmetic operations, $+, \times, -, \div$ and so on. Starting with any arithmetic expression, arithmetic rules can be applied until there is a single number to which no further rules apply (except erasure). This number is called the **value of an expression**.

An algebraic expression defined is an arrangement of symbols that – unlike an arithmetic expression – includes one or more letters. An equation with a letter on one side and an expression on the other offers the option – within a given context – of substituting that expression for an occurrence of that letter in some other expression. Except for the occurrences of letters, algebraic expressions conform to the same formation and transformation rules as arithmetic expressions. But an algebraic expression cannot be evaluated. That is, un- less arithmetic expressions are substituted for all variables in the expression. If that is done the result is an arithmetic expression, so it has a value. In other words, the value of an algebraic expression varies depending on what arithmetic expressions are substituted for its letters. Indeed, this basic combination of substitution and evaluation leads inevitably to the abstract idea of a function.

In summary, the Ground for any Foundation of Mathematics is the possibility of copying, adding, deleting, substituting, and rearranging of expressions, especially lists.

2.2.3   *Diagrams Rule by Diagram Rules*

The following is a *Foundation for Mathematics* as required for this book.

Natural language xpressions such as "we write," "we choose to write," "we usually write," "we sometimes simply write," and so on, are common in mathematical writing. A declaration in the Discourse that a described diagram "exists" is equivalent to asserting the right but not the obligation to draw the diagram as described on the Surface. Roughly speaking, in this Foundation of Mathematics, *everything is a diagram.*

In any context and for any expressions A and B a diagram of the form $A = B$ is called an *equation*. This equation signifies that A may be substituted for B in whatever expression B is a sub-expression, or B may be substituted for A in whatever expression A is a sub-expression.

Obviously, $A = A$ changes nothing, and $A = B$ signifies the same substitution possibilities as $B = A$. By the definition of the substitution algorithm, If $A = B$ and $B = C$ then $A = C$.

In any context and for any expressions A and B there exists at most one diagram $A := B$. If this diagram exists it is called the *defining equation for A* in the given context. In a defining equation the left side – usually but not always a symbol – is considered to be an abbreviation for the right side, although without necessarily restricting which side may substitute for the other.

2.2.4   *Dot & Arrow*

Aside 2.2.3. "Connecting the dots" means that, given various items, understanding "the big picture" means finding the links between them. I remember a fun thing to do as a kid was to draw lines from one numbered dot to another numbered dot in the order of the numbers and gradually see a picture emerge. Long before that, ancient farmers, poets, and sailors connected dots with lines to create astral constellations.

Diagrams of dots and connections represented by (straight or curved) line segments (with or without arrowheads) abound. A road map has dots for locations and curved lines representing roads between them. Wiring and plumbing diagrams for a house are also pictures of dots and their connections. And so on. The ubiquity of such diagrams naturally gave rise to mathematical abstractions and the most basic abstraction is called a "graph" (more precisely, a "directed graph"), which is a certain kind of mathematical structure represented by a drawing of some dots and some arrows with tails and heads at dots. Usually the dots and arrows are labeled with mathematical *expressions*.

There is a distinction between a graph in which the connections between points are not assumed to have any particular directionality – like two-way roads in a road map – and a directed graph, where exactly one end of each connection has an arrowhead or other means to distinguish it from the other end. An arrowhead is very common, since it is a visual metaphor for the directed flight of an arrow. By "graph" I always mean directed graph.

"Change is the fountainhead of all dialectical concepts," but "qualitative change eludes arithmomorphic schematization" ([Georgescu-Roegen (1971)] p. 63). (In systems engineering the stable, persistent state of a system is called "steady-state," and the transitory intermediates between steady-states are called "transients.") I think graphs of labeled dots and arrows schematize the unitary notion of continuous change, but only at the price of insistence on a distinction: dots represent rest and arrows represent motion. This is a graphic distinction between *Being* and *Becoming*.[1]

The literature of mathematics and computer science is not entirely consistent on what to call these two representations. Figure 2.1 exhibits common locations.

I prefer the descriptive terms "labeled dot" and "labeled arrow" since these are items that are visible on the page. The dot is not usually drawn, only its label is. But the arrow is definitely drawn, with its label alongside. The labels on dots and arrows are mathematical expressions.

A profound conceptual breakthrough is that many graphs are naturally endowed with a multiplicative structure over and above mere dots and connecting arrows.

Look, if you can drive from New York City, New York to New Haven,

---

[1] Ilya Prigogine relegates classical and quantum mechanics to the physics of "being," and for him, modern thermodynamics assumes the mantle of the physics of "becoming" [Prigogine (1980)].

Rest (*Being*)	Motion (*Becoming*)
vertex	arc
node	arrow
object	arrow
vertex	edge
vertex	arrow
node	link
node	arc
state	transition
steady-state	transient

Fig. 2.1   Alternative Terminologies for elements of a directed graph.

Connecticut and then from New Haven to Boston, Massachusetts – that is two arrows end-to-end – you can certainly drive straight through from New York to Boston, which is one arrow "composed" of two. Nothing mysterious about that, it is common sense. The algebraic formalization of this "composition" of arrows makes precise two obvious aspects of this Composition Law.

First, if you have three arrows end-to-end, then composing the first two to yield one arrow which is then composed with the third, you get exactly the same result as composing the first arrow with the composition of the last two. This must resonate with something analogous you know from arithmetic which is that if you have three numbers, to get their product you will get exactly the same result if you multiply the first two and multiply by the third, or if you multiply the first number by the result of multiplying the last two. In high school one is taught to call this the "Associative Law of Multiplication",

$$(r * s) * t = r * (s * t).$$

Second, with that analogy in mind you know also that multiplying a number times 1, or in the other order if you multiply 1 times the same number, in both cases the result is just the number – it retains its identity.

$$1 * r = r$$

$$r * 1 = r.$$

These are the "Identity Laws for Multiplication". Interestingly enough, there is a second analogy, since one is also taught the "Associative Law of

Addition" and that 0 is an identity for addition:

$$(r + s) + t = r + (s + t)$$
$$0 + r = r$$
$$r + 0 = r .$$

Likewise for arrows, except every dot is given its own 1 called its "identity arrow." So the composition of the identity arrow of a dot with any arrow with tail at the dot just yields the same arrow, unchanged, and likewise, composing any arrow with the identity arrow at its head also yields just the same arrow.

Example 2.1. In high school one may learn about points and vectors in a plane – these could represent physical forces, or velocities. From one point it is easy to draw several arrows away from the point

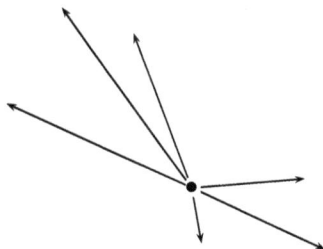

or towards the point

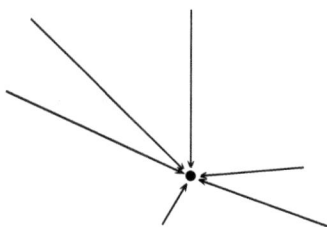

To simplify discourse it is better to use symbols to name points, and in diagrams points are actually replaced by their names. So, for any two vectors $p \to q$ and $q \to r$ with the tip end of one at the back end of the other as in

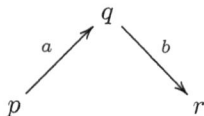

the sum vector is constructed as in

$$(2.1)$$

For any three vectors in a chain

$$(2.2)$$

which is the Associative Law for Vector Summation learned in high school

$$(a + b) + c = a + (b + c) \, .$$

Assuming that for each point p there exists a vector 0_p of length 0 from p to p, this vector sum operation obviously satisfies for any vector $p \xrightarrow{a} q$ the Additive Identity Laws

$$0_p + a = a$$

$$a + 0_q = a$$

Thus, points and vectors in a plane constitute a directed graph with a Summation Law satisfying the Associative and Identity Laws. Moreover, for any two points there exists exactly one vector from the first point to the second point.

By 1940 mathematicians – already well-accustomed to using letters (and combinations of symbols) for mathematical objects – started using arrows to represent mappings between objects ([Mac Lane (1971)] p. 29). In 1945 a full-blown algebraic theory of "categories," the arrows between them called "functors," and even called "natural transformations" appeared [Eilenberg and Mac Lane (1986)].

Roughly speaking, a category is a directed graph with the additional structure that every dot has an identity arrow, two arrows that are connected end-to-end can be composed to yield a third arrow, and this law of composition satisfies some common-sense associativity and identity conditions.

Aside 2.2.4. The terminology "dot" and "arrow" I use to describe a directed graph is not quite standard. These terms are descriptive of the actual marks made on paper. In category theory the dots are called "objects" and arrows are called "morphisms." The reason for the word "objects" is that when Eilenberg and Mac Lane invented category theory almost all their examples of categories were categories of well-known structured mathematical objects such as vector spaces, groups, topological spaces, ordered sets – and the transformations connecting these objects.

Originally, Eilenberg and Mac Lane used the word "mapping" [Eilenberg and Mac Lane (1986)] for an arrow between dots, but that was replaced by the word *morphism*. The word "morphism" for arrow in a category seems to be abstracted from "homomorphism" (Greek $o\mu o\zeta$ (homos) meaning "same" and $\mu o\rho\phi\eta$ (morphe) meaning "shape"). Long before category theory was invented "homomorphism" meant "structure preserving map between two algebraic structures" [Waerden (1931, 1937, 1940, 1949, 1953)].

I reserve the terminology of dots and arrows for directed graphs, and objects and morphisms for categories.

Axiom 2.2.1. $\boxed{\text{One}}$

There exists a diagram **1** and in Discourse the locations "A in B" or "A belongs to B" or "A is a selection in B" are synonymous and short for "there exists on the Surface a diagram $\mathbf{1} \xrightarrow{A} B$." As an abbreviation, $A \in B := \mathbf{1} \xrightarrow{A} B$.

Remark 2.2. It is possible that both $A \in B$ and $A \in C$ without $B = C$. Unpacking the abbreviations, this implies that possibly $B \xleftarrow{A} \mathbf{1} \xrightarrow{A} C$ with $B \neq C$. Such a diagram may raise the eyebrows of the experienced purist who demands that in a diagram any two arrows with the same label must have exactly the same head and tail. In this book, however, that convention about arrows and their labels is not adopted unless contravened in a specific context.

Axiom 2.2.2. **Identity**

For every diagram D there exists a diagram $D \xrightarrow{1_D} D$ called the **identity arrow** of D.

$$D \quad \bigcirc \quad 1_D$$

Fig. 2.2   The identity arrow of a diagram on the Surface.

2.3   Foundation: Category & Functor

> Category theory asks of every type of Mathematical object: "What are the morphisms?" It suggests that these morphisms should be described at the same time as the objects.
>
> [Mac Lane (1971)]

Aside 2.3.1. The most important category – if only by virtue of its ubiquity in mathematics – is that of "sets" and their "functions." Like any extremely important concept, it is not so easy to pin down exactly what is a set, let alone a function. In the beginning, Georg Cantor wrote "By an "aggregate" we are to understand any collection into a whole M of definite and separate objects m of our intuition or our thought." So, right away even though you probably get the idea, you have every right to be wary when a fundamental concept is introduced by reference to your own thoughts. Subsequent mathematicians managed to hide their thoughts behind an increasingly elaborate formal apparatus called "axiomatic set theory." This is a formal language about a "membership relation" and an "empty set," with a handful of carefully wrought axioms. These axioms guarantee the existence of certain sets provided some other set or sets exist to begin with, and before you know it, a hierarchy of sets exists based solely on the empty set, which exists. There is an axiom that says if some set exists, then the set whose sole element is that set also exists. Hence, there is a set whose sole element is that set, and so on. This implies that there exists an infinity of sets. It seems like we get something from nothing, but that is not so in this case because the axioms assert the existence of new sets based on existing sets, and the empty set exists. A particular kind of set is called a relation, and a particular kind of relation between two sets is called a function, and once that definition is given, it is easy to say what is the category of sets and functions.

Axiomatic set theory is an axiomatic theory, which means it is a theory, and so you probably want to know, what is a theory? A mathematical theory is a formal system, so what then is a formal system? Ultimately when you track down these things you get to a point where the question is, what is a mathematical language? And at that point you get to the interface between natural language, which nearly everyone learns as a child, and the logical use of symbols, which is a very delicate subject requiring great *mathematical* maturity. This book does not go that way.

Writers of mathematics are strongly advised to avoid technical terms and "especially the creation of new ones, whenever possible" [Steenrod *et al.* (1973)]. On the other hand, I think it is necessary to carefully avoid interference by possibly misleading preconceptions, especially those inherent in terminological conventions. Matters are exacerbated by the fact that actually there are several theories of sets and functions. The "set theory" in this book is extremely conservative. Consequently I had the impulse to substitute for the standard terminology of "element," "set," and "function" the words "item," "sort," and "map." But I will not do that. It is better not to muddy the water by using terminology that could confuse the beginner who wishes to read other mathematical science works, nor to adopt conventions that could raise eyebrows if not technical questions among the experts. My attitude is that instead of assuming you know about Cantorian set theory [Cantor (1915,1955)], or some version of naive set theory [Halmos (1960)], or axiomatic set theory [Suppes (1960)], I will insist that dots may be labeled by expressions that stand for **sets**, that there may be labeled arrows between sets which I shall call **maps**, and that there is a category 𝕊et of sets and maps with certain common sense properties. One convention I adopt is to distinguish between a generic map between sets, and map defined by a specified formula, which I will call a **function**.

The usual formal preparation for defining a new kind of mathematical object depends on being prepared to interpret a language with terms such as "primitive symbol," "proper and improper symbol," "denumerable sequence," "variable" and the like. Therefore, a rigorous definition of "category" seems to require a minimal conception of "naive set theory." This procedure places set theory of some kind ahead of category theory. However, I believe this ranking is not necessary, so in this book the Ground for introducing new objects is entirely formal but in terms of marks on paper – the Surface – according to guidelines declared in Discourse.

Remark 2.3. Diagrams in a category are like diagrams of points and vectors in a plane. However, unlike points and vectors in a plane, a category diagram need not have any arrows between two given objects, nor need a category diagram have only one morphism between two given objects. All category diagrams of objects and morphisms are drawn in a plane, with the understanding that lengths and directions of category diagram morphisms are irrelevant. In a given Context the same diagram may be re-drawn with its objects shifted around so long as the morphisms connecting them are retained. It is like a rickety apparatus made of freely jointed telescopic tubes. Not only that, if there are objects with the same label in a diagram, then the diagram may be re-drawn with the two objects merged into one, provided, again, that all connecting morphisms are retained.

The basic idea of category theory is that mathematical objects are related to one another, so that diagrams may depict relations with morphisms between symbols for objects, and the morphisms are labeled with symbols representing particular relations.

2.3.1 *Category*

Definition 2.4. An expression $\mathbb{C}$ is a **category** if

Composition Law for any diagram $1 \xrightarrow{D} \mathbb{C}$ such that

$$D = \quad \begin{array}{ccc} & Y & \\ {\scriptstyle a}\nearrow & & \searrow {\scriptstyle b} \\ X & & Z \end{array}$$

(2.3)

there exists a diagram

$$
\begin{array}{ccc}
 & Y & \\
a \nearrow & & \searrow b \\
X & \xrightarrow{\quad ba \quad} & Z
\end{array}
\qquad \mathbb{C}\,;
\tag{2.4}
$$

Associative Law for any diagram in $\mathbb{C}$

$$
\begin{array}{ccc}
Y & & W \\
\uparrow a & \searrow b & \uparrow c \\
X & & Z
\end{array}
\tag{2.5}
$$

there exists a diagram

$$
\begin{array}{ccc}
Y & \xrightarrow{\ cb\ } & W \\
\uparrow a & & \uparrow c \\
X & \xrightarrow{\ ba\ } & Z
\end{array}
\qquad \mathbb{C}\,;
\tag{2.6}
$$

and there exists an equation $c(ba) = (cb)a$, and

Identity Laws for any diagrams $A \xrightarrow{f} B$ and $C \xrightarrow{g} A$ in $\mathbb{C}$ there exist diagrams

$$
\begin{array}{ccc}
 & A & \\
1_A \nearrow & & \searrow f \\
A & \xrightarrow{\ f\ } & B
\end{array}
\quad \mathbb{C}
\qquad
\begin{array}{ccc}
 & A & \\
g \nearrow & & \searrow 1_A \\
C & \xrightarrow{\ g\ } & A
\end{array}
\quad \mathbb{C}\,.
\tag{2.7}
$$

and there exist equations $f1_A = f$ and $1_A g = g$.

In Eq. (2.4) the label ba for the arrow $X \to Z$ is called the **composition** of $X \xrightarrow{a} Y$ followed by $Y \xrightarrow{b} Z$, and also a and b are **composable** because the head of a equals the tail of b. Other notations for ba are $b \circ a = ba$ and $a \overrightarrow{\circ} b = ba$ when it is convenient to denote the composition by a symbol that reads from left to right in the same direction as the morphisms. Equation (2.6) resulting from the composition of three arrows in Eq. (2.5) is the **Associative Law for Composition**. The diagrams Eq. (2.7) are the **Identity Laws for Composition**.

Definition 2.5. If $1 \xrightarrow{x} X$ and $X \xrightarrow{f} Y$ are composable morphisms in a category then an alternative notation for their composition is $f(x) := f \circ x$.

2.3.2   Functor

Definition 2.6. If $\mathbb{C}$ and $\mathbb{D}$ are categories then a diagram $\mathbb{C} \xrightarrow{F} \mathbb{D}$ is a **functor** if

(1) x in $\mathbb{C}$ implies $F(x)$ in $\mathbb{D}$,

(2) $x \xrightarrow{f} y$ in $\mathbb{C}$ implies $F(x) \xrightarrow{F(f)} F(y)$ in $\mathbb{D}$,

(3) $x \xrightarrow{1_x} x$ in $\mathbb{C}$ implies $F(1_x) = 1_{F(x)}$ in $\mathbb{D}$,

(4)

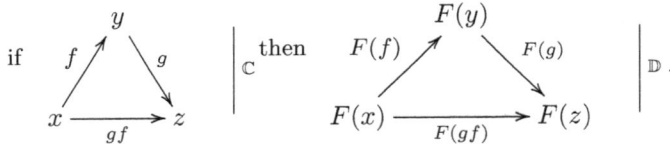

Hence $F(g \circ_{\mathbb{C}} f) = F(g) \circ_{\mathbb{D}} F(f)$ by (4), a Distributive Law.

Theorem 2.7. *There exists a category* $\mathbb{C}at$ *– the* **category of categories** *whose objects are categories and whose morphisms are functors.*

2.3.3   Isomorphism

Definition 2.8. For any $A \xrightarrow{f} B$ of a category $\mathbb{C}$ say $A \xrightarrow{f} B$ is an **isomorphism from** A **to** B if there exists $B \xrightarrow{g} A$ in $\mathbb{C}$ such that

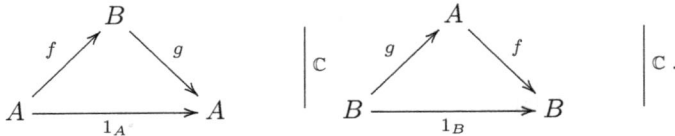

Theorem 2.9. *Every identity arrow* $A \xrightarrow{1_A} A$ *is an isomorphism. If* $\mathbb{C} \xrightarrow{F} \mathbb{D}$ *is a functor and* $A \xrightarrow{f} B$ *is an isomorphism in* $\mathbb{C}$ *then* $F(A) \xrightarrow{F(f)} F(B)$ *is an isomorphism in* $\mathbb{D}$.

2.4   Examples of Categories & Functors

2.4.1   *Finite Set*

Axiom 2.4.1. $\boxed{\text{Void Set}}$

There exists a set $\emptyset$ such that there is no diagram $\mathbf{1} \xrightarrow{x} \emptyset$.

No doubt, for high school students and mathematical science teachers the most intuitive concept of set is that of *finite* set. In the textbooks on set theory the concept of finite set is a special case of the concept of set. In this book the most primitive mathematical notion is that of list on the Surface. Two lists are considered to represent the exact same finite set if one list is merely a rearrangement of the other.

Axiom 2.4.2. $\boxed{\text{Finite Set Axiom}}$

(1) There exists a diagram $\mathbb{F}in$ and $\mathbf{1}$ in $\mathbb{F}in$.
(2) If $A_1, \ldots, A_n$ are distinct expressions (in Discourse or on Surface) then there is a diagram $\{\, A_1 \cdots A_n \,\}$ and it is in $\mathbb{F}in$.
(3) If $\{\, A_1 \cdots A_n \,\}$ and $\{\, B_1 \cdots B_n \,\}$ in $\mathbb{F}in$ and the list $[\, B_1 \cdots B_n \,]$ is a rearrangement of the list $[\, A_1 \cdots A_n \,]$ then there is a diagram $\{\, B_1 \cdots B_n \,\} = \{\, A_1 \cdots A_n \,\}$.
(4) If $A := \{\, A_1 \cdots A_n \,\}$ in $\mathbb{F}in$ then there exists $\mathbf{1} \xrightarrow{A_j} A$ for $1 \leq j \leq n$. Moreover, if $\mathbf{1} \xrightarrow{x} A$ then there exists j such that $1 \leq j \leq n$ and $x = A_j$.

Axiom 2.4.3. $\boxed{\text{Finite Map}}$

Let $X := \{\, A_1 \cdots A_m \,\}$ and $Y := \{\, B_1 \cdots B_n \,\}$ in $\mathbb{F}in$. Then $X \xrightarrow{f} Y$ is in $\mathbb{F}in$ if and only if

(1) for every i such that $1 \leq i \leq m$ there exists a unique j such that $1 \leq j \leq n$ and

$$A_i \xmapsto{\ f\ } B_j \; :=$$

(2) $\mathbf{1} \xrightarrow{r} f$ if and only if there are i such that $1 \leq i \leq m$ and j such that $1 \leq j \leq n$ and $A_i \xmapsto{f} B_j$.

Theorem 2.10. *There exists a category* $\mathbb{F}in$ *whose objects are the finite sets and whose morphisms are the finite maps.*

Axiom 2.4.4. $\boxed{\text{List}}$

For any list $[\, A_1 \cdots A_n \,]$ of sets $A_1, \ldots, A_n$ there exists a set $A_1 \times \cdots \times A_n$ such that there exists a diagram $\mathbf{1} \xrightarrow{L} A_1 \times \cdots \times A_n$ if and only if there exist diagrams $\mathbf{1} \xrightarrow{x_i} A_i$ for $i = 1, \ldots, n$ such that $L = [\, x_1 \cdots x_n \,]$. For each $i = 1, \ldots, n$ there exists a function

$$A_1 \times \cdots \times A_n \xrightarrow{\pi_i} A_i$$

such that if $\mathbf{1} \xrightarrow{[\, x_1 \cdots x_n \,]} A_1 \times \cdots \times A_n$ then $[\, x_1 \cdots x_n \,] \xmapsto{\pi_i} x_i$. For any diagram

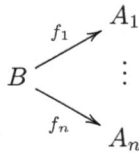

there exists a unique diagram

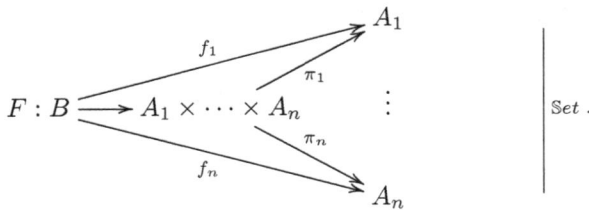

This diagram implies that $\pi \circ F = f_i$ for $i = 1, \cdots, n$. The unique map F determined by $f_1, \cdots, f_n$ is denoted by $(\, f_1 \cdots f_n \,)$.

2.4.2   *Set*

By analogy with finite sets and finite maps, sets are the objects of the category $\mathbb{S}et$ in which the morphisms are the maps:

Axiom 2.4.5. $\boxed{\text{Set}}$

There exists a category $\mathbb{S}et$ such that

(1) $\mathbf{1}$ in $\mathbb{S}et$.

(2) For every directed graph $\mathbb{G}$ there exists $\mathbf{1} \xrightarrow{\mathrm{Dot}(\mathbb{G})} \mathbb{S}et$ and there exists $\mathbf{1} \xrightarrow{\mathrm{Arr}(\mathbb{G})} \mathbb{S}et$ such that $\mathbf{1} \xrightarrow{A} \mathrm{Dot}(\mathbb{G})$ if and only if A is a dot of $\mathbb{G}$, and $\mathbf{1} \xrightarrow{u} \mathrm{Arr}(\mathbb{G})$ if and only if u is an arrow of $\mathbb{G}$.

(3) Let X and Y in $\mathbb{S}et$. Then $X \xrightarrow{f} Y$ in $\mathbb{S}et$ if and only if

  (a) for every $\mathbf{1} \xrightarrow{k} X$ there exists a unique diagram

$$k \xmapsto{\ f\ } fk \ :=$$

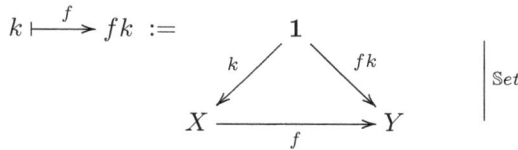

  (b) $\mathbf{1} \xrightarrow{r} f$ if and only if there is $\mathbf{1} \xrightarrow{k} X$ such that $k \xmapsto{\ f\ } fk$.

Definition 2.11. If $x \xmapsto{\ f\ } y$ in $\mathbb{S}et$ say x **maps to** y **by** f, and also write $y = f(x) := fx$.

Definition 2.12. For a diagram $X \xrightarrow{F} Y$ in $\mathbb{S}et$ call X the **domain** and Y the **codomain** of F.

Aside 2.4.1. This opaque technical language goes against my urge towards intelligible terminology. To ease my conscience I might just call the domain the "tail", and the codomain the "head" of the arrow F. That being said, the words "tail" and "head" are descriptive of marks on the Surface and reserved for directed graph diagrams, but the word "domain" is historically well-established for referring to the "domain of definition" of a function. And more recently there is a well-established habit of tacking the prefix "co-" onto any term for which there is a logical "dual" or "opposite" of some kind: a co-head is a tail.

Definition 2.13. A map $A \xrightarrow{u} B$ in $\mathbb{S}et$ is an **inclusion** if for any diagram

$$(2.8)$$

$\mathbb{S}et$

there exists an equation

$$a = x \ .$$

Definition 2.14. If $1 \xrightarrow{x} X$ and $X \xrightarrow{f} Y$ are composable maps in $\mathbb{S}et$ then an alternative notation for their composition is $f(x) := f \circ x$ for any $1 \xrightarrow{x} X$.

Therefore, if $A \xrightarrow{u} B$ is an inclusion then $u(a) = u \circ a = a$ for any $a \in A$.

Axiom 2.4.6. $\boxed{\text{Map Equality}}$

If

$$A \underset{g}{\overset{f}{\rightrightarrows}} B$$

then $f = g$ if and only if $f(a) = g(a)$ for all $a \in A$.

Theorem 2.15. *If*

$$A \underset{g}{\overset{f}{\rightrightarrows}} B$$

and f, g are both inclusions, then $f = g$.

Proof.   Since both f, g are inclusions, $f(a) = a = g(a)$ for any $a \in A$. The conclusion follows from the Map Equality Axiom.   □

Definition 2.16. If there exists an inclusion from A to B then the unique inclusion from A to B is denoted by $A \hookrightarrow B$.

Theorem 2.17. *Denote an inclusion $A \hookrightarrow B$ by u. Then for any diagram*

$\mathbb{S}et$

there exists an equation $f = g$.

Proof. Calculate $f(x) = u(f(x)) = u \circ f(x) = u \circ g(x) = u(g(x)) = g(x)$ for any $x \in X$. The conclusion follows from the Map Equality Axiom. $\square$

Axiom 2.4.7. $\boxed{\text{Equalizer Map}}$

For every diagram

$$X \underset{g}{\overset{f}{\rightrightarrows}} Y$$

in $\mathbb{S}et$ there exists a unique diagram

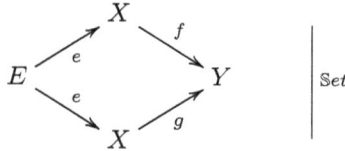

such that for every diagram

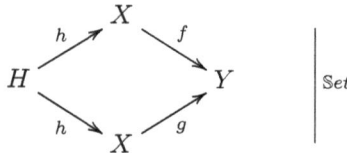

there exists a unique diagram

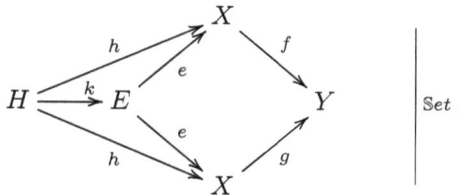

Axiom 2.4.8. $\boxed{\text{Complement}}$

For any inclusion $A \hookrightarrow B$ there exists an inclusion $D \hookrightarrow B$ such that $A \cap D = \emptyset$ and $A \cup D = B$.

Axiom 2.4.9. ⎡Finite Intersection & Union⎤

For any sets X and Y there exists a diagram

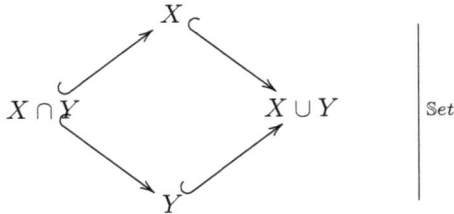

such that if there exists a diagram

(2.9)

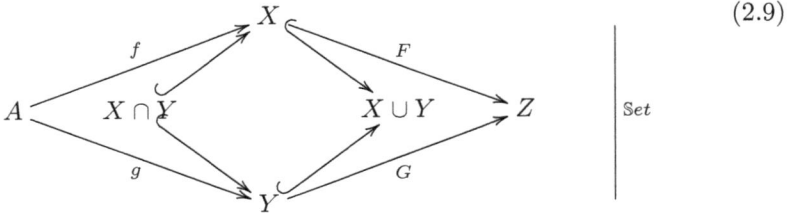

then there exists a unique diagram

(2.10)

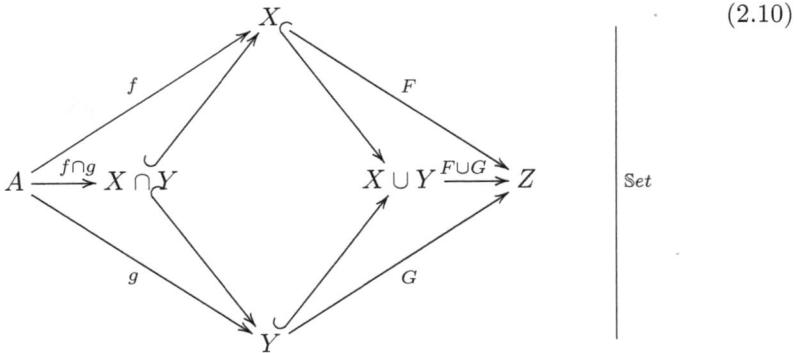

The set $X \cap Y$ is called the **intersection** and $X \cup Y$ the **union** of X and Y.

The existence of diagram Eq. (2.9) is equivalent to the equations $f(a) = g(a)$ for all $1 \xrightarrow{a} A$, so with Eq. (2.10) the Axiom declares that if maps

f and g from A to X and Y agree on all of A then they factor through a map $f \cap g$ from A to the common part of X and Y, their intersection. Dually, the Axiom declares that if maps F and G to a set Z agree on the intersection – as in $F|_{X \cap Y} = G|_{X \cap Y}$ – then they combine to form a map $F \cup G$ from the union to Z.

Theorem 2.18. *If there exist two inclusions $C \hookrightarrow B$ and $D \hookrightarrow B$ such that $A \cap C = \emptyset$ and $A \cup C = B$ and $A \cap D = \emptyset$ and $A \cup D = B$, then $C = D$.*

Proof. Say $1 \xrightarrow{x} C$ then $1 \xrightarrow{x} B$ but $1 \xrightarrow{x} A$ is impossible, so $1 \xrightarrow{x} D$. Hence, $C \hookrightarrow D$. By symmetry, $D \hookrightarrow C$. Therefore, $C = D$. □

Axiom 2.4.10. | Set Subtraction |

Given any inclusion $A \hookrightarrow B$ there exists a unique inclusion satisfying the conditions of the theorem. It is denoted by $B \backslash A \hookrightarrow B$.

Theorem 2.19. *For any inclusion $A \hookrightarrow B$ and any two distinct sets X and Y there exists a unique arrow $A \cup B \backslash A \to \{ X\, Y \}$ such that*

$$
\begin{array}{ccccc}
A & \hookrightarrow & A \cup B \backslash A & \hookleftarrow & B \\
\downarrow & & \downarrow & & \downarrow \\
1 & \hookrightarrow & \{ X\, Y \} & \hookleftarrow & 1
\end{array}
\qquad \Big|\, \text{Set}
$$

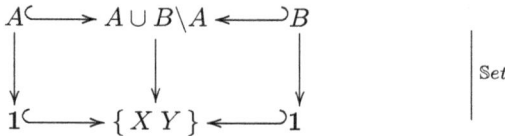

Theorem 2.20. $\emptyset \hookrightarrow A$ *for any set.*

Proof. Without arrows $1 \to \emptyset$ the conclusion is vacuously true by the definition of material implication. □

Theorem 2.21. *If $A \xrightarrow{f} X$ and $A \xrightarrow{g} X$ are inclusions then $f = g$.*

Proof. For any $\xrightarrow{x} A$ there is a diagram $fx = x = gx$ hence a diagram $fx = gx$. Therefore, $f = g$. □

Therefore, if there is an inclusion $A \xrightarrow{f} X$ it is drawn with a hook and without a label as in $A \hookrightarrow X$.

Example 2.22. For any set A the identity map $A \xrightarrow{1_A} A$ is an inclusion.

Definition 2.23. Let X and Y belong to $\mathbb{S}et$. The diagram $X = Y$ is a diagram abbreviation defined by

$$X = Y := \begin{bmatrix} X \hookrightarrow Y \\ Y \hookrightarrow X \end{bmatrix}$$

Definition 2.24. A map $A \xrightarrow{f} X$ in $\mathbb{S}et$ is called a **surjection** if for any map $1 \xrightarrow{x} X$ there exists at least one map $1 \xrightarrow{a} A$ such that

(2.11)

in which case the arrow is drawn with a double head as in $A \xrightarrow{f} X$.

The union of a set of sets has two aspects. On one hand it is the "smallest" set that includes all those sets. On the other, any element of the union must be an element of one of those sets. In conventional axiomatic set theory these two aspects are conflated by a theorem. In this book they are distinguished.

Axiom 2.4.11. $\boxed{\textbf{Union}}$

For any set X there exists a set $\bigcup X$ such that

(1)
$$\text{if } 1 \xrightarrow{A} X \text{ then } A \hookrightarrow \bigcup X$$

(2) if Z is a set such that

$$\text{if } 1 \xrightarrow{A} X \text{ then } A \hookrightarrow Z$$

then

$$\bigcup X \hookrightarrow Z .$$

(3) $1 \xrightarrow{u} \bigcup X$ if and only if there exist diagrams
$$1 \xrightarrow{u} A$$
$$1 \xrightarrow{A} X .$$

Similarly, if W is a set and X_a is a set for each arrow $1 \xrightarrow{a} W$, then there exists a set

$$\bigcup_{1 \xrightarrow{a} W} X_a$$

such that

(1)

$$\text{if } 1 \xrightarrow{a} W \text{ then } X_a \hookrightarrow \bigcup_{1 \xrightarrow{a} W} X_a$$

(2) if Z is a set such that

$$\text{if } 1 \xrightarrow{a} W \text{ then } X_a \hookrightarrow Z$$

then

$$\bigcup_{1 \xrightarrow{a} W} X_a \hookrightarrow Z \ .$$

(3) $1 \xrightarrow{u} \bigcup_{1 \xrightarrow{a} W} X_a$ if and only if there exist diagrams

$$1 \xrightarrow{a} W$$
$$1 \xrightarrow{u} X_a \ .$$

([Kelley (1955)] p. 255)([Lawvere and Rosebrugh (2003)] pp. 247–248)

Remark 2.25. The symbol a is bound in the context of the expression $\bigcup_{1 \xrightarrow{a} W} X_a$.

The intersection of the sets in a set of sets is the "largest" set that is included in all of those sets. Again, there is an aspect of the intersection that says an element belongs to the intersection if it belongs to every set.

Axiom 2.4.12. $\boxed{\text{Intersection}}$

(1) if $1 \xrightarrow{A} X$ then $\bigcap X \hookrightarrow X$;

(2) if Z is a set such that if

$$1 \xrightarrow{A} X \text{ then } Z \hookrightarrow X$$

then

$$Z \hookrightarrow \bigcap X \; ;$$

(3) $1 \xrightarrow{k} \bigcap X$ if and only if $1 \xrightarrow{k} A$ for all sets A such that

$$1 \xrightarrow{k} A$$

$$1 \xrightarrow{A} X \; .$$

Theorem 2.26. *The category of finite sets is a subcategory of the category of sets.*

2.4.3 *Exponentiation of Sets*

Theorem 2.27. *For every set A there exists a functor* $\mathrm{Set} \xrightarrow{A \times (\text{-})} \mathrm{Set}$ *such that on a set X the value is $A \times X$ and on a map $X \xrightarrow{f} Y$ the value is the map $A \times f := 1_A \times f : A \times X \to A \times Y$.*

A fundamental intuition about maps of sets is that a map $A \times X \xrightarrow{f} Y$ corresponds to a map that assigns to each element of X a map from A to Y. Formalizing this intuition brings out an analogy that every high school mathematics teacher should endeavor to appreciate. The first step is an axiom which suggests the analogy by using the same notation for the set of maps from X to Y that is used for the exponential y^x of a number y raised to the xth power. The analogy extends to Set the rule $y^{a*x} = (y^a)^x$ for numbers.

Axiom 2.4.13. **Exponential**

For any sets X and Y there exists a unique set denoted by Y^X such that there exists a diagram $X \xrightarrow{f} Y$ if and only if there exists a diagram $1 \xrightarrow{f} Y^X$.

Given an item $1 \xrightarrow{a} A$ the value of a map $A \xrightarrow{f} X$ has been defined by $f(a) := f \circ a$. Another aspect of the intuition about maps of sets is that there must exist an "evaluation map" that assigns to $a \in A$ and $f \in X^A$ the value $f(a)$.

Axiom 2.4.14. $\boxed{\text{Evaluation}}$

For every set A

(1) for every set X there exists a diagram

$$A \times X^A \xrightarrow{\varepsilon_X^A} X \; ;$$

(2) for every set Y and diagram

$$A \times Y \xrightarrow{f} X$$

there exists a unique diagram

This axiom determines a bijective correspondence between $X^{A \times Y}$ and $(X^A)^Y$, thus completing the analogy with $x^{a*y} = (x^a)^y$ for numbers. The special case $Y = 1$ implies $\varepsilon_X^A(a, f) = f(a)$ for any element $a \in A$ and map $A \times 1 = A \xrightarrow{f} X$, thus fulfilling the intuition that evaluation is a map.

2.4.4 *Pointed Set*

Perhaps the simplest mathematical structure more complicated than a set is a set together with a selected item belonging to the set.

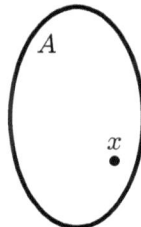

Fig. 2.3 A pointed set $1 \xrightarrow{x} A$.

Definition 2.28. A diagram $1 \xrightarrow{x} A$ where A is a set is called a **pointed set with point x and underlying set** A.

If $1 \xrightarrow{y} B$ is a pointed set then a diagram

$$F := \qquad (2.12)$$

where f is a set map is called a **pointed set map from** $1 \xrightarrow{x} A$ **to** $1 \xrightarrow{y} B$ **with underlying map** f.

Axiom 2.4.15. $\boxed{\text{Pointed set}}$

If $X := 1 \xrightarrow{x} A$ is a pointed set then there exists a diagram $1 \xrightarrow{X} \mathbb{S}et_*$, that is, X belongs to $\mathbb{S}et_*$. If also $Y := 1 \xrightarrow{y} B$ belongs to $\mathbb{S}et_*$ and there is a diagram F as in Eq. (2.12), then there exists a diagram $X \xrightarrow{F} Y$ which belongs to $\mathbb{S}et_*$.

Theorem 2.29. *For any chain diagram in* $\mathbb{S}et_*$

$$\qquad (2.13)$$

there exists in $\mathbb{S}et_*$ *a diagram*

$$\qquad (2.14)$$

Moreover, for every $1 \xrightarrow{x} A$ *in* $\mathbb{S}et_*$ *there exists in* $\mathbb{S}et_*$ *a diagram*

$$\qquad (2.15)$$

Proof. If $y = fx$ and $z = gy$ then $z = g(fx) = (gf)x$ by substitution and the Associative Law in $\mathbb{S}et$. The equation $x = 1_A x$ is the definition of 1_A. $\qquad\square$

Definition 2.30. The diagram Eq. (2.14) is called **composition diagram** and Eq. (2.15) the **identity diagram**.

Theorem 2.31. *Composition and identity diagrams of pointed set maps satisfy the Associative and Identity Laws:*

$$(HG)F = H(GF)$$
$$1_Y F = F$$
$$F1_X = F\,.$$

Therefore, $\mathbb{S}et_*$ is a category, so $1 \xrightarrow{\;\mathbb{S}et_*\;} \mathbb{C}at$.

Proof. These equations follow directly from the corresponding equations for diagrams of sets and maps. $\qquad\square$

Theorem 2.32. *There is a forgetful functor from the category $\mathbb{S}et_*$ of pointed sets to the category $\mathbb{S}et$ of sets, namely by forgetting about the distinguished element.*

2.4.5 *Directed Graph*

Directed graphs are exceedingly more ubiquitous than pointed sets in mathematics, computer science, chemistry, and biology.

Definition 2.33. A **directed graph** $\mathbb{G}$ is a diagram

$$D \xleftarrow{\;\text{tail}\;} A \xrightarrow{\;\text{head}\;} D$$

of sets called **dots** $D = \mathrm{Dot}(\mathbb{G})$ and **arrows** $A = \mathrm{Arr}(\mathbb{G})$ and two maps, **tail** assigning to an arrow a its tail dot, $\mathrm{tail}(a)$, and its head dot, $\mathrm{head}(a)$. A **directed graph map** is a diagram of the form

$$
\begin{array}{ccc}
\mathrm{Dot}(\mathbb{G}) \xleftarrow{\;\text{tail}\;} \mathrm{Arr}(\mathbb{G}) \xrightarrow{\;\text{head}\;} \mathrm{Dot}(\mathbb{G}) \\
\big\downarrow{\scriptstyle d} \qquad\quad \big\downarrow{\scriptstyle a} \qquad\quad \big\downarrow{\scriptstyle d} \\
\mathrm{Dot}(\mathbb{H}) \xleftarrow{\;\text{tail}\;} \mathrm{Arr}(\mathbb{H}) \xrightarrow{\;\text{head}\;} \mathrm{Dot}(\mathbb{H})
\end{array}
\qquad \mathbb{S}et\,.
$$

Theorem 2.34. *There exists a category $\mathbb{G}ph$ of directed graphs and their maps.*

Axiom 2.4.16. | **Directed Graph** |

For every directed graph arrow a with tail x and head y there exists a diagram

$$x \xrightarrow{a} y \, .$$

For every set X there exists a directed graph $\text{Gph}(X)$ such that $\text{Dot}(\text{Gph}(X)) := X$ and $1 \xrightarrow{u} \text{Arr}(\text{Gph}(X))$ if and only if $u = 1_x$ for some $1 \xrightarrow{x} X$, and $\text{tail}(1_x) = \text{head}(1_x) = x$ for all $1 \xrightarrow{x} X$.

Theorem 2.35. *There exists an inclusion functor from the category* $\mathbb{S}\text{et}$ *to the category* $\mathbb{G}\text{ph}$ *of directed graphs.*

Theorem 2.36. *There exists a forgetful functor from the category* $\mathbb{G}\text{ph}$ *of directed graphs to the category* $\mathbb{S}\text{et}$ *of sets, namely by forgetting about the set of arrows and retaining only the set of dots.*

Theorem 2.37. *There exists a forgetful functor from the category* $\mathbb{C}\text{at}$ *of categories to the category* $\mathbb{G}\text{ph}$ *of directed graphs, namely by forgetting about composition and retaining only the set of objects and the set of morphisms.*

Aside 2.4.2. The professionals' eyebrows shoot up because in the traditional foundations of mathematics there is a sharp distinction between "sets" and "classes" adopted to resolve Russell's Paradox about the set of sets that are not members of themselves. The root of this difficulty is the idea that a set may be defined in terms of a "property" specified by a logical formula. This book intends to sidestep such issues by adopting the Ground distinction between Discourse and Surface, and does not deploy properties specified by formulas. Then again, there is no injunction against developing the formal apparatus of mathematical logic and elaborating mathematical theories, provided this is done on a Surface according to the Ground Rules of Discourse.

2.4.6 *Dynamic System*

Another mathematical structure slightly more complicated than a set is a set together with a map to itself. Lightly paraphrasing ([Lawvere and Schanuel (1997)] p. 137), "the idea is that A is the set of possible *states,* either of a natural system or of a machine, and that the given map d represents the evolution of states, either the natural evolution in one unit of time of the system left to itself, or the change of internal state that will occur as a result of pressing a button (or other control) on the outside of the machine once."

Definition 2.38. Let A be a set. A **dynamic system** is a diagram of the form

$$A \xrightarrow{\quad d \quad} A \qquad (2.16)$$

where A is called the **underlying set** of the dynamic system and the label d is its **dynamic**.

If

$$B \xrightarrow{\quad e \quad} B \qquad (2.17)$$

is also a dynamic system then a **dynamic system map** is a diagram of the form

$$
\begin{array}{ccc}
A & \xrightarrow{\ d\ } & A \\
\downarrow{\scriptstyle f} & & \downarrow{\scriptstyle f} \\
B & \xrightarrow{\ e\ } & B
\end{array}
\qquad \text{Set} \qquad (2.18)
$$

Composition of dynamics system maps is induced by composition of set maps in the sense that for any chain diagram

$$
\begin{array}{ccc}
A & \xrightarrow{\ d\ } & A \\
\downarrow{\scriptstyle f} & & \downarrow{\scriptstyle f} \\
B & \xrightarrow{\ e\ } & B \\
\downarrow{\scriptstyle g} & & \downarrow{\scriptstyle g} \\
C & \xrightarrow{\ h\ } & C
\end{array}
\qquad \text{Set} \qquad (2.19)
$$

there exists a diagram

$$
\begin{array}{ccc}
A & \xrightarrow{\ d\ } & A \\
\downarrow{\scriptstyle gf} & & \downarrow{\scriptstyle gf} \\
C & \xrightarrow{\ h\ } & C
\end{array}
\qquad \text{Set} . \qquad (2.20)
$$

Moreover, for every dynamic system $A \xrightarrow{d} A$ there exists an identity diagram

$$
\begin{array}{ccc}
A & \xrightarrow{d} & A \\
\downarrow{\scriptstyle 1_A} & & \downarrow{\scriptstyle 1_A} \\
A & \xrightarrow{d} & A
\end{array}
\qquad \Big| \; Set \; .
\tag{2.21}
$$

Theorem 2.39. *Composition and identity diagrams of dynamic system maps satisfy the Associative and Identity Laws. Therefore the dynamic systems and their maps form a category.*

Definition 2.40. The category of dynamic systems and their maps is denoted by $\mathbb{D}yn$. Hence, $1 \xrightarrow{\mathbb{D}yn} \mathbb{C}at$.

Theorem 2.41. *There exists a forgetful functor from the category of dynamical systems to the category of sets, namely by forgetting the dynamic.*

2.4.7 Initialized Dynamic System

Blending the idea of a pointed set with the idea of a dynamic system yields the the idea of a dynamic system with a distinguished element that is considered to be the initial element in a (possibly infinite) sequence of elements obtained by iterating the dynamic.

Definition 2.42. Let A be a set. An **initialized dynamic system** is a diagram of the form

$$
1 \xrightarrow{\bullet_A} A \xrightarrow{d} A
\tag{2.22}
$$

where $A \xrightarrow{d} A$ is a dynamic system and the selected element $\bullet_A$ is the **initial condition** of the dynamic.

If

$$
1 \xrightarrow{\bullet_B} B \xrightarrow{e} B
\tag{2.23}
$$

is also an initialized dynamic system then an **initialized dynamic system map** is a diagram of the form

$$
\begin{array}{ccccc}
1 & \xrightarrow{\bullet_A} & A & \xrightarrow{d} & A \\
\Big\| & & \downarrow{\scriptstyle f} & & \downarrow{\scriptstyle f} \\
1 & \xrightarrow{\bullet_B} & B & \xrightarrow{e} & B
\end{array}
\qquad \Big| \; Set
\tag{2.24}
$$

Composition of initialized dynamics system maps is induced by composition of set maps in the sense that for any chain diagram

$$
\begin{array}{ccccc}
1 & \xrightarrow{\bullet_A} & A & \xrightarrow{d} & A \\
\| & & \downarrow{f} & & \downarrow{f} \\
1 & \xrightarrow{\bullet_B} & B & \xrightarrow{e} & B \\
\| & & \downarrow{g} & & \downarrow{g} \\
1 & \xrightarrow{\bullet_C} & C & \xrightarrow{h} & C
\end{array}
\qquad \mathbb{Set}
\qquad (2.25)
$$

there exists a diagram

$$
\begin{array}{ccccc}
1 & \xrightarrow{\bullet_A} & A & \xrightarrow{d} & A \\
\| & & \downarrow{gf} & & \downarrow{gf} \\
1 & \xrightarrow{\bullet_C} & C & \xrightarrow{h} & C
\end{array}
\qquad \mathbb{Set} .
\qquad (2.26)
$$

Moreover, for every dynamic system $A \xrightarrow{d} A$ there exists an identity diagram

$$
\begin{array}{ccccc}
1 & \xrightarrow{\bullet_A} & A & \xrightarrow{d} & A \\
\| & & \downarrow{1_A} & & \downarrow{1_A} \\
1 & \xrightarrow{\bullet_A} & A & \xrightarrow{d} & A
\end{array}
\qquad \mathbb{Set} .
\qquad (2.27)
$$

Theorem 2.43. *Composition and identity diagrams of dynamic system maps satisfy the Associative and Identity Laws. Therefore the dynamic systems and their maps form a category.*

Definition 2.44. The category of initialized dynamic systems and their maps is denoted by $i\mathbb{Dyn}$. Hence, $1 \xrightarrow{i\mathbb{Dyn}} \mathbb{Cat}$.

Theorem 2.45. *There exists a forgetful functor from the category of initialized dynamical systems to the category of dynamical systems, namely by forgetting the initial condition.*

Intuitively, the sequence generated by the initial condition $1 \xrightarrow{\bullet_A} A$ in Eq. (2.22) is

$$\bullet_A$$
$$d \circ \bullet_A$$
$$d \circ d \circ \bullet_A$$
$$d \circ d \circ d \circ \bullet_A$$
$$\cdots\cdots\cdots\cdots$$

Axiom 2.4.17. $\boxed{\textbf{Natural Number}}$

There exists an initialized dynamic system $1 \xrightarrow{0} \mathbb{N} \xrightarrow{s} \mathbb{N}$ in $i\mathbb{D}yn$ such that for any initialized dynamic system Eq. (2.22) there exists a unique initialized dynamic system map

$$
\begin{array}{ccccc}
1 & \xrightarrow{\;0\;} & \mathbb{N} & \xrightarrow{\;s\;} & \mathbb{N} \\
\| & & \downarrow{\scriptstyle s_A} & & \downarrow{\scriptstyle s_A} \\
1 & \xrightarrow{\;\bullet_A\;} & A & \xrightarrow{\;d\;} & A
\end{array}
\qquad \Big| \quad Set\,.
\tag{2.28}
$$

An element of the set $\mathbb{N}$ is called a **natural number** and the initial natural number 0 is called, of course, **zero**. The dynamic $\mathbb{N} \xrightarrow{s} \mathbb{N}$ is called **successor**, and the map $\mathbb{N} \xrightarrow{s_A} A$ is the **trajectory** of the initial condition $\bullet_A$ due to the dynamic d.

Rigorously, the sequence generated by the initial condition $1 \xrightarrow{\bullet_A} A$ in Eq. (2.22) is

$$s_A(0) = \bullet_A$$
$$s_A(s(n)) = d(s_A(n))\,.$$

Definition 2.46. If x is an object of a category $\mathbb{C}$ such that for every object y of $\mathbb{C}$ there exists a unique morphism $x \to y$ in $\mathbb{C}$, then x is called an **initial object** of $\mathbb{C}$.

Accordingly, the natural numbers are an initial object in the category of initialized dynamic systems.

2.4.8 *Magma*

After the ideas of adding structure to a set by selecting an item, as in a pointed set, or of adding structure by selecting a self map, another idea is to add structure that combines two items of a set to yield a set. This is the basic idea, for example, of adding any two numbers, or multiplying them, or of adding any two vectors with tail ends at the same point, and so on.

Definition 2.47. Let A be a set. A **magma** is a diagram of the form

$$A \times A \xrightarrow{\;*\;} A \qquad (2.29)$$

where A is called the **underlying set** and $*$ is called the **binary operator** of the magma.

If

$$B \times B \xrightarrow{\;\circ\;} B \qquad (2.30)$$

is a magma then a **magma map** is a diagram of the form

$$
\begin{array}{ccc}
A \times A & \xrightarrow{\;*\;} & A \\
{\scriptstyle f \times f}\big\downarrow & & \big\downarrow{\scriptstyle f} \\
B \times B & \xrightarrow{\;\circ\;} & B
\end{array}
\quad \text{Set} .
\qquad (2.31)
$$

Composition of magma maps is induced by composition of set maps in the sense that for any chain diagram

$$
\begin{array}{ccc}
A \times A & \xrightarrow{\;*\;} & A \\
{\scriptstyle f \times f}\big\downarrow & & \big\downarrow{\scriptstyle f} \\
B \times B & \xrightarrow{\;\circ\;} & B \\
{\scriptstyle g \times g}\big\downarrow & & \big\downarrow{\scriptstyle g} \\
C \times C & \xrightarrow{\;\odot\;} & C
\end{array}
\quad \text{Set}
\qquad (2.32)
$$

there exists a diagram

$$
\begin{array}{ccc}
A \times A & \xrightarrow{\;*\;} & A \\
{\scriptstyle gf \times gf}\big\downarrow & & \big\downarrow{\scriptstyle gf} \\
C \times C & \xrightarrow{\;\odot\;} & C
\end{array}
\quad \text{Set} .
\qquad (2.33)
$$

Moreover, for every magma $A \times A \overset{*}{\to} A$ there exists an identity diagram

$$
\begin{array}{ccc}
A \times A & \overset{*}{\longrightarrow} & A \\
{\scriptstyle 1_A \times 1_A} \big\downarrow & & \big\downarrow {\scriptstyle 1_A} \qquad \big| \;\; \textit{Set}\,. \\
A \times A & \overset{*}{\longrightarrow} & A
\end{array}
\tag{2.34}
$$

Theorem 2.48. *Composition and identity diagrams of magma maps satisfy the Associative and Identity Laws. Therefore the magmas and their maps form a category.*

Definition 2.49. The category of magmas and their maps is denoted by $\mathbb{M}gm$. Hence, $1 \xrightarrow{\;\mathbb{M}gm\;} \mathbb{C}at$.

Theorem 2.50. *There exists a forgetful functor from the category of magmas to the category of sets, namely by forgetting the binary operator to leave only the underlying set.*

2.4.9 *Semigroup*

A semigroup is a magma whose binary operator satisfies the Associative Law. In other words there is no structural difference between a magma and a semigroup but a conditional difference: if three items x, y and z are combined in a semigroup as in $(x * y) * z$ or as in $x * (y * z)$ then the result is the same:

$$
(x * y) * z = x * (y * z)\,.
$$

Definition 2.51. A **semigroup** is a magma $A \times \overset{*}{\to} A$ such that there exists a diagram

$$
\begin{array}{ccc}
(A \times A) \times A & \rightleftarrows & A \times (A \times A) \\
{\scriptstyle * \times 1_A} \big\downarrow & & \big\downarrow {\scriptstyle 1_A \times *} \qquad \big\downarrow \textit{Set} \\
A \times A \overset{*}{\longrightarrow} A & \overset{*}{\longleftarrow} & A \times A
\end{array}
\tag{2.35}
$$

where the top row represents the associativity of list formation.

A semigroup map between semigroups is just a magma map, and so the composition and identity diagrams for semigroups are the same as the diagrams for magma maps.

Theorem 2.52. *Composition and identity diagrams of semigroup maps automatically satisfy the Associative and Identity Laws. Therefore the semigroups and their maps form a category.*

Definition 2.53. The category of semigroups and their maps is denoted by $\mathbb{S}gr$. Hence, $1 \xrightarrow{\mathbb{S}gr} \mathbb{C}at$.

Theorem 2.54. *There exists an inclusion functor from the category of semigroups to the category of magmas.*

2.4.10 *Monoid*

The notion of a monoid (a semigroup with identity) plays a central role in category theory [Mac Lane (1971)].

A new level of complication arises by blending the idea of semigroup with the idea of pointed set. A monoid is both a semigroup and a pointed set with the same underlying set and in which the selected "point" satisfies the Identity Law with respect to the binary operator of the semigroup.

Definition 2.55. A **monoid** is a diagram $M := A \times A \xrightarrow{*} A \xleftarrow{e} 1$ such that $A \times A \xrightarrow{*} A$ is a semigroup, $1 \xrightarrow{e} A$ is a pointed set, and there exists a diagram

$$A \xrightarrow{(1_A\ e)} A \times A \xleftarrow{(e\ 1_A)} A \qquad \qquad (2.36)$$

with 1_A, $*$, 1_A to A, $\quad Set\ .$

The semigroup $A \times A \xrightarrow{*} A$ is called the **underlying semigroup** of the monoid, and the item e selected by the pointed set $1 \xrightarrow{e} A$ is called the **identity** of the monoid. If $N := B \times B \xrightarrow{\circ} B \xleftarrow{f} 1$ is a monoid then a diagram

$$\begin{array}{ccc} A \times A & \xrightarrow{\ *\ } & A \\ {\scriptstyle g\times g}\downarrow & & \downarrow{\scriptstyle g} \\ B \times B & \xrightarrow{\ \circ\ } & B \end{array} \quad \xleftarrow{e} 1 \qquad \xleftarrow{f} \qquad Set \qquad (2.37)$$

is a **monoid map** if the left square is a semigroup map and the right triangle is a pointed set map.

Theorem 2.56. *The definitions of composition and identity diagrams for monoids follow the same pattern exhibited by pointed sets, magmas, and semigroups. Likewise for the theorem that states that these diagrams satisfy the Associative and Identity Laws. Therefore the monoids and their maps form a category.*

Definition 2.57. The category of monoids and their maps is denoted by $\mathbb{M}on$. Hence, $1 \xrightarrow{\text{Mon}} \mathbb{C}at$.

Theorem 2.58. *There exists a forgetful functor from the category of monoids to the category of semigroups, namely by forgetting the identity element of the monoid and retaining only the underlying semigroup.*

Theorem 2.59. *If there are two monoids with the same underlying set A as in*

$$M_1 = A \times A \xrightarrow{*} A \xleftarrow{e} 1$$
$$M_2 = A \times A \xrightarrow{\square} A \xleftarrow{f} 1$$

and the binary operations $*, \square$ are related by the Exchange Law

$$(a * b)\square(c * d) = (a\square c) * (b\square d)$$

then $* = \square$, $e = f$, and the one monoid $M_1 = M_2$ is a commutative monoid.

Proof. First, $(e * f)\square(f * e) = (e\square f) * (f\square e)$, hence $f = f\square f = e * e = e$, so the two monoid identity elements coincide. Calculuate

$$
\begin{aligned}
x\square y &= (e * x)\square(y * e) \\
&= (e\square y) * (x\square e) \\
&= (f\square y) * (x\square f) \\
&= y * x \\
&= (y\square f) * (f\square x) \\
&= (y * f)\square(f * x) \\
&= (y * e)\square(e * x) \\
&= y\square x \ .
\end{aligned}
$$

$\square$

Remark 2.60. This often-cited theorem is due to B. Eckmann and P. Hilton [Eckmann and Hilton (1962)]. It is included for its intrinsic interest, but also because it illustrates the power of the Exchange Law.

2.4.11 *Group*

A group seems to blend the idea of monoid with the idea of dynamical system, in the sense that a group has an associative binary operator and an identity item, plus a self-map of the underlying set. However, that self-map – the dynamic – simply "inverts" items, so that the inverse of the inverse of an item is the item itself. Much more important is the relationship between the dynamic and the binary operator.

Definition 2.61. A **group** is a diagram

$$A \times A \xrightarrow{*} A \xleftarrow{i} A \xleftarrow{e} 1$$

such that $A \times A \xrightarrow{*} A \xleftarrow{e} 1$ is a monoid, $A \xrightarrow{i} A$ is a dynamical system, and there exists a diagram

$$\begin{array}{ccc}
A \xrightarrow{(i\,1_A)} A \times A \xleftarrow{(1_A\,i)} A \\
\downarrow \qquad\quad \downarrow * \qquad\quad \downarrow \\
1 \xrightarrow{\quad e \quad} A \xleftarrow{\quad e \quad} 1
\end{array} \qquad\qquad \text{Set}\,. \tag{2.38}$$

The monoid $A \times A \xrightarrow{*} A \xleftarrow{e} 1$ is called the **underlying monoid** of the group. The map $A \xrightarrow{i} A$ is called the **inversion operator** of the group.

Theorem 2.62. *The definitions of group map – usually called a "group homomorphism" – composition, and identity diagrams for groups follow the same pattern exhibited by pointed sets, magmas, semigroups, and monoids. Likewise for the theorem that states that these diagrams satisfy the Associative and Identity Laws. Therefore the groups and their maps form a category.*

Definition 2.63. The category of groups and their maps is denoted by $\mathbb{Grp}$. Hence, $1 \xrightarrow{\text{Grp}} \mathbb{Cat}$.

Theorem 2.64. *There exists a forgetful functor from the category of groups to the category of monoids, namely by forgetting the inversion operator and retaining only the underlying monoid.*

2.4.12 *Commutative Group*

A commutative group is a group that satisfies the condition that the order in which two items are combined – by the binary operator of the underlying monoid – is irrelevant. Thus, the definition of commutative involves no added structure, but does require an added condition.

Definition 2.65. A **commutative group** is a group

$$A \times A \xrightarrow{*} A \xleftarrow{i} A \xleftarrow{e} 1$$

for which there exists a diagram

$$A \times A \underset{\tau}{\overset{\tau}{\rightleftarrows}} A \times A \qquad\qquad (2.39)$$

$$\diagdown_{*} \qquad \diagup_{*}$$

$$A$$

$$\bigg|\; Set\;.$$

where τ is the operator that reverses the order of two items in a list.

Remark 2.66. The condition Eq. (2.39) could be a requirement on any magma binary operator $*$, giving rise thus to the concepts of "commutative magma", "commutative semigroup", and "commutative monoid."

Definition 2.67. The category of commutative groups and their maps is denoted by $c\mathbb{G}rp$. Hence, $1 \xrightarrow{c\mathbb{G}rp} \mathbb{C}at$.

Theorem 2.68. *There exists an inclusion functor from the category of commutative groups to the category of groups.*

2.4.13   *Ring*

Higher levels of complexity are broached by combining multiple magmas upon the same underlying set. For example, fractions are added *and* multiplied by high school students.

Definition 2.69. A **ring** is a diagram

$$1 \xrightarrow{0} A \xleftarrow{+} A \times A \xrightarrow{*} A \xleftarrow{1} A$$

where the underlying structure $1 \xrightarrow{0} A \xleftarrow{+} A \times A$ is a commutative group and the underlying structure $A \times A \xrightarrow{*} A \xleftarrow{1} A$ is a monoid, such that there exists a diagram expressing the Distributive Law of $*$ over $+$, namely that for any items a, b, c and d of A there exist equations

$$a * (b + c) = a * b + a * c \qquad\qquad \text{Left Distributive Law}\,,$$
$$(a + b) * c = a * c + b * c \qquad\qquad \text{Right Distributive Law}\,.$$

The ring is a **commutative ring** if the underlying monoid is a commutative monoid.

Definition 2.70. The category of rings and their maps is denoted by $\mathbb{R}ng$, and that of commutative rings by $c\mathbb{R}ng$. Hence, $1 \xrightarrow{\mathbb{R}ng} \mathbb{C}at$ and $1 \xrightarrow{c\mathbb{R}ng} \mathbb{C}at$.

Theorem 2.71. *There exist two forgetful functors from the category of rings, namely to the category of commutative groups and to the category of monoids.*

Aside 2.4.3. Those with a highly refined taste for the abstract will certainly enjoy [Beck (1969)] on the topic of Distributive Law. In any case the theory of rings is a very broad area of mathematical research. For this book I introduced the concept of ring in anticipation of defining the concept of field.

2.4.14 *Field*

The real numbers, $\mathbb{R}$, are fundamental for calculation in this book. This is because they can be added, subtracted, multiplied, and divided (except for 0).

Definition 2.72. A **field** is a commutative ring

$$1 \xrightarrow{0} A \xleftarrow{\pm} A \times A \xrightarrow{*} A \xleftarrow{1} A$$

together with a dynamical system

$$A \backslash 0 \xrightarrow{i} A \backslash 0$$

such that

$$A \backslash 0 \times A \backslash 0 \xrightarrow{*\backslash 0} A \backslash 0 \xleftarrow{1} 1$$

is a commutative group. An item $1 \xrightarrow{x} A$ is called a **number**.

Definition 2.73. The category of fields and their maps is denoted by $\mathbb{F}ld$, hence $1 \xrightarrow{c\mathbb{R}ng} \mathbb{C}at$.

Theorem 2.74. *There exists an inclusion functor from the category of fields to the category of commutative rings. There are also two forgetful functors to the category of groups, namely by forgetting the multiplicative group, and by forgetting the additive group.*

It must be appreciated, first of all, that the set subtraction $A\backslash 0$ is the first occurrence of subtraction in the sequence of definitions starting with' that of magma. Second, $*\backslash 0$ is the restriction to $A\backslash 0 \times A\backslash 0$ of the $*$ binary operator of the ring. Third, that $0 \neq 1$ is implied by the assumption $1 \xrightarrow{1} A\backslash 0$. Fourth, although not used in this book a field map would be a a ring map that is also a dynamical systems map.

2.4.15   *Vector Space over a Field*

Definition 2.75. A **vector space** $\mathbb{V}$ **over a field** is a diagram

$$1 \xrightarrow{0} A \xleftarrow{+} A \times A \xrightarrow{*} A \xleftarrow{1} A \tag{2.40}$$

$$A\backslash 0 \xrightarrow{-} A\backslash 0 \tag{2.41}$$

$$V \times V \xrightarrow{+} V \xleftarrow{-} V \xleftarrow{\overrightarrow{0}} 1 \tag{2.42}$$

$$A \times V \xrightarrow{\cdot} V \tag{2.43}$$

such that Eqs. (2.40)–(2.41) define a field of **scalars** with underlying set A, Eq. (2.42) defines a commutative group of **vectors**vectors with identity vector $\overrightarrow{0}$ and inverse operator $-$, Eq. (2.43) defines **scalar multiplication of vectors**, and for scalars $r, s \in A$ and vectors $\overrightarrow{a}, \overrightarrow{b} \in V$ there exist equations

$$1 \cdot \overrightarrow{a} = \overrightarrow{a}$$
$$r \cdot (s \cdot \overrightarrow{a}) = (r * s) \cdot \overrightarrow{a}$$
$$r \cdot (\overrightarrow{a} + \overrightarrow{b}) = r \cdot \overrightarrow{a} + r \cdot \overrightarrow{b}$$
$$(r + s) \cdot \overrightarrow{a} = r \cdot \overrightarrow{a} + s \cdot \overrightarrow{a} .$$

Take care to note that $+$ in Eq. (2.40) for the field of scalars is distinct from the addition for the group of vectors in Eq. (2.42).

Example 2.76. The real numbers $\mathbb{R}$ are a vector space over themselves.

Theorem 2.77. *For a vector space $\mathbb{V}$ over $\mathbb{R}$ and $\overrightarrow{a} \in V, r \in \mathbb{R}$ the following equations exist:*

$$0 \cdot \overrightarrow{a} = \overrightarrow{0} \tag{2.44}$$

$$r \cdot \overrightarrow{0} = \overrightarrow{0} \tag{2.45}$$

$$r \cdot (-\overrightarrow{a}) = -(r \cdot \overrightarrow{a}) \tag{2.46}$$

$$(-1) \cdot \overrightarrow{a} = -\overrightarrow{a} \tag{2.47}$$

Proof. Calculate $0 \cdot \overrightarrow{a} + \overrightarrow{a} = 0 \cdot \overrightarrow{a} + 1 \cdot \overrightarrow{a} = (0+1) \cdot \overrightarrow{a} = 1 \cdot \overrightarrow{a} = \overrightarrow{a}$, so subtracting $\overrightarrow{a}$ from both sides yields Eq. (2.44).

Calculate $r \cdot \overrightarrow{0} + r \cdot \overrightarrow{0} = r \cdot (\overrightarrow{0} + \overrightarrow{0}) = r \cdot \overrightarrow{0} = r \cdot \overrightarrow{0} + \overrightarrow{0}$, so subtracting $r \cdot \overrightarrow{0}$ from both sides yields Eq. (2.45).

Calculate $r \cdot \overrightarrow{a} + r \cdot (-\overrightarrow{a}) = r(\overrightarrow{a} + (-\overrightarrow{a})) = r \cdot \overrightarrow{0} = \overrightarrow{0}$, so subtracting $r \cdot \overrightarrow{a}$ from both sides yields Eq. (2.46).

Calculate $0 = 0 \cdot \overrightarrow{a} = (1 + (-1)) \cdot \overrightarrow{a} = 1 \cdot \overrightarrow{a} + (-1) \cdot \overrightarrow{a} = \overrightarrow{a} + (-1) \cdot \overrightarrow{a}$, so subtracting $\overrightarrow{a}$ from both sides yields Eq. (2.47). $\qquad\square$

2.4.16 *Ordered Field*

Finally, an ordered field such as the numbers corresponding to the points of "the real line" is a field with the addition structure of a selected sub-set of positive numbers.

Definition 2.78. An **ordered field** is a field

$$1 \xrightarrow{0} A \xleftarrow{+} A \times A \xrightarrow{*} A \xleftarrow{1} 1$$

$$A \backslash 0 \xrightarrow{i} A \backslash 0$$

together with the added structure of inclusions $N \hookrightarrow A \backslash 0 \hookleftarrow P$ such that $A \backslash 0 = N \cup P$ and there exist diagrams

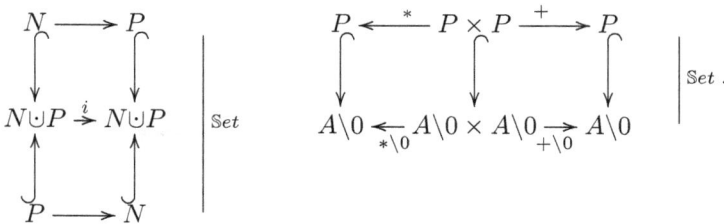

Definition 2.79. The category of ordered fields and their maps is denoted by $o\mathbb{F}ld$, hence $1 \xrightarrow{o\mathbb{F}ld} \mathbb{C}at$.

Theorem 2.80. *There exists an inclusion functor from the category of ordered fields to the category of fields. There exists a forgetful functor from the category of ordered fields to the category of totally ordered sets.*

2.4.17 *Topology*

Recall from Axiom 2.4.11 for the category of sets that for any set X there exists a set $\bigcup X$ such that

$$\text{if } 1 \xrightarrow{A} X \text{ then } A \hookrightarrow \bigcup X$$

and, if Z is a set such that

$$\text{if } 1 \xrightarrow{A} X \text{ then } A \hookrightarrow Z$$

then

$$\bigcup X \hookrightarrow Z \,.$$

Definition 2.81. A **topology** is a set T such that

(1) if A is in T then $\bigcup A$ is in T ;
(2) if $U_1, \ldots, U_n$ are in T then $U_1 \cap \cdots \cap U_n$ is in T .

The set $\bigcup T$ is in T since $T \hookrightarrow T$ and is called the **space** of T. It is also said that $\bigcup T$ is a **topological space with topology** T.

Definition 2.82. If Z is a topology then a map $\bigcup T \xrightarrow{f} \bigcup Z$ in $\mathbb{S}et$ is called a **continuous map from (the space of)** T **to (the space of)** Z if for every V in Z there exists an inverse image diagram

$$
\begin{array}{ccc}
U & \longrightarrow & V \\
\big\uparrow & & \big\uparrow \\
\big\downarrow & & \big\downarrow \\
\bigcup T & \xrightarrow[f]{} & \bigcup Z
\end{array}
\qquad \Big| \ \mathbb{S}et
$$

such that U is in T.

Axiom 2.4.18. $\boxed{\text{Topology}}$

There exists a diagram $\mathbb{T}op$ such that if T is a topology then $1 \xrightarrow{T} \mathbb{T}op$ and if $\bigcup T \xrightarrow{f} \bigcup Z$ is a continuous map then $T \xrightarrow{f} Z$ is in $\mathbb{T}op$.

Definition 2.83. The category of topological spaces and their maps is denoted by $\mathbb{T}op$, hence $1 \xrightarrow{\mathbb{T}op} \mathbb{C}at$.

Theorem 2.84. *Composition in* $\mathbb{T}op$ *is justified by the universal property of inverse image diagrams.*

2.5 Constructions

A functor is a diagram $\mathbb{C} \xrightarrow{F} \mathbb{C}$ in $\mathbb{C}at$. For example, since a semigroup is just a magma without additional structure and satisfies a condition (Associative Law), for every diagram $\mathbf{1} \xrightarrow{S} \mathbb{S}gr$ there exists a diagram $\mathbf{1} \xrightarrow{S} \mathbb{M}gm$, and a diagram $S \xrightarrow{f} T$ in $\mathbb{S}gr$ is automatically a diagram in $\mathbb{M}gm$. Hence there is an **inclusion functor**

$$\mathbb{S}gr \hookrightarrow \mathbb{M}gm .$$

This inclusion may be considered to "forget" the condition that the Associative Law holds in a semigroup, but without forgetting the binary operator structure.

On the other hand given a magma $M := A \times A \xrightarrow{*} A$ forgetting the binary operator $*$ but retaining the underlying set A, and likewise for a magma map

$$
\begin{array}{ccc}
A \times A & \xrightarrow{\;*\;} & A \\
{\scriptstyle f \times f}\downarrow & & \downarrow{\scriptstyle f} \\
B \times B & \xrightarrow{\;\circ\;} & B
\end{array}
\qquad \Big| \;\mathbb{S}et
\tag{2.48}
$$

forgetting everything but the underlying map $A \xrightarrow{f} B$ determines a **forgetful functor**

$$\mathbb{M}gm \xrightarrow{U} \mathbb{S}et .$$

Indeed, there is such a forgetful functor for every category introduced above, defined just by retaining underlying sets and maps in every case, except in the case of a topology. For a topology T the underlying set is taken to be the space $\bigcup T$, and for a continuous map $T \xrightarrow{f} Z$ the underlying map of sets is $\bigcup T \xrightarrow{f} \bigcup Z$.

Aside 2.5.1. To me an inclusion or forgetful functor is mildly interesting, but nowhere near as interesting as a *construction functor*. This is a functor that builds a new structure out of one or more given structures.

2.5.1 *Magma Constructed from a Set*

Let A be a set. By the List Axiom and the Union Axiom there exists a sequence of sets built up from A:

$$A_0 := A$$
$$A_1 := A_0 \cup A_0 \times A_0$$
$$A_2 := A_1 \cup A_1 \times A_1$$
$$\vdots$$
$$A_{n+1} := A_n \cup A_n \times A_n$$
$$\vdots$$

By the Union Axiom there exists a set which is the union of $A_0, A_1, A_2, \ldots$. (If A is the empty set then each set in the sequence is empty and the union is empty.)

Definition 2.85. Define

$$\mathrm{Mgm}(A) := \bigcup_{n \geq 0} A_n \ .$$

and call elements of A the **generators** of $\mathrm{Mgm}(A)$.

Theorem 2.86. *For any x and y in $\mathrm{Mgm}(A)$ the list $[x\ y]$ is also in $\mathrm{Mgm}(A)$.*

Proof. Without loss of generality, let $x \in A_m$ and $y \in A_n$ for some $m \leq n$. Then either $m = n$ and both x and y are in A_n so that $[x\ y] \in A_{n+1}$, or $A_m \subset A_{m+1} \subset \cdots \subset A_n$, and again both x and y are in A_n so that $[x\ y] \in A_{n+1} \subset \mathrm{Mgm}(A)$. $\qquad\square$

Definition 2.87. For any set A define the magma

$$\mathrm{Mgm}(A) \times \mathrm{Mgm}(A) \xrightarrow{\ *_A\ } \mathrm{Mgm}(A) \tag{2.49}$$

by $x *_A y := [x\ y]$.

Theorem 2.88. *There exists a functor $\mathbb{S}\mathrm{et} \xrightarrow{\ \mathrm{Mgm}\ } \mathbb{M}\mathrm{gm}$ defined on an object A to be Eq. (2.49), and on morphisms $A \xrightarrow{f} B$ in $\mathbb{S}\mathrm{et}$ by $\mathrm{Mgm}(f)(x) := f(x)$ for generators $x \in A$, and if $[x\ y] \in \mathrm{Mgm}(A)$ by*

$$\mathrm{Mgm}(f)([x\ y]) := [\mathrm{Mgm}(f)(x)\ \mathrm{Mgm}(f)(y)] \ .$$

Theorem 2.89.

(1) For every set A there exists a diagram

$$A \xhookrightarrow{\eta_A} \text{Set}(\text{Mgm}(A)) \; ;$$

(2) for every magma M and diagram

$$A \xrightarrow{f} \text{Set}(M)$$

there exists a unique diagram

2.5.2 Category Constructed from a Directed Graph

A directed graph $\mathbb{G}$ automatically generates a category $\text{Cat}(\mathbb{G})$. This is for "free" in the sense of "no cost" because the category is constructed from freely available materials, just the dots and arrows of the graph. The basic idea is that the dots of the directed graph become the objects of the category, and by the Identity Axiom each object acquires its identity morphism. The morphisms of the category are the *paths* of composable arrows in the directed graph, including possibly the new identity arrows, which become the identity morphisms of the category. The domain of a path is the tail of its first arrow and the codomain is the head of the last arrow. If the domain and codomain of the path happen to be the same object then the path is a *cycle*. "Composable" just means that the head of one arrow in the path matches the tail of the next arrow in the path – except of course, maybe, for the last arrow in the path. Thus, the Law of Composition for two paths in which the codomain of one matches the domain of the other is merely to join the two paths into one longer path. Therefore, any morphism of the category factors into the composition of all its successive arrows. In other words, the directed graph is buried in the category it generates. The only qualification in this story is that the composition of a morphism with an identity morphism must be the morphism itself. In other words, any path of arrows is equal to the path obtained by omitting any identity arrows that may occur in it. Here are formal details of this construction.

Let $\mathbb{G}$ be the directed graph $D \xleftarrow{\text{tail}} A \xrightarrow{\text{head}} D$. By the Directed Graph Axiom for the set D of dots there exists a directed graph $\text{Gph}(D)$ whose arrows $\text{Arr}(\text{Gph}(D))$ are the identity arrows of the dots in D. Define

$$\text{Arr}_0 := \text{Arr}(\text{Gph}(D))$$
$$\text{Arr}_1 := \text{Arr}_0 \cup A .$$

Thus, Arr_1 consists of the arrows of $\mathbb{G}$ supplemented by the identity arrows of its dots. The set A_2 of pairs of composable arrows in Arr_1 is constructed by appealing to the Equalizer Axiom with regard to parallel morphisms in $\mathbb{S}et$, and is given by

$$A_2 \xhookrightarrow{\quad\text{equalizer}\quad} \text{Arr}_1 \times \text{Arr}_1 \begin{array}{c} \xrightarrow{\pi_1} \text{Arr}_1 \searrow^{\text{head}} \\ \searrow_{\pi_2} \text{Arr}_1 \quad\nearrow_{\text{tail}} \end{array} D$$

$$\text{Arr}_2 := A_2 \cup \text{Arr}_1 .$$

This way Arr_2 includes the original arrows of $\mathbb{G}$, the identity arrows of its dots, and the composable pairs of all those arrows. Continuing, let

$$A_{n+1} \xhookrightarrow{\quad\text{equalizer}\quad} \text{Arr}_n \times \text{Arr}_n \begin{array}{c} \xrightarrow{\pi_1} \text{Arr}_1 \searrow^{\text{head}} \\ \searrow_{\pi_2} \text{Arr}_1 \quad\nearrow_{\text{tail}} \end{array} D$$

$$\text{Arr}_{n+1} := A_{n+1} \cup \text{Arr}_n .$$

By construction,

$$A \hookrightarrow \text{Arr}_1 \hookrightarrow \text{Arr}_2 \cdots \hookrightarrow \text{Arr}_n \hookrightarrow \cdots .$$

Define the set of morphisms of $\text{Cat}(\mathbb{G})$ by

$$\text{Arr}(\text{Cat}(\mathbb{G})) := \bigcup_{n \geq 0} \text{Arr}_n .$$

The Law of Composition in $\text{Cat}(\mathbb{G})$ may be defined formally but it is best to draw diagrams. Suppose

$$a = x_0 \xrightarrow{a_1} x_1 \xrightarrow{a_2} x_2 \to \cdots \xrightarrow{a_m} x_m$$
$$b = y_0 \xrightarrow{b_1} y_1 \xrightarrow{b_2} y_2 \to \cdots \xrightarrow{b_n} y_n$$

are two paths. Then a and b are composable if $x_m = y_0$, and their composition is

$$a = x_0 \xrightarrow{a_1} x_1 \xrightarrow{a_2} x_2 \to \cdots \xrightarrow{a_m} x_m = y_0 \xrightarrow{b_1} y_1 \xrightarrow{b_2} y_2 \to \cdots \xrightarrow{b_n} y_n$$

with the understanding that if any of the arrows a_i or b_j is an identity arrow then it may be omitted and its (equal) ends merged into a single object.

Theorem 2.90. *There exists a functor* $\mathbb{G}ph \xrightarrow{Cat} \mathbb{C}at$ *which assigns* $Cat(\mathbb{G})$ *to* $\mathbb{G}$ *and to a directed graph map* $\mathbb{G} \xrightarrow{f} \mathbb{H}$ *the functor* $Cat(\mathbb{G}) \xrightarrow{Cat(f)} Cat(\mathbb{H})$ *defined on a morphism*

$$a = x_0 \xrightarrow{a_1} x_1 \xrightarrow{a_2} x_2 \to \cdots \xrightarrow{a_m} x_m$$

by

$$Cat(f)(a) = f(x_0) \xrightarrow{f(a_1)} f(x_1) \xrightarrow{f(a_2)} f(x_2 \to \cdots \xrightarrow{f(a_m)} f(x_m) \,.$$

Besides getting $Cat(\mathbb{G})$ for free – basically just by tacking together arrows of $\mathbb{G}$ – there is another "free" aspect to this result. That is, apart from the trivial identifications that ensure composition with identity arrows satisfies the Identity Laws, the morphisms of $Cat(\mathbb{G})$ are free of any non-trivial equations. The only way two morphisms of $Cat(\mathbb{G})$ might be equal is if they differ at most by appearance of identity arrows along the paths.

Theorem 2.91.

(1) For every directed graph $\mathbb{G}$ there exists a diagram of directed graphs

$$\mathbb{G} \xhookrightarrow{\eta_{\mathbb{G}}} \mathrm{Gph}(\mathrm{Cat}(\mathbb{G})) \; ;$$

(2) for every category $\mathbb{C}$ and diagram of directed graphs

$$\mathbb{G} \xrightarrow{f} \mathrm{Gph}(\mathbb{C})$$

there exists a unique diagram

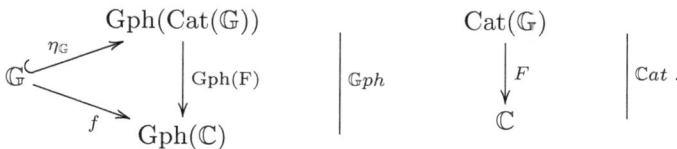

This theorem says there is a bijective correspondence between $\mathbb{G}$-shaped diagrams in $\mathbb{C}$ and functors from the category freely generated by $\mathbb{G}$ to $\mathbb{C}$.

Remark 2.92.

> *Adjoint functors* **are, at least from a mathematical perspective, the greatest achievement of category theory thus far: it essentially unifies all known mathematical constructs of a variety of areas of mathematics such as algebra, geometry, topology, analysis and combinatorics within a single mathematical concept. The restriction of adjoint functors to posetal categories ... play an important role in computer science when reasoning about computational processes** [Coecke and Éric Oliver Paquette (2009)].

It is no accident that there is a common thread connecting the key diagram in the Evaluation Axiom for sets and the key diagrams in these theorems about generating objects such as magmas and categories from simpler objects such as sets and directed graphs. These three exemplify the deepest concept in category theory called "adjointness" [Lawvere (1969)]. This book stops short of exploring "adjoint functor" but any student or teacher of mathematical science will certainly encounter this circle of ideas in applications of category theory to mathematics, physics, and computer science [Mac Lane (1971)].

2.5.3 *Category Constructed from a Topological Space*

Definition 2.93. Let X be a topological space, x and y points of X, and $0 < r \in \mathbb{R}$. A **path of duration** r **in** X **from** x **to** y is a continuous map $[0, r] \xrightarrow{\alpha} X$ such that $\alpha(0) = x$ and $\alpha(r) = y$.

Theorem 2.94. *There exists a functor* $\mathbb{Top} \xrightarrow{\text{Pth}} \mathbb{Cat}$ *such that* $\text{Obj}(\text{Cat}(X)) = X$ *and* $\text{Mor}(\text{Cat}(X)(x,y))$ *is the set of paths of some duration from* x *to* y. *The composition of a path* α *from* x *to* y *with duration* r *and* β *from* y *to* z *with duration* s *is the path* $\alpha \overrightarrow{\partial} \beta$ *with duration* $r + s$ *defined to have value* $\alpha(t)$ *for argument* t *such that* $0 \leq t \leq r$, *and* $\beta(t - r)$ *for* $r \leq t \leq r + s$. *The identity morphism at* x *is the constant path of duration 0 with value* x *[Brown (2006)].*

The forgetful functor from $\mathbb{Cat}$ to $\mathbb{Gph}$ supplies for each topological space X the directed graph $\text{Gph}(\text{Cat}(X))$ so that corresponding to a path $[0, r] \xrightarrow{\alpha} X$ from x to y there exists an arrow $x \xrightarrow{\alpha} y$.

Chapter 3

Calculus as an Algebra of Infinitesimals

Aside 3.0.1. Miss Keit was my sixth grade teacher in PS.139, Brooklyn, New York. She had invited a woman to talk to the class about mathematics (this was in the early 1950s). The lady asked if someone could spell the word, "infinitesimal." I did it, how could you go wrong? Years later, while in high school, a customer in my mother's ceramic art supply studio in Greenwich Village, New York offered to tutor me in calculus. (Whenever she learned a customer was somehow involved in engineering or science she would introduce me to them.) The young gentleman very, very patiently wrote out a calculus notebook for me, all the while striving to help me understand Δx and Δy. Sorry to say, I did not get it. But I did not forget that I did not get it, and late in my high school career during a summer vacation I encountered "A Course in Mathematical Analysis, Volume I" by Édouard Goursat. I tried to discover algebraic axioms for a theory of infinitesimals based on his definition on page 19, "Any *variable* quantity which approaches zero as a limit is called an *infinitely small quantity*, or simply an *infinitesimal*. The condition that the quantity be variable is essential, for a constant, however small, is not an infinitesimal unless it is zero." I failed then, but I never stopped trying.

In college I enjoyed mastering *epsilonics*. But that was not nearly as satisfying as the revelation provided by Abraham Robinson, who gave us a (hyper)real theory of infinitesimals.

3.1   Real & Hyperreal

> The hyperreal numbers can be algebraically manipulated just like the real numbers.
>
> ———————————————
>
> [Keisler (2002)]

Axiom 3.1.1. | **Real Numbers** |

There exists a set $\mathbb{R}$ with the structure of an ordered field.

Axiom 3.1.2. | **Hyperreal Numbers** |

There exists a set $\mathbb{H}$ with the structure of an ordered field and an inclusion $\mathbb{R} \hookrightarrow \mathbb{H}$ that preserves ordered field structure.

Axiom 3.1.3. | **Infinitesimal Numbers** |

There exists an inclusion $\mathbb{I} \hookrightarrow \mathbb{H}$ and there exists ε in $\mathbb{I}$ such that if

$$\varepsilon \text{ in } \mathbb{I} \text{ and } \varepsilon > 0$$
$$r \text{ in } \mathbb{R} \text{ and } r > 0$$

then

$$\varepsilon < r\,.$$

Axiom 3.1.4. | **Transfer** |

For every real map $\mathbb{R}^n \xrightarrow{f} \mathbb{R}$ there exists a diagram

$$
\begin{array}{ccc}
\mathbb{R}^n & \xrightarrow{\ f\ } & \mathbb{R} \\
\big\downarrow & & \big\downarrow \\
\mathbb{H}^n & \xrightarrow[f^{\mathbb{H}}]{} & \mathbb{H}
\end{array}
\qquad \Big| \ \mathbb{O}fld
$$

such that for every commutative diagram involving real maps f there exists a commutative diagram obtained by substituting $f^{\mathbb{H}}$ for its corresponding f.

Theorem 3.1. *There exists H in $\mathbb{H}$ such that $r < H$ for any $r > 0$ in $\mathbb{R}$.*

Proof. By definition of ordered field for any x, y, a in $\mathbb{R}$ if $0 < x < y$ and $a > 0$ then $0 < \frac{x}{a} < \frac{y}{a}$. In particular, if $a = xy$ then $0 < \frac{1}{y} < \frac{1}{x}$. By the Transfer Axiom for any X, Y in $\mathbb{H}$ if $0 < X < Y$ then $0 < \frac{1}{Y} < \frac{1}{X}$. In particular, the conclusion follows by defining $H := \frac{1}{\varepsilon}$ after selecting $X = \varepsilon > 0$ in $\mathbb{I}$ (which exists by the Infinitesimal Axiom) and $Y = \frac{1}{r}$ for any $r > 0$ in $\mathbb{R}$. $\qquad\square$

Corollary 3.2. *There exists a set $\infty \hookrightarrow \mathbb{H}$ such that $\frac{1}{\varepsilon}$ in ∞ if and only if ε in $\mathbb{I}$.*

Axiom 3.1.5. $\boxed{\textbf{Standard Part}}$

There is a map $\mathbb{H}\backslash\infty \xrightarrow{\text{st}} \mathbb{R}$ such that $\text{st}(x) \approx x$ for any x in $\mathbb{H}$.

Theorem 3.3. *If $x \in \mathbb{R}$ then $\text{st}(x) = x$.*

Proof. $\text{st}(x) \approx x$ for $x \in \mathbb{R}$ implies $\text{st}(x) - x \in \mathbb{I}$ so $\text{st}(x) - x < r$ for any $0 < r \in \mathbb{R}$. Hence, $\text{st}(x) - x = 0$. $\qquad\square$

Definition 3.4. A hyperreal number is called **finite** if it is strictly between two real numbers; **positive infinite** if it is greater than every real number; and **negative infinite** if less than every real number.

Theorem 3.5. *Let ε, δ in $\mathbb{I}$, let b, c be finite but not infinitesimal hyperreal numbers, and let H, K be infinite hyperreal numbers.*

(1) The only infinitesimal real number is 0 and every real number is finite;

(2) $-\varepsilon$ is infinitesimal, $-b$ is finite but not infinitesimal, and $-H$ is infinite;

(3) if $\varepsilon \neq 0$ then $\frac{1}{\varepsilon}$ is infinite, $\frac{1}{b}$ is finite but not infinitesimal, and $\frac{1}{H}$ is infinitesimal;

(4) $\varepsilon + \delta$ is infinitesimal, $b + \varepsilon$ is finite but not infinitesimal, $b + c$ is finite but possibly infinitesimal, $H + \varepsilon$ and $H + b$ are infinite;

(5) $\delta\varepsilon$ and $b\varepsilon$ are infinitesimal, bc is finite but not infinitesimal, and Hb, HK are infinite;

(6) $\varepsilon/b, \varepsilon/H$ and b/H are infinitesimal; b/c is finite but not infinitesimal; if $\varepsilon \neq 0$ then b/ε, H/ε and H/b are infinite.

Proof. By the Transfer Axiom. □

Theorem 3.6.

(1) *Every hyperreal number which is between two infinitesimals is infinitesimal.*
(2) *Every hyperreal number which is between two finite hyperreal numbers is finite.*
(3) *Every hyperreal number which is greater than some positive infinite number is positive infinite.*
(4) *Every hyperreal number which is less than some negative infinite number is negative infinite.*

Definition 3.7. Two hyperreal numbers b and c are **infinitely close** and there exists a diagram $b \approx c$, if $b - c$ is in $\mathbb{I}$.

Theorem 3.8.

(1) *If ε in $\mathbb{I}$ then $b \approx b + \varepsilon$.*
(2) *b is in $\mathbb{I}$ if and only if $b \approx 0$.*
(3) *If b and c in $\mathbb{R}$ and $b \approx c$ then $b = c$.*

Theorem 3.9. *Let a, b and c in $\mathbb{H}$. Then*

(1) *$a \approx a$;*
(2) *if $a \approx b$ then $b \approx a$;*
(3) *if $a \approx b$ and $b \approx c$ then $a \approx c$.*

Theorem 3.10. *Let b be a finite hyperreal number. Then*

(1) *$\text{st}(b)$ is in $\mathbb{R}$;*
(2) *$b = \text{st}(b) + \varepsilon$ for some ε om $\mathbb{I}$;*
(3) *if b is in $\mathbb{R}$ then $\text{st}(b) = b$.*

Theorem 3.11. *Let a and b be finite hyperreal numbers. Then*

(1) *$\text{st}(-a) = a\,\text{st}(a)$;*
(2) *$\text{st}(a + b) = \text{st}(a) + \text{st}(b)$;*
(3) *$\text{st}(ab) = \text{st}(a) - \text{st}(b)$;*
(4) *$\text{st}(ab) = \text{st}(a)\,\text{st}(b)$;*
(5) *if $\text{st}(b) \neq 0$ then $\text{st}(a/b) = \text{st}(a)/\text{st}(b)$.*

3.2 Variable

What is the difference between a mathematical variable and a physical variable? How does a computer program variable differ from a mathematical variable?

3.2.1 *Computer Program Variable*

Computers are complex physical objects which receive physical inputs – from keyboard, mouse, and so on – and produce physical outputs – visual displays, motion controls, and so on – according to plans – computer programs – stored in memory. A "computer memory" consists of memory locations. A "memory location" is a physical system that may persist in more than one alternative physical state. Operation of a computer entails many billions of state changes of memory locations per second.

A computer program does not refer directly to hardware memory locations. Instead it is composed in a language that refers to "variables" of a "virtual machine" that sits atop a tower of virtual machines at successive lower levels which ultimately refer to hardware memory locations.

3.2.2 *Mathematical Variable*

Mathematicians communicate their ideas using expressions. Even a single symbol is considered to be an expression, albeit a very simple one. Among the most fundamental operations upon an expression is to *substitute* a different symbol for one or more symbols in the expression. Expressions are most frequently written on paper or on a blackboard – or, these days – typed to a computer screen. In any given context, substitution for certain symbols may be dis-allowed by fiat. Within that context, these unchanging symbols are called "constants." Some symbols are even constant regardless of the context. For example, the numerical digits $0, 1, 2, 3, 4, 5, 6, 7, 8, 9$ are constant throughout all mathematical contexts. Symbols for which substitution *is* allowed are mathematical variables.

A blackboard on which expressions are composed and transformed according to mathematical rules may be considered analogous to a computer memory. That is, the physical state of a region of the blackboard changes when it is erased and something else is written at that region. In that sense there is some analogy between mathematical variables and computer program variables.

3.2.3   *Physical Variable*

Aside 3.2.1. Physical variables are the measurable changes in nature. They are represented by mathematical variables – with a difference. My first idea on the difference between a mathematical variable and a physical variable was that every physical variable is represented by a mathematical variable *together with a specific means of measurement.* I was thinking of things like distance, measured with a ruler, and time, measured with a clock.

Every physical measurement yields a positive integer multiple of a *unit of measurement.* Corresponding to the two basic types of motion in 3-dimensional space – translation and rotation – are the two basic units of measurement: distance units and angle units. Time, for example, is displayed in terms of angle units on the face of a clock. Even measurements that do not at first appear to be in terms of distance and angle, such as mass or charge or viscosity, are ultimately measured using instruments that change a distance or an angle in correspondence with changes in the variable measured. Bottom line: a physical variable is associated with an *instrument* that yields changes in distance or angle that can be counted based on a specified unit. This instrument and unit characterize the variable.

In any mathematical expression involving physical variables it must always be understood that the arithmetic operations on the variables carry with them the physical dimensions of the variables. High-end physicists have even employed dimensional analysis to generate hypotheses about Nature.[1]

This was my whole understanding of what makes a physical variable different from a mathematical variable, until I read the following by eminent mathematical physicist, Freeman Dyson.

> **We now take it for granted that electric and magnetic fields are abstractions not reducible to mechanical models. To see that this is true, we need only look at the units in which the electric and magnetic fields are supposed to be measured. The conventional unit of electric field-strength is the square-root of a joule per cubic meter. A joule is a unit of energy and a meter is a unit of length, but a square-root of a joule is not a unit of anything tangible. There is no way we can imagine measuring directly the**

---

[1]The so-called *Planck length* may be derived using a dimensional analysis argument.

square-root of a joule. The unit of electric field-strength is a mathematical abstraction, chosen so that the square of a field-strength is equal to an energy-density that can be measured with real instruments. The unit of energy density is a joule per cubic meter, and therefore we say that the unit of field-strength is the square-root of a joule per cubic meter. This does not mean that an electric field-strength can be measured with the square-root of a calorimeter. It means that an electric field-strength is an abstract quantity, incommensurable with any quantities that we can measure directly.[2]

The second connection between Maxwell theory and quantum mechanics is a deep similarity of structure. Like the Maxwell theory, quantum mechanics divides the universe into two layers. The first layer contains the wave-functions of Schrödinger, the matrices of Heisenberg and the state-vectors of Dirac. Quantities in the first layer obey simple linear equations. Their behaviour can be accurately calculated. But they cannot be directly observed. The second layer contains probabilities of particle collisions and transmutations, intensities and polarisations of radiation, expectation-values of particle energies and spins. Quantities in the second layer can be directly observed but cannot be directly calculated. They do not obey simple equations. They are either squares of first-layer quantities or products of one first-layer quantity by another. In quantum mechanics just as in Maxwell theory, Nature lives in the abstract mathematical world of the first layer, but we humans live in the concrete mechanical world of the second layer. We can describe Nature only in abstract mathematical language, because our verbal language is at home only in the second layer. Just as in the case of the Maxwell theory, the abstract quality of the first-layer quantities is revealed in the units in which they are expressed. For example, the Schrödinger wave-function is expressed in a unit which is the square root of

---

[2] "Why is Maxwell's Theory so hard to understand?" an essay by Professor Freeman J. Dyson, FRS, Professor Emeritus, Institute of Advanced Study, Princeton, USA

an inverse cubic meter. **This fact alone makes clear that the wave-function is an abstraction, forever hidden from our view. Nobody will ever measure directly the square root of a cubic meter.**[3]

So, I have to change my story: although every physical variable must have a dimensional unit, not every dimensional unit is a unit of measurement. Then again, even a physical variable without a unit of measurement must be related by some calculation – e.g., squaring – to a physical variable that does have a unit of measurement.

3.3   Right, Left & Two-Sided Limit

Let $X \hookrightarrow \mathbb{R}$, $Y \hookrightarrow \mathbb{R}$ and $X \xrightarrow{f} Y$.

Definition 3.12. For c in X and L in $\mathbb{R}$ say L is a **limit of $f(x)$ as x approaches c from the right** provided that $f(x) \approx L$ for every hyperreal x such that $c < x$ and $c \approx x$.

Theorem 3.13. *If L and L' are limits of $f(x)$ as x approaches c from the right then $L = L'$.*

Proof. $L \approx f(x) \approx L'$ implies that L, L' are infinitely close real numbers, hence must be equal.   □

In words, if there is a limit from the right, then there is exactly one. If there is a limit from the right then it is denoted by

$$\lim_{c \leftarrow x} f(x) \, .$$

The symbol x is bound in this expression context. Note also that the strict inequality $c < x$ implies that $x - c$ is never 0.

The notion of left limit is defined by substituting $>$ for $<$:

Definition 3.14. For c in X and L in $\mathbb{R}$ say L is a **limit of $f(x)$ as x approaches c from the left** provided that $f(x) \approx L$ for every hyperreal x such that $c > x$ and $c \approx x$.

The proof that a limit from the left is unique if it exists is by substitution of $>$ for $<$ in the above proof, and if the limit from the left exists it is

---

[3]ibid.

denoted by

$$\lim_{x \to c} f(x) .$$

Definition 3.15. If left and right limits of $f(x)$ at c both exist and are equal then say **the limit of** $f(x)$ **at** c **exists**. The common value of the left and right limits is denoted by

$$\lim_{x \,@\, c} f(x) := \lim_{c \leftarrow x} f(x) = \lim_{x \to c} f(x).$$

The symbol "@" is pronounced "at."

3.4 Continuity

Let $X \hookrightarrow \mathbb{R}$, $Y \hookrightarrow \mathbb{R}$ and $X \xrightarrow{f} Y$.

Definition 3.16. For c in X say $f(x)$ is *continuous at c from the right* if $f(c)$ is the limit of $f(x)$ as x approaches c from the right, that is, if

$$\lim_{c \leftarrow x} f(x) = f(c).$$

Likewise, say $f(x)$ is **continuous at** c **from the left** if

$$\lim_{x \to c} f(x) = f(c).$$

Say $f(x)$ is **continuous at** c if it continuous from both sides, that is, if

$$\lim_{x \,@\, c} f(x) = f(c).$$

Finally, say $X \xrightarrow{f} Y$ is **continuous** if $f(x)$ is continuous at c for all c in X.

3.5 Differentiable, Derivative & Differential

Let $X \hookrightarrow \mathbb{R}$, $Y \hookrightarrow \mathbb{R}$ and $X \xrightarrow{f} Y$.

Definition 3.17. For c in X say $f(x)$ **is differentiable at** c if the two-sided limit of $\frac{f(x)-f(c)}{x-c}$ as x approaches c exists. If it exists then

$$f'(c) := \lim_{x \,@\, c} \frac{f(x) - f(c)}{x - c}$$

is called the **derivative of** $f(x)$ **at** c. If $f'(c)$ exists for all c in X then $f(x)$ is called **differentiable** and the map $X \xrightarrow{f'} Y$ that assigns $f'(c)$ to c is called the **derivative** of $X \xrightarrow{f} Y$.

Theorem 3.18. *If $X \xrightarrow{f} Y$ is differentiable then it is continuous.*

Proof. Let $c \in X$. Hence, if $x \in \mathbb{H}$ and $x \approx c$, then $\frac{f(x)-f(c)}{x-c} \in \mathbb{H}$ and $\frac{f(x)-f(c)}{x-c} \approx f'(c)$, so $f(x) - f(c) \approx f'(c) \cdot (x - c) \approx 0$. Therefore, $f(x) \approx f(c)$. $\qquad\square$

Theorem 3.19. *Let $X \hookrightarrow \mathbb{R}$, $Y \hookrightarrow \mathbb{R}$ and $X \xrightarrow{f} Y$. If f is differentiable at c then*

$$f'(c) = \mathrm{st}\left(\frac{f(c+\epsilon) - f(c)}{\epsilon} \right)$$

for any choice of $\epsilon \in \mathbb{I}$.

Proof. Let $\varepsilon \in \mathbb{I}$ and $x := c + \varepsilon$. By definition of derivative,

$$f'(c) \approx \frac{f(x) - f(c)}{x - c} = \frac{f(c+\varepsilon) - f(c)}{\varepsilon} \ .$$

The result follows since $\mathrm{st}(f'(c)) = f(c)$. $\qquad\square$

$$
\begin{array}{cc}
[U] & [V] \\
X \xrightarrow{\ \ f\ \ } Y \\
x & y \\
& y = f(x)
\end{array}
\qquad
\begin{array}{cc}
[U] & [V] \\
X \xrightarrow{\ \ y\ \ } Y \\
x & y \\
& y = y(x)
\end{array}
$$

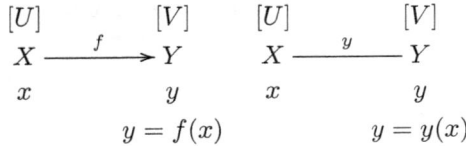

Fig. 3.1 U, V are measurement units associated with the sets of values X, Y for variables x, y. On the left, f is a map from X to Y, and $y = f(x)$ means that application of f to a given value of x yields a value for the variable y. On the right the same information is conveyed except that one letter less is needed, namely y stands ambiguously for both a map f and a variable.

Figure 3.1 exhibits a function f relating variable x with unit of measurement $[U]$ taking its values in set X to the variable y measured in units $[V]$ and values in Y. A standard abuse of notation is to write $y = y(x)$ without explicit mention of f, which is an abuse because one could forget that y *depends on x via f*. Thus,

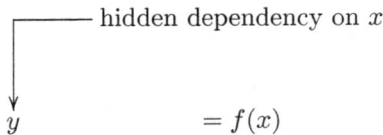

$$
\begin{array}{l}
\quad\rule[0.3em]{2em}{0.4pt}\ \text{hidden dependency on } x \\
\Big\downarrow \\
y \qquad\qquad\qquad = f(x)
\end{array}
$$

A deeper problem with $y = y(x)$ together with the definition of an equation – as a declaration of the right but not the obligation to substitute either side of the equation for the other in a given context –is that it may lead to endless silliness, such as $y = y(x) = y(x)(x) = y(x)(x)(x) = \cdots$. One may attempt to formulate rules for preventing this sort of thing, or one may simply consider the abuse to be harmless if tempered by common sense. That is my strategy.

The differential notation, which has been in use longer than any other, is due to Leibniz. Although it is by no means indispensable, it possesses certain advantages of symmetry and of generality which are convenient, especially in the study of maps of several variables. This notation is founded upon the use of infinitesimals....Any variable quantity which approaches zero as a limit is called an *infinitely small quantity*, or simply an *infinitesimal*. The condition that the quantity be variable is essential, for a constant, however small, is not an infinitesimal unless it is zero [Édouard Goursat (1959)].

Definition 3.20. For a variable x with unit of measurement $[U]$ taking its values in X the **differential** dx denotes a variable with the same unit of measurement as x and taking its values in $\mathbb{I}$.

Since $X \xrightarrow{f} Y$ differentiable implies f is continuous, $f(x) - f(c)$ is infinitesimal if $x - c$ is infinitesimal. This is the same as saying $f(x + dx) - f(x)$ is infinitesimal for any value of dx in $\mathbb{I}$. Define the differential of y by

$$dy = dy(x, dx) := f'(x)dx \ .$$

The beauty of this definition is that although both dx and dy are *variables* with infinitesimal values, the *ratio* of dy to dx is a real number. This only works because of the abuse of notation that hides the dependency of dy on both x and dx.

$$\boxed{} \text{ hidden dependency on } x \text{ and } dx \qquad (3.1)$$

$$dy \qquad\qquad = f'(x) \cdot dx$$

Further variations on notation are summarized in Eq. (3.2), and a convention in case the unit of measurement of the independent variable is time [**TME**] is reviewed in Fig. 3.3.

$$
\begin{array}{ccc}
[U] & & [V] \\
X & \xrightarrow{\ f\ } & Y \\
x & & y \\
& y = f(x) &
\end{array}
\qquad \mapsto \qquad
\begin{array}{ccc}
[U] & & [V]/[U] \\
X & \xrightarrow{\ f'\ } & \mathbb{R} \\
x & & \frac{dy}{dx} \\
& & \frac{dy}{dx} = f'(x)
\end{array}
$$

Fig. 3.2   Standard diagram of a differentiable function on the left leads to diagram with differential notation on the right.

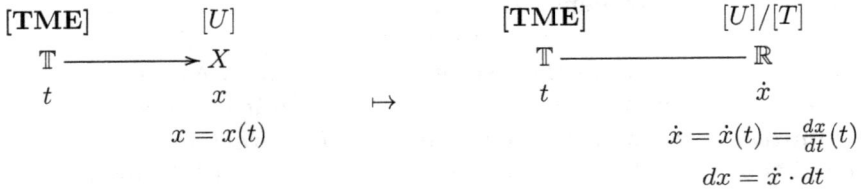

$$
\begin{array}{ccc}
[\mathbf{TME}] & & [U] \\
\mathbb{T} & \xrightarrow{\hspace{1cm}} & X \\
t & & x \\
& x = x(t) &
\end{array}
\qquad \mapsto \qquad
\begin{array}{ccc}
[\mathbf{TME}] & & [U]/[T] \\
\mathbb{T} & \xrightarrow{\hspace{1cm}} & \mathbb{R} \\
t & & \dot{x} \\
& & \dot{x} = \dot{x}(t) = \frac{dx}{dt}(t) \\
& & dx = \dot{x} \cdot dt
\end{array}
$$

Fig. 3.3   A differentiable function of time t on the left leads to the diagram on the right where the derivative with respect to time is indicated with a dot over the dependent variable x.

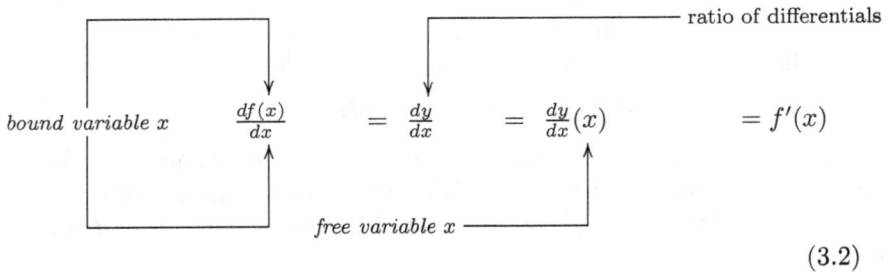

$$
\textit{bound variable } x \qquad \frac{df(x)}{dx} \; = \; \frac{dy}{dx} \; = \; \frac{dy}{dx}(x) \qquad = f'(x)
$$

ratio of differentials

free variable x

$$(3.2)$$

3.5.1   *Partial Derivative*

Recall 3.5.1. Given a list $[\,x_1 \cdots x_n\,]$ of length n and a position $i, 1 \le i \le n$ then a new list is easy to form by omitting the item at position i. This new list of length $n-1$ is denoted by $[\,x_1 \cdots \hat{x} \cdots x_n\,]$. Omission at position i is a map $X^n \to X^{n-1}$ and is "complementary" to projection $X^n \xrightarrow{\pi_i} X$ onto the item at position i.

Definition 3.21. If $y = y(x_1, \ldots, x_n) = f(x_1, \ldots, x_n)$, or in a diagram if

$$\mathbb{R}^n \longrightarrow \mathbb{R}$$
$$(x_1, \ldots, x_n) \qquad y = y(x_1, \ldots, x_n)$$

then for any choice of an item-and-omission $(x_1, \ldots, \hat{x}_i, \ldots, x_n)$ as above, a new map of one independent variable x_i is defined by

$$y_{x_1 \cdots \hat{x}_i \cdots x_n}(x_i) := y(x_1, \ldots, x_i, \ldots, x_n).$$

Assuming this map is differentiable for any choice of item-and-omission, the **partial derivative of** $y = y(x_1, \ldots, x_n)$ **with respect to** x_i is alternatively notated and defined by

$$\partial_i y = \partial_i y(x_1, \ldots, x_n) \tag{3.3}$$
$$= \frac{\partial y}{\partial x_i}$$
$$= \frac{\partial y}{\partial x_i}(x_1, \ldots, x_n)$$
$$= \frac{\partial y}{\partial x_i}\bigg|_{x_1 \cdots \hat{x}_i \cdots x_n}$$
$$:= \frac{dy_{x_1 \cdots \hat{x}_i \cdots x_n}}{dx_i}(x_i)$$
$$= f'_{x_1 \cdots \hat{x}_i \cdots x_n}(x_i) \ .$$

In thermodynamics, as in all branches of physics, a symbol denotes a *physical quantity*, irrespective of how it is related to other quantities. Thus, in the equation of state

$$\theta = \theta(P, v) \ , \tag{3.4}$$

it may become necessary to perform a change of variables, replacing, for example, the specific volume v by the refractive index n. We would still write

$$\theta = \theta(P, n) \tag{3.5}$$

retaining the symbol θ for temperature, regardless of the fact that the mathematical forms [above] cannot, in general, be identical. In mathematics, as explained, two different symbols would be used to emphasize this fact. In forming partial derivatives in thermodynamics we would write

$$\left(\frac{\partial \theta}{\partial P}\right)_v \text{ and } \left(\frac{\partial \theta}{\partial v}\right)_P \text{ for equation (3.4)}$$

and

$$\left(\frac{\partial \theta}{\partial P}\right)_n \text{ and } \left(\frac{\partial \theta}{\partial n}\right)_P \text{ for equation (3.5)},$$

placing the emphasis on the physical quantity, θ in this example, rather than on the functional relation which links it to the properties chosen as independent [Kestin (1979)].

3.6 Curve Sketching Reminder

The following definitions and theorems are adapted from ([Keisler (2002)] Chapter 3). Let $[a, b] \hookrightarrow \mathbb{R}$ be a closed interval and $[a, b] \xrightarrow{f} \mathbb{R}$.

Definition 3.22.

(1) f is **constant** on $[a, b]$ if $f(s) = f(t)$ for all $s, t \in [a, b]$;
(2) f is **increasing** on $[a, b]$ if $f(s) < f(t)$ for all $s < t \in [a, b]$;
(3) f is **decreasing** on $[a, b]$ if $f(s) > f(t)$ for all $s < t \in [a, b]$.

Theorem 3.23. *If f is continuous on $[a, b]$ and differentiable on (a, b) and $y = f(x)$ then*

(1) $\frac{dy}{dx}(x) = 0$ for all $x \in (a, b)$ if and only if f is constant on $[a, b]$;

(2) $\frac{dy}{dx}(x) > 0$ for all $x \in (a, b)$ if and only if f is increasing on $[a, b]$;

(3) $\frac{dy}{dx}(x) < 0$ for all $x \in (a, b)$ if and only if f is decreasing on $[a, b]$.

Definition 3.24. The graph of $y = f(x)$ is **concave upward** on $[a, b]$ if

$$f(t) < \frac{(t - s)f(u) + (u - t)(fs)}{u - s}$$

for all $s < t < u \in [a, b]$. In words this condition says that $f(t)$ is below the chord of the graph connecting $(s, f(s))$ to $(u, f(u))$.

The graph of $y = f(x)$ is **concave downward** on $[a, b]$ if

$$f(t) > \frac{(t-s)f(u) + (u-t)(fs)}{u-s}$$

for all $s < t < u \in [a, b]$. In words this condition says that $f(t)$ is above the chord of the graph connecting $(s, f(s))$ to $(u, f(u))$.

Theorem 3.25. *If f is continuous on $[a, b]$ and twice differentiable on (a, b) and $y = f(x)$ then*

(1) $\dfrac{d^2y}{dx^2} > 0$ *if and only if f is concave upward on $[a, b]$ if and only if* $\dfrac{dy}{dx}$ *is increasing on (a, b);*

(2) $\dfrac{d^2y}{dx^2} < 0$ *if and only if f is concave downward on $[a, b]$ if and only if* $\dfrac{dy}{dx}$ *is decreasing on (a, b).*

Definition 3.26. Let $t \in (a, b)$.

(1) f has a **local maximum** at t if there exists a diagram $t \in (a_0, b_0) \hookrightarrow (a, b)$ such that $f(t) \geq f(x)$ for $x \in (a_0, b_0)$;

(2) f has a **local minimum** at t if there exists a diagram $t \in (a_0, b_0) \hookrightarrow (a, b)$ such that $f(t) \leq f(x)$ for $x \in (a_0, b_0)$.

3.7   Integrability

Let $[a, b] \xrightarrow{f} Y$.

Definition 3.27. For $0 < \Delta x$ in $\mathbb{R}$ and $n = \lfloor \frac{b-a}{\Delta x} \rfloor$, the number

$$\sum_{j=0}^{j=n-1} f(a + j\Delta x)\Delta x + f(a + n\Delta x)(b - a - n\Delta x)$$

is denoted by $\overset{b}{\underset{a}{S}} f(x)\Delta x$. Say $f(x)$ is **integrable** if the limit of $\overset{b}{\underset{a}{S}} f(x)\Delta x$

exists as Δx approaches 0 from the right. If $\lim_{0 \leftarrow \Delta x} \overset{b}{\underset{a}{S}} f(x)\Delta x$ exists then

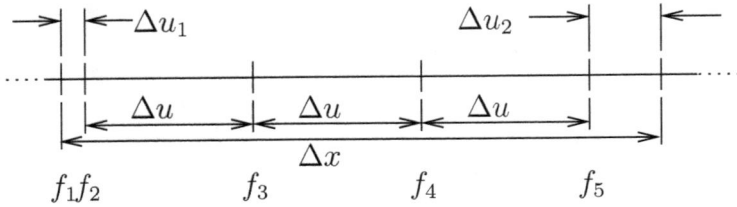

Fig. 3.4 Subdivisions.

it is called the **integral of** $f(x)$ **over** $[a, b]$ and this limit is denoted by $\int_a^b f(x)dx$.

Theorem 3.28. *If ε is a positive real number then there is a positive real number δ such that if $0 < \Delta x < \delta$ and $0 < \Delta u < \delta$, then*

$$\left| \underset{a}{\overset{b}{S}} f(x)\Delta x - \underset{a}{\overset{b}{S}} f(u)\Delta u \right| < \varepsilon. \tag{3.6}$$

Proof. Let $\varepsilon > 0$. Since $[a, b] \xrightarrow{f} Y$ is uniformly continuous, there exists $\delta > 0$ such that $|x - y| < \delta$ implies $|f(x) - f(y)| < \varepsilon_1 := \frac{\varepsilon}{b-a}$. Assume $0 < \Delta u < \Delta x < \delta$. Then the difference

$$\underset{a}{\overset{b}{S}} f(x)\Delta x - \underset{a}{\overset{b}{S}} f(u)\Delta u \tag{3.7}$$

may be re-written as a sum of terms with factor Δx each of which is a sum of terms with factor Δu – or smaller – terms.

$$f_1 \Delta x - f_1 \Delta u_1 - f_2 \Delta u - f_3 \Delta u - f_4 \Delta u - f_5 \Delta u_2$$
$$= f_1(\Delta u_1 + 3\Delta u + \Delta u_2) - f_1 \Delta u_1 - f_2 \Delta u - f_3 \Delta u - f_4 \Delta u - f_5 \Delta u_2$$
$$= (f_1 - f_1)\Delta u_1 + (f_1 - f_2)\Delta u + (f_1 - f_3)\Delta u + (f_1 - f_4)\Delta u + (f_1 - f_5)\Delta u_2.$$

Thus, by the Triangle Inequality the absolute value of (3.7) is not greater than a sum of terms including

$$|f_1 - f_1|\Delta u_1 + |f_1 - f_2|\Delta u + |f_1 - f_3|\Delta u + |f_1 - f_4|\Delta u + |f_1 - f_5|\Delta u_2.$$

However, each of these absolute values is bounded by ε_1. Therefore, the absolute value of (3.7) is not greater than a sum of terms

$$\varepsilon_1 \Delta u_1 + \varepsilon_1 \Delta u + \varepsilon_1 \Delta u + \varepsilon_1 \Delta u + \varepsilon_1 \Delta u_2 = \varepsilon_1 \Delta x.$$

Summing these $\varepsilon_1 \Delta x$ yields

$$\left| \sum_{a}^{b} f(x)\Delta x - \sum_{a}^{b} f(u)\Delta u \right| < \varepsilon_1 (b - a) = \varepsilon. \qquad (3.8)$$

$\square$

Theorem 3.29. *If $[a, b] \xrightarrow{f} Y$ is continuous then it is integrable.*

Proof. Since $[a, b] \xrightarrow{f} Y$ is bounded (image of a continuous map on a compact space is compact hence bounded) it follows that for any Δx the sum

$$\lim_{0 \leftarrow \Delta x} \sum_{a}^{b} f(x)\Delta x$$

– which depends only on Δx – is bounded, regardless of the choice of Δx. Thus, the hyperreal extension of $\displaystyle\lim_{0 \leftarrow \Delta x} \sum_{a}^{b} f(x)\Delta x$ is also bounded. Hence, its values for different infinitesimals dx are limited hyperreals, so their standard parts are finite real numbers. By Theorem 3.28 any two such values are equal, so that common value is

$$\lim_{0 \leftarrow \Delta x} \sum_{a}^{b} f(x)\Delta x \ .$$

$\square$

Aside 3.7.1. I was compelled to write up the above proof by a suspiciously brief proof in the otherwise superb book ([Henle and Kleinberg (1979)], p. 60) on infinitesimal calculus.

Project 3.7.1. Use the Transfer Axiom to translate the above "epsilonics" proof into a hyperreal proof [Goldblatt (1998)][Tao (2008)].

3.8   Algebraic Rules for Calculus

3.8.1   *Fundamental Rule*

Starting with a value and changing it by successively incrementing or decrementing it yields a result that equals the starting value plus the sum of the changes, whether they are negative or positive. For a high school student this banal calculation is an application of the Associative Law of Addition. The plausibility of the *Fundamental Theorem of Calculus* is rooted in this minor observation about change. Actually, there is a deeper symmetry about change, expressed in the following theorem.

Theorem 3.30. *(Fundamental Theorem of Calculus) If* $y = y(x)$ *and* a, b *in* $\mathbb{R}$ *then*

$$\int_a^b \frac{dy}{dx} \cdot dx = y(b) - y(a)\,, \tag{3.9}$$

$$\frac{d}{dx} \int_a^x y \cdot dx = y(x)\,. \tag{3.10}$$

3.8.2   *Constant Rule*

Theorem 3.31. *If* a *in* $\mathbb{R}$ *and* x *is a variable then*

$$\frac{da}{dx} = 0 \ \ or \ \ da = 0\,.$$

3.8.3   *Addition Rule*

Theorem 3.32. *If* $u = u(x)$ *and* $v = v(x)$ *then*

$$\frac{d}{dx}(u + v) = \frac{du}{dx} + \frac{dv}{dx} \ \ or \ \ d(u + v) = du + dv\,.$$

3.8.4   *Product Rule*

Theorem 3.33. *If* $u = u(x)$ *and* $v = v(x)$ *then*

$$\frac{d}{dx}(u \cdot v) = u \cdot \frac{dv}{dx} + \frac{du}{dx} \cdot v$$

$$or \ \ d(u \cdot v) = u \cdot dv + du \cdot v\,.$$

3.8.5 Scalar Product Rule

Theorem 3.34. *If* $\vec{u} = \vec{u}(x)$ *and* $\vec{v} = \vec{v}(x)$ *then*

$$\frac{d}{dx}\langle \vec{u} \mid \vec{v}\rangle = \langle u \mid \frac{d\vec{v}}{dx}\rangle + \langle \frac{d\vec{u}}{dx} \mid \vec{v}\rangle .$$

3.8.6 Chain Rule

Theorem 3.35. *Assume there exists a diagram*

$$\mathbb{T} \longrightarrow U \times V \longrightarrow Y$$

$$t \qquad u \quad v \qquad\qquad\qquad y = y(u,v)$$

$$(u,v) = (u,v)(t)$$

$$u = u(t)$$

$$v = v(t) \qquad\qquad y = y(u,v) = y(u,v)(t) = y(u(t),v(t)) .$$

Then

$$\frac{dy(u,v)}{dt}(t) \approx \frac{dy}{dv}(u(t),v(t))\dot{v}(t) + \frac{dy}{du}(u(t),v(t))\dot{u}(t) .$$

Proof.

$$\frac{dy(u,v)}{dt}(t) \approx \frac{y(u,v)(t+\varepsilon) - y(u,v)(t)}{\varepsilon}$$

$$= \frac{y(u(t+\varepsilon),v(t+\varepsilon)) - y(u(t),v(t))}{\varepsilon}$$

$$= \frac{y(u(t+\varepsilon),v(t+\varepsilon)) - y(u(t+\varepsilon),v(t)) + y(u(t+\varepsilon),v(t)) - y(u(t),v(t))}{\varepsilon}$$

$$= \frac{y(u(t+\varepsilon),v(t+\varepsilon)) - y(u(t+\varepsilon),v(t))}{\varepsilon} + \frac{y(u(t+\varepsilon),v(t)) - y(u(t),v(t))}{\varepsilon}$$

$$= \frac{y(u(t),v(t+\varepsilon)) - y(u(t),v(t))}{\varepsilon} + \frac{y(u(t+\varepsilon),v(t)) - y(u(t),v(t))}{\varepsilon}$$

$$\approx \frac{y(u(t),v(t)+\varepsilon\dot{v}(t))) - y(u(t),v(t))}{\varepsilon}$$

$$+ \frac{y(u(t)+\varepsilon\dot{u}(t)),v(t)) - y(u(t),v(t))}{\varepsilon}$$

$$= \frac{y(u(t),v(t)+\varepsilon\dot{v}(t))) - y(u(t),v(t))}{\varepsilon\dot{v}(t)}\dot{v}(t)$$

$$+ \frac{y(u(t)+\varepsilon\dot{u}(t)),v(t)) - y(u(t),v(t))}{\varepsilon\dot{u}(t)}\dot{u}(t)$$

$$\approx \frac{dy}{dv}(u(t),v(t))\dot{v}(t) + \frac{dy}{du}(u(t),v(t))\dot{u}(t). \qquad\qquad \square$$

3.8.7 *Exponential Rule*

Theorem 3.36. *If $y = e^x$ then*

$$\frac{dy}{dx} = e^x .$$

Proof. See Appendix A.   □

3.8.8 *Change-of-Variable Rule*

Theorem 3.37. *If $z = z(y)$ and $y = y(x)$ then*

$$\int_{y_0}^{y_1} z(y)dy = \int_{x_0}^{x_1} z(y(x))\frac{dy}{dx}dx$$

where $y(0) = y(x_0)$ and $y_1 = y(x_1)$.

3.8.9 *Increment Rule*

Theorem 3.38.

$$f(x + \epsilon) = f(x) + \epsilon \cdot f'(x) + \epsilon\delta \tag{3.11}$$

$$f(\overrightarrow{x} + \overrightarrow{\epsilon}) = f^{(1)}(\overrightarrow{x})\epsilon_1 + \cdots + f^{(n)}(\overrightarrow{x})\epsilon_n + \epsilon_1 \cdots \epsilon_n\delta \tag{3.12}$$

3.8.10 *Quotient Rule*

Theorem 3.39. *If $u = u(x)$ and $v = v(x)$ then*

$$\frac{d}{dx}\frac{u}{v} = \frac{1}{v}\left(\frac{du}{dx} - \frac{u}{v}\frac{dv}{dx}\right) \ \ or \ d(u/v) = (vdu - udv)/v^2 .$$

3.8.11 *Intermediate Value Rule*

Theorem 3.40. *If $[a, b] \xrightarrow{f} \mathbb{R}$ is continuous and $y \in [f(a), f(b)]$ then there exists $x \in [a, b]$ such that $y = f(x)$.*

Proof. ([Lang (2002)] p. 237).   □

Corollary 3.41. *If $[a, b] \xrightarrow{f} \mathbb{R}$ is continuous then there exists a map $[f(a), f(b)] \xrightarrow{g} \mathbb{R}$ such that $g(y) \in [a, b]$ and $y = f(g(y))$.*

Proof. For each $y \in [f(a), g(b)]$ choose (by virtue of the Axiom of Choice) $x \in [a, b]$ such that $y = f(x)$. Since the choice of x depends on y define $g(y) := x$, hence $y = f(x) = f(g(y))$. □

Remark 3.42. The opposite equation $x = g(f(x))$ may not hold for all $x \in [a, b]$ since it may be that $y = f(x_1) = f(x_2)$ for some $x_1 \neq x_2$ in $[a, b]$, so that $x_1 = g(f(x_1))$ but $g(f(x_2)) = g(f(x_1)) = x_1 \neq x_2$.

Aside 3.8.1. During the oral examination for my doctoral degree at Dalhousie University in 1973 I was asked a seemingly elementary question about a proof regarding continuous maps. I asked for a moment to marshal my thoughts, but had to announce, "I just marshalled the empty set." This was a tad amusing – perhaps – but very embarrassing. What I needed to have replied was that the Axiom of Choice is required. Ever since then I am very conscious of the occasional need for the Axiom of Choice, even in seemingly elementary contexts.

3.8.12 Mean Value Rule

Theorem 3.43. *If $[a, b] \xrightarrow{f} \mathbb{R}$ is continuous, and f is differentiable on (a, b), then there exists $c \in (a, b)$ such that*

$$f'(c) = \frac{f(b) - f(a)}{b - a} \ .$$

Proof. ([Lang (2002)] pp. 88–89). □

3.8.13 Monotonicity Rule

Theorem 3.44. *Let $[a, b] \xrightarrow{f} \mathbb{R}$ be continuously differentiable on (a, b). If $f'(x) \geq 0$ for all $x \in (a, b)$ and $f'(x) > 0$ (respectively, $f'(x) < 0$) for some $x \in (a, b)$, then f is strictly increasing (respectively, decreasing) on $[a, b]$.*

Proof. ([Lang (2002)] p. 92). □

3.8.14 Inversion Rule

Theorem 3.45. *If $[a, b] \xrightarrow{f} \mathbb{R}$ is continuously differentiable on (a, b) and $f'(x) \neq 0$ for $x \in (a, b)$, then there exists a map $[f(a), f(b)] \xrightarrow{g} \mathbb{R}$ such that* *g is differentiable on* $(f(a), f(b))$, *g is inverse to* f, *and*

$$g'(y) = \frac{1}{f'(x)} \cdot$$

if $y = f(x)$.

Proof. By the Intermediate Value Rule Corollary there exists $[f(a), f(b)] \xrightarrow{g} \mathbb{R}$ such that $y = f(g(y))$ for $y \in [f(a), f(b)]$, and by the Monotonicity Rule $x = g(f(x))$ for $x \in [a, b]$.

Let $y = f(x)$ and $x = g(y)$. For any infinitesimal $\varepsilon \neq 0$ let $\delta := g(y + \varepsilon) - g(y)$ (where g also denotes its own hyperreal extension). Then

$$f(x) + \varepsilon = y + \varepsilon = f(g(y + \varepsilon)) = f(g(y) + \delta) = f(x + \delta)$$

so $\varepsilon = f(x + \delta) - f(x)$. By definition of derivative,

$$f'(x) \approx \frac{f(x + \delta) - f(x)}{\delta} = \frac{\varepsilon}{g(y + \varepsilon) - g(y)} \cdot$$

Since $f'(x) \neq 0$,

$$\frac{1}{f'(x)} \approx \frac{g(y + \varepsilon) - g(y)}{\varepsilon} \cdot$$

since the choice of ε is arbitrary, the right-hand side equals $g'(x)$. $\qquad\square$

Theorem 3.46. *The following are true:*

(1) *If there exists a map* $A \times C \xrightarrow{f} B$ *then for every* $y \in C$ *there exists a map* $A \xrightarrow{f_y} B$ *such that* $f_y(a) = f(a, y)$;

(2) *if for every* $y \in C$ *there exists a map* $A \xrightarrow{f_y} B$ *then there exists a map* $A \times C \xrightarrow{f} B$ *such that* $f(a, y) = f_y(a)$;

(3) *if there exists a map* $A \times C \xrightarrow{f} B$ *then there exits a map* $C \xrightarrow{\tilde{f}} B^A$ *such that* $\tilde{f}(y)(a) = f(a, y)$;

(4) *if there exists a map* $C \xrightarrow{\tilde{f}} B^A$ *then there exists a map* $A \times C \xrightarrow{f} B$ *such that* $f(a, y) = \tilde{f}(y)(a)$.

Theorem 3.47. *The following are equivalent:*

(1) *For every* $y \in C$ *there exists a diagram*

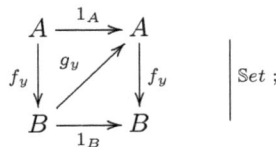

$$
\begin{array}{ccc}
A & \xrightarrow{1_A} & A \\
f_y \downarrow & \nearrow{\scriptstyle g_y} & \downarrow f_y \\
B & \xrightarrow[1_B]{} & B
\end{array}
\qquad \text{Set} \ ;
$$

(2) there exists a diagram

$$
\begin{array}{ccc}
A \times C & \xrightarrow{\;1_{A \times C}\;} & A \times C \\
\end{array}
\qquad (3.13)
$$

$$
(\widetilde{f}\,\pi_C) \downarrow \quad \overset{(\widetilde{g}\,\pi_C)}{\nearrow} \quad \downarrow (\widetilde{f}\,\pi_C) \qquad \text{Set}\,.
$$

$$
\begin{array}{ccc}
B \times C & \xrightarrow[\;1_{B \times C}\;]{} & B \times C
\end{array}
$$

Proof. $\widetilde{f}(a,y) :=: f_y(a)$. $\qquad\qquad\qquad\qquad\qquad\qquad\qquad\qquad\square$

Remark 3.48. In these theorems the role of y is considered to be a "parameter" or an "index" – with values in C – of a family of maps between A and B.

Definition 3.49. Let the map $A \times C \xrightarrow{f} B$ be represented by $z = z(a, y)$, which is to say, $z = f(a, y)$ for $a \in A$ and $y \in C$. Then $z = z(a, y)$ is **solvable for** a if there exists a map $B \times C \xrightarrow{g} A$ such that Eq. (3.13) is true. Put another way, $a = g(z, y)$ for $z \in B$ and $y \in C$, and

$$
z = z(a(z, y), y)
$$
$$
a = a(z(a, y), y)\,.
$$

Theorem 3.50. *(Solvability) If*

$$
\begin{array}{ccc}
\mathbb{R} \times \mathbb{R} & \longrightarrow & \mathbb{R} \\
a \quad y & & z = z(a, y)
\end{array}
$$

is continuously differentiable with respect to a and $\dfrac{\partial z}{\partial a} > 0$, then $z = z(a, y)$ is solvable for a and

$$
\frac{\partial a}{\partial z} = \frac{1}{\frac{\partial z}{\partial a}}\,.
$$

Proof. For every $y \in \mathbb{R}$ the map defined by $z_y = z_y(a) := z(a, y)$ satisfies the conditions of the Inversion Rule. $\qquad\qquad\qquad\qquad\qquad\square$

3.8.15 Cyclic Rule

Theorem 3.51. *If $T = T(U, V)$, $U = U(T, V)$ and $\left(\dfrac{\partial U}{\partial T}\right)_V > 0$ then*

$$\left(\frac{\partial T}{\partial V}\right)_U = -\frac{\left(\frac{\partial U}{\partial V}\right)_T}{\left(\frac{\partial U}{\partial T}\right)_V}.$$

Proof. For arbitrary differentials dT, dU and dV,

$$dU = \left(\frac{\partial U}{\partial T}\right)_V dT + \left(\frac{\partial U}{\partial V}\right)_T dV \qquad (3.14)$$

$$dT = \left(\frac{\partial T}{\partial U}\right)_V dU + \left(\frac{\partial T}{\partial V}\right)_U dV. \qquad (3.15)$$

Dividing (3.14) throughout by $\left(\dfrac{\partial U}{\partial T}\right)_V$ yields

$$\frac{dU}{\left(\frac{\partial U}{\partial T}\right)_V} = dT + \frac{\left(\frac{\partial U}{\partial V}\right)_T}{\left(\frac{\partial U}{\partial T}\right)_V} dV,$$

and substitution for dT using (3.15) yields

$$\frac{dU}{\left(\frac{\partial U}{\partial T}\right)_V} = \left(\frac{\partial T}{\partial U}\right)_V dU + \left[\left(\frac{\partial T}{\partial V}\right)_U + \frac{\left(\frac{\partial U}{\partial V}\right)_T}{\left(\frac{\partial U}{\partial T}\right)_V}\right] dV,$$

hence

$$\left[\frac{1}{\left(\frac{\partial U}{\partial T}\right)_V} - \left(\frac{\partial T}{\partial U}\right)_V\right] dU = \left[\left(\frac{\partial T}{\partial V}\right)_U + \frac{\left(\frac{\partial U}{\partial V}\right)_T}{\left(\frac{\partial U}{\partial T}\right)_V}\right] dV. \qquad (3.16)$$

But the coefficient of dU on the left-hand side of Eq. (3.16) vanishes by the Inversion Rule and the hypothesis $\left(\dfrac{\partial U}{\partial T}\right)_V > 0$. Therefore, the coefficient of $dV \neq 0$ on the right side also vanishes, which is equivalent to the desired result. $\square$

3.8.16   *Homogeneity Rule*

Definition 3.52. A function $f : \mathbb{R}^n \to \mathbb{R}$ is **homogeneous of order** $n \geq 0$ if for any $\lambda \in \mathbb{R}$ and $(x_1, \ldots, x_n) \in \mathbb{R}^n$ it is true that

$$f(\lambda x_1, \ldots, \lambda x_n) = \lambda^n f(x_1, \ldots, x_n) . \tag{3.17}$$

Example 3.53. Any linear functional $A : \mathbb{R}^n \to \mathbb{R}$ is homogeneous of order 1.

Theorem 3.54. *If* $f : \mathbb{R}^n \to \mathbb{R}$ *is a homogeneous function of order 1 then*

$$f(x_1, \ldots, x_n) = \frac{\partial f}{\partial x_1} x_1 + \cdots + \frac{\partial f}{x_n} x_n . \tag{3.18}$$

Proof.   The derivatives of the two sides of Eq. (3.17) with respect to λ are equal, so for $n = 1$ by the Constant Rule and the Chain Rule

$$f(x_1, \ldots, x_n) = \frac{d}{d\lambda} f(\lambda x_1, \ldots, \lambda x_n)$$

$$= \frac{\partial f}{\partial x_1} \frac{d\lambda x_1}{d\lambda} + \cdots + \frac{\partial f}{\partial x_n} \frac{d\lambda x_n}{d\lambda}$$

$$= \frac{\partial f}{\partial x_1} x_1 + \cdots + \frac{\partial f}{x_n} x_n . \qquad \square$$

3.9   Three Gaussian Integrals

Three so-called Gaussian integrals appear in most expositions of statistical mechanics, are easy to prove, and so are gathered together here.

Theorem 3.55. *The following equations exist:*

(1) $\int\limits_{-\infty}^{+\infty} e^{-x^2} dx = \sqrt{\pi};$

(2) $\int\limits_{-\infty}^{+\infty} e^{-ax^2} dx = \sqrt{\frac{\pi}{a}};$

(3) $\int\limits_{-\infty}^{+\infty} x^2 e^{-ax^2} dx = \frac{\sqrt{\pi}}{2a\sqrt{a}};$

Proof.

(1) Let $G := \int\limits_{-\infty}^{+\infty} e^{-x^2} dx$. Then

$$G^2 = \int\limits_{-\infty}^{+\infty} e^{-x^2} dx \int\limits_{-\infty}^{+\infty} e^{-y^2} dy$$

$$= \int\limits_{-\infty}^{+\infty} e^{-(x^2+y^2)} dx dy$$

$$= \int\limits_{0}^{2\pi} \int\limits_{0}^{\infty} e^{-r^2} r\,dr\,d\theta$$

(if $r^2 := x^2 + y^2$ so that the infinitesimal arclength at radius r and angle $d\theta$ is $r d\theta$)

$$= 2\pi \int\limits_{0}^{\infty} e^{-r^2} dr$$

$$= 2\pi \int\limits_{0}^{\infty} e^{-u} \frac{du}{2} = \pi$$

(if $u := r^2$ so that $du = 2r dr$) .

(2) Substitute $u := \sqrt{a}x$ in (1).

(3)

$$\int\limits_{-\infty}^{+\infty} x^2 e^{-ax^2} dx = -\int\limits_{-\infty}^{+\infty} -x^2 e^{-ax^2} dx$$

$$= -\int\limits_{-\infty}^{+\infty} \frac{d}{da} e^{-ax^2} dx$$

$$= -\frac{d}{da} \int\limits_{-\infty}^{+\infty} e^{-ax^2} dx$$

$$= -\frac{d}{da} \sqrt{\frac{\pi}{a}} = \frac{\sqrt{\pi}}{2a\sqrt{a}} \text{ by (2) .} \qquad \square$$

3.10 Three Differential Equations

The existence and uniqueness of a solution for any ordinary differential equation with given initial condition is a standard theorem of mathematics ([Arnold (1978)], Chapter 4). Everybody should know how to solve (at least) three differential equations.

Theorem 3.56. *Let b and y_0 in $\mathbb{R}$. The initial value problem*

$$\frac{dy}{dx} = by$$

$$y(0) = y_0$$

has the unique solution

$$y(x) := y_0 e^{bx}.$$

Proof. To begin with, $y(0) = y_0 e^{b \cdot 0} = y_0$. Calculate

$$\frac{d}{dx} y_0 e^{bx} = y_0 \frac{d}{dx} e^{bx} \qquad \text{by the Product and Constant Rules}$$

$$= y_0 e^{bx} \frac{d}{dx} bx \qquad \text{by the Exponential Rule and the Chain Rule}$$

$$= by \qquad \text{by the Product and Constant Rules.} \qquad \square$$

Theorem 3.57. *Let a, b and y_0 in $\mathbb{R}$. The initial value problem*

$$\frac{dy}{dx} = a + by$$

$$y(0) = y_0$$

has the unique solution

$$y(x) := \left(\frac{a}{b} + y_0\right) e^{bx} - \frac{a}{b}.$$

Proof. First, $y(0) = \left(\frac{a}{b} + y_0\right) e^{b \cdot 0} - \frac{a}{b} = y_0$. Second,

$$\frac{d}{dx} \left(\frac{a}{b} + y_0\right) e^{bx} - \frac{a}{b} = \left(\frac{a}{b} + y_0\right) b e^{bx}$$

and

$$a + by = a + b \left(\left(\frac{a}{b} + y_0\right) e^{bx} - \frac{a}{b} \right)$$

$$= a + (a + by_0) e^{bx} - a . \qquad \square$$

Theorem 3.58. *Let a in $\mathbb{R}$ and $\mathbb{R} \xrightarrow{b} \mathbb{R}$. The initial value problem*

$$\frac{dx}{dt}(t) = ax(t) + b(t)$$

$$x(0) = x_0$$

has the unique solution $\mathbb{R} \xrightarrow{x} \mathbb{R}$ defined by

$$x(t) := x_0 e^{at} + \int_0^t b(s) e^{a(t-s)} ds.$$

Proof.   ([Zwanzig (2001)], Appendix I)

Suppose the solution has the form $x(t) := e^{at} y(t)$ for some $\mathbb{R} \xrightarrow{y} \mathbb{R}$. Then

$$ax(t) + b(t) = \frac{dx}{dt} = e^{at}\frac{dy}{dt} + ae^{at}y(t) = e^{at}\frac{dy}{dt} + ax$$

hence

$$\frac{dy}{dt} = b(t)e^{-at}$$

and so

$$y(t) = x(0) + \int_0^t b(s)e^{-as}ds \ .$$

Therefore, the solution is given by

$$x(t) := e^{at}y(t) = e^{at}\left(x(0) + \int_0^t b(s)e^{-as}ds \right)$$

$$= x(0)e^{at} + e^{at}\int_0^t b(s)e^{-as}ds$$

$$= x(0)e^{at} + \int_0^t b(s)e^{at}e^{-as}ds$$

$$= x(0)e^{at} + \int_0^t b(s)e^{at-as}ds$$

$$= x(0)e^{at} + \int_0^t b(s)e^{a(t-s)}ds \ .$$

$\square$

3.11 Legendre Transform

High school students study a circle of ideas involving the equation

$$y = m \cdot x + b$$

for a line with slope m and y-intercept b. At each point x where a dependent variable $y = y(x)$ is differentiable there corresponds an equation

$$y(x) = m(x) \cdot x + b(x)$$

where by definition $m(x) := \dfrac{dy}{dx}(x)$ and $b(x)$ is the y-intercept of the tangent line to the graph of y with respect to x at x. Equivalently,

$$y(x) - b(x) = m(x) \cdot x \ .$$

The key observation leading to the definition of the Legendre Transform of $y = y(x)$ is that if it happens to be the case that $m = m(x)$ is invertible, so that $x = x(m)$ and

$$x = x(m(x))$$
$$m = m(x(m))$$

then, also equivalently,

$$y(x(m)) - b(x(m)) = m \cdot x(m) \ .$$

The variable $\widetilde{b}(m) := b(x(m))$ – which is nothing but the varying y-intercept of the aforementioned tangent line – depends on the slope m of that tangent line. Thus, given the $1 - 1$ correspondence between values of x and values of m, $\widetilde{b}$ is to m as y is to x, and $\widetilde{b} = \widetilde{b}(m)$ *is* the Legendre Transform of $y = y(x)$. More can be said. Calculate

$$
\begin{aligned}
\frac{d\widetilde{b}}{dm} &= \frac{d}{dm}(y(x(m)) - m \cdot x(m)) \\
&= \frac{dy}{dx}(x(m))\frac{dx}{dm} - m \cdot \frac{dx}{dm} - x \\
&= m(x(m))\frac{dx}{dm} - m \cdot \frac{dx}{dm} - x \\
&= m \cdot \frac{dx}{dm} - m \cdot \frac{dx}{dm} - x \\
&= -x(m) \ ,
\end{aligned}
$$

$$
\frac{d^2\widetilde{b}}{dm^2} = -\frac{dx}{dm} = -\frac{1}{\frac{dm}{dx}} = -\frac{1}{\frac{d^2y}{dx^2}} \ .
$$

Therefore, if $\frac{d^2y}{dx^2} < 0$ – that is, if the graph of $y = y(x)$ is convex upward – then $\frac{d^2\widetilde{b}}{dm^2} > 0$ is convex downward. In particular, if $m(x_*) = \frac{dy}{dx}(x_*) = 0$ – so that y has a maximum at x_* – then

$$\widetilde{b}(0) = \widetilde{b}(m(x_*)) = y(x(m(x_*))) = y(x_*) .$$

Theorem 3.59. *Given $X, M \hookrightarrow \mathbb{R}$ and maps*

$$\mathbb{R} \longleftarrow X \underset{\longleftarrow}{\overset{\longrightarrow}{\rule{0pt}{0pt}}} M \longrightarrow \mathbb{R}$$
$$y \qquad x \qquad m \qquad b$$

such that

$$m = m(x(m)) \tag{3.19}$$
$$x = x(m(x)) \tag{3.20}$$

the following are logically equivalent:

$$b(m) = y(x(m)) - m \cdot x(m) \qquad and \qquad \frac{db}{dm} = -x \tag{3.21}$$

$$y(x) = m(x) \cdot x + b(m(x)) \qquad and \qquad \frac{dy}{dx} = m. \tag{3.22}$$

Proof.   By symmetry of notation it suffices to prove (3.21) implies (3.22).

$$
\begin{aligned}
x \cdot m(x) + b(m(x)) &= m(x) \cdot x + [y(x(m(x))) - m(x) \cdot x(m(x))] && \text{by (3.21)}\\
&= m(x) \cdot x + y(x) - m(x) \cdot x && \text{by (3.20)}\\
&= y(x) && \text{cancellation.}
\end{aligned}
$$

hence $y(x) = m(x) \cdot x + b(m(x))$ so

$$
\begin{aligned}
\frac{dy}{dx} &= \frac{d}{dx}(m(x) \cdot x + b(m(x)))\\
&= x\frac{dm}{dx} + \frac{dx}{dx}m(x) + \frac{db}{dm}\frac{dm}{dx} && \text{Product and Chain rules}\\
&= x\frac{dm}{dx} + m(x) - x\frac{dm}{dx} && \text{by (3.21)}\\
&= m(x) && \text{cancellation.} \qquad \square
\end{aligned}
$$

A standard geometric interpretation of this result is illustrated in ([Tester and Modell (2004)] pp. 144–5). With respect to an application of this method for switching independent variables, Saunders Mac lane wrote, "It has taken me over fifty years to understand the derivation of Hamilton's

equations." His explanation of the "trick" is a technically profound explo-
ration ([Mac Lane (1986)], pp. 282–289). But the fairly simple algebraic
observation above leads by itself to an algorithm for exchanging some independent variables in a map of several variables for the partial derivatives
of the map with respect to those variables. This capability is important in
thermodynamics because some theoretically useful independent variables –
such as entropy – are not practically measurable, while the corresponding
partial derivative – such as temperature – is readily amenable to measurement.

Let the variable

$$y = y(u, v) \tag{3.23}$$

depend on vectors u and v of independent variables. Define new variables
by

$$\xi_u := \frac{\partial y}{\partial u}(u, v) \tag{3.24}$$

$$\xi_v := \frac{\partial y}{\partial v}(u, v) \tag{3.25}$$

so that

$$dy = \langle \xi_u | du \rangle + \langle \xi_v | dv \rangle , \tag{3.26}$$

where it should be understood throughout that by definition if $u = (u_1, \ldots, u_m)$, say, then $du = (du_1, \ldots, du_m)$ and $\frac{\partial y}{\partial u} := (\frac{\partial y}{u_1}, \ldots, \frac{\partial y}{u_m})$ is
the gradient map of y, with independent variables (u, v).

The following algorithm yields a variable

$$\tilde{y}_u = \tilde{y}_u(\xi_u, v) \tag{3.27}$$

such that

$$d\tilde{y}_u = -u d\xi_u + \xi_v dv . \tag{3.28}$$

Furthermore, if the algorithm is performed with input Eq. (3.27) then the
output is Eq. (3.23). Invertibility of the algorithm means, in other words,
"no information is lost in transformation."

The algorithm has three steps. First, given Eq. (3.23) define

$$\tilde{y}_u := y - \xi_u \cdot u . \tag{3.29}$$

At this stage $\tilde{y}_u = \tilde{y}_u(u, v)$, in other words there is as yet no change in the independent variables from y to $\tilde{y}_u$. Second, solve Eq. (3.24) for u in terms of ξ_u and v, yielding

$$u = u(\xi_u, v) \ . \tag{3.30}$$

Third, use Eq. (3.30) to substitute $u(\xi_u, v)$ for u in Eq. (3.29), yielding the **Legendre transform** of $y = y(u, v)$ **with respect to** u:

$$\tilde{y}_u(\xi_u, v) = y(u(\xi_u, v), v) - \xi_u \cdot u(\xi_u, v) \ , \tag{3.31}$$

where the independent variables are now given by $\tilde{y}_u = \tilde{y}_u(\xi_u, v)$. Finally, based on Eq. (3.26) calculate

$$d\tilde{y}_u = dy - \xi_u \cdot du - u \cdot d\xi_u \tag{3.32}$$
$$= \xi_u \cdot du + \xi_v \cdot dv - \xi_u \cdot du - u \cdot d\xi_u \tag{3.33}$$
$$= -u \cdot d\xi_u + \xi_v \cdot dv \ , \tag{3.34}$$

which proves Eq. (3.28).

3.12 Lagrange Multiplier

Recall 3.12.1. The *infimum* and *supremum* operations have the following algebraic properties:

(1) $-\sup(X) = \inf(-X)$;
(2) $\sup(a + X) = a + \sup(X)$;
(3) if $\overline{X} := -\sup(X)$ then a in $X \Rightarrow a \leq \sup(X) \Rightarrow -a \geq -\sup(X) \Leftrightarrow -a \geq \overline{X} \Leftrightarrow a \leq -\overline{X}$.

for $a \in \mathbb{R}$ and $X \subset \mathbb{R}$.

Definition 3.60. Given a map $X \xrightarrow{f} \mathbb{R}$ and a true-or-false condition $P(x)$ specified for each $x \in X$, define

$$\max f(x)|_{P(x)} := \{\, x^* \text{ in } X \mid P(x^*) \text{ and if } P(x) \text{ then } f(x) \leq f(x^*) \,\}$$

$$\min f(x)|_{P(x)} := \{\, x^* \text{ in } X \mid P(x^*) \text{ and if } P(x) \text{ then } f(x^*) \leq f(x) \,\}$$

Write $\max f(x)|_x$ (or $\min f(x)|_x$) when there is no condition for x to satisfy – which is the same as saying the condition is merely that x be an item of X. Note that x is bound in these expression contexts.

Remark 3.61. These sets formalize the idea of a value (x^*) at which a map (f) is an extreme value, but qualified by the condition that the value satisfy a condition P, or "constraint." The definition in itself does not guarantee that the sets are non-void, nor if they are non-void that they contain exactly one item.

Carefully note that x is a bound symbol in expression contexts such as $\max f(x)|_{P(x)}$.

Theorem 3.62. *Given*

$$\mathbb{R} \times \mathbb{R} \times \mathbb{R} \longrightarrow \mathbb{R}$$
$$x \quad y \quad z \qquad\qquad u$$

then for any (y_0, z_0) in $\mathbb{R} \times \mathbb{R}$ the following are equivalent:

(1) (x^*, y_0, z_0) *in* $\max u(x, y_0, z_0)|_x$;

(2) $\left(\dfrac{\partial u}{\partial x}\right)(x^*, y_0, z_0) = 0$ *and* $\left(\dfrac{\partial^2 u}{\partial x^2}\right)(x^*, y_0, z_0) < 0$.

Definition 3.63. Let A be a set and let $A \xrightarrow{f} \mathbb{R}$ be a map. The maximum value reached by $f(x)$ for x in A may or may not exist, and if it does exist there may be more than one argument x at which it is reached. If $\bot$ denotes "undefined" then $\operatorname*{Max}_{x \text{ in } A} f(x)$ is defined to be the maximum value reached by $f(x)$ for x in A if that value is reached, or otherwise $\operatorname*{Max}_{x \text{ in } A} f(x) = \bot$.

The set of arguments x in A at which $f(x)$ reaches a maximum value is denoted by $\operatorname*{ArgMax}_{x \text{ in } A} f(x)$. By definition,

$$\operatorname*{Max}_{x \text{ in } A} f(x) = \bot \text{ if and only if } \operatorname*{ArgMax}_{x \text{ in } A} f(x) = \emptyset . \tag{3.35}$$

Definition 3.64. Assume $A \subset \mathbb{R}$. For a differentiable map $A \xrightarrow{f} \mathbb{R}$, an argument x in A is a **critical point** of $f(x)$ if $f'(x) = 0$ ([Buck (1978)], p. 133). The set of critical points of f is denoted by $\operatorname*{Crit}_{x \text{ in } A} f(x)$.

Theorem 3.65. *Assume $A \subset \mathbb{R}$. For a differentiable map $A \xrightarrow{f} \mathbb{R}$,*

$$\operatorname*{ArgMax}_{x \text{ in } A} f(x) \hookrightarrow \operatorname*{Crit}_{x \text{ in } A} f(x) . \tag{3.36}$$

This standard result of the infinitesimal calculus says that an argument where a differentiable map is at a maximum value is an argument where the derivative of the map is zero. It does not say there exists such an argument.

Entirely dual definitions and a theorem hold for the minima of maps. That is, $\underset{x \text{ in } A}{\text{ArgMin}} f(x)$ denotes the possibly empty set of arguments at which $f(x)$ reaches a minimum value, and also

$$\underset{x \text{ in } A}{\text{ArgMin}} f(x) \hookrightarrow \underset{x \text{ in } A}{\text{Crit}} f(x) . \tag{3.37}$$

Therefore, if $\underset{x \text{ in } A}{\text{ArgExt}} f(x) := \underset{x \text{ in } A}{\text{ArgMax}} f(x) \cup \underset{x \text{ in } A}{\text{ArgMin}} f(x)$ it follows that

$$\underset{x \text{ in } A}{\text{ArgExt}} f(x) \hookrightarrow \underset{x \text{ in } A}{\text{Crit}} f(x) . \tag{3.38}$$

A more general result holds for maps depending on multiple arguments. Let $A \subset \mathbb{R}^n$ and let $A \xrightarrow{f} \mathbb{R}$ be differentiable with respect to each of its arguments. In this case an argument $x = (x_1, \ldots, x_n)$ in A is a critical point of $y = f(x)$ if

$$\partial_i f(x_1, \ldots, x_n) := \frac{\partial y}{\partial x_i}(x_1, \ldots, x_n) = 0, \quad i = 1, \ldots, n . \tag{3.39}$$

Roughly speaking, the Method of Lagrange Multipliers is an algorithm for determining arguments of a differentiable map at which extrema are reached, but also constrained to belong to a subset of the domain of definition of the map which is defined by one or more differentiable maps also defined on that domain.

Definition 3.66. Let $A \xrightarrow{g} \mathbb{R}$ be a differentiable map and define a **zero of** $g(x)$ **in** A to be an argument x in A such that $g(x) = 0$. Let $\text{Zero}(g) := \underset{x \text{ in } A}{\text{Zero}}(g(x))$ denote the set of zeros of $g(x)$ in A. (Zeros of a map g are the same as **roots** of the equation $g(x) = 0$.)

The formal justification of the Method of Lagrange Multipliers is

Theorem 3.67. *Let $A \subset \mathbb{R}^n$ and let $A \xrightarrow{f} \mathbb{R}$ and $A \xrightarrow{g} \mathbb{R}$ be differentiable with respect to each of their arguments. Then*

$$\underset{\text{Zero}(g)}{\text{ArgExt}} f(x) = \underset{(x,\lambda) \text{ in } A \times \mathbb{R}}{\text{ArgExt}} (f(x) - \lambda g(x)) \tag{3.40}$$

$$\hookrightarrow \underset{(x,\lambda) \text{ in } A \times \mathbb{R}}{\text{Crit}} (f(x) - \lambda g(x)) , \tag{3.41}$$

where the map $A \times \mathbb{R} \xrightarrow{f - \lambda g} \mathbb{R}$ is defined pointwise by $(f - \lambda g)(x, \lambda) = f(x) - \lambda g(x)$.

Therefore, a necessary condition for an extreme value of $f(x)$ where x is constrained to be a zero of $g(x)$ is that $n+1$ partial derivatives of $f(x) - \lambda g(x)$ be zero. This is a problem of solving for $n+1$ unknowns $(x_1, \ldots, x_n, \lambda)$ given a system of $n+1$ equations, which may or may not be possible. In summary, the Method of Lagrange Multipliers reduces the problem of constrained *extrema* to a problem of solving a system of equations ([Buck (1978)], p. 539).

Chapter 4

Algebra of Vectors

4.1 Introduction

This chapter is about calculations within and upon finite-dimensional vector spaces. For example, the real numbers are a 1-dimensional vector space of dimension 1. Within a vector space the items are scalars and vectors – real numbers are both, as illustrated in Fig. 4.1. In general vectors are

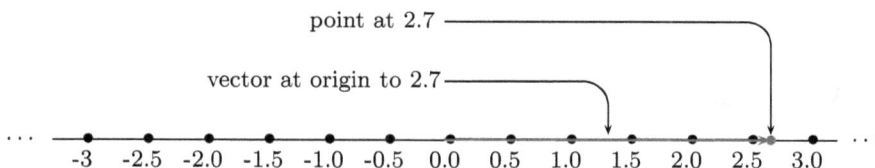

Fig. 4.1 The head of a vector at the origin of the 1-dimensional vector space of real numbers is a point.

"higher-dimensional" than scalars. A scalar multiplies a vector, Fig. 4.2, and vectors can be added to vectors yielding vectors, Fig. 4.3.

Also, two vectors in a finite-dimensional vector space can be multiplied to yield a scalar.

As suggested in Example (2.1) a vector space is also endowed with a categorical structure in which the objects are the vectors, but vectors are identified with points (their heads), and the morphisms are vectors situated between points. All of this is algebraic structure *within* a vector space.

Vector operators are maps between vector spaces, and vector spaces together with their operators form yet another category with a rich algebraic structure arising from the fact that vector spaces themselves can be

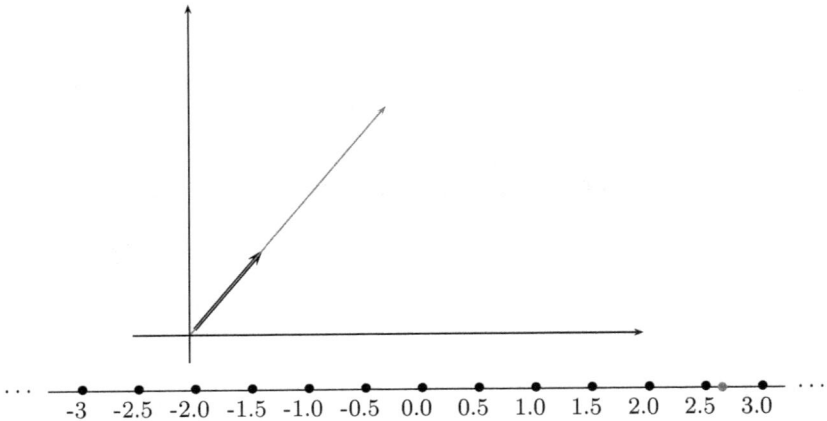

Fig. 4.2 A vector of magnitude 1 in a 2-dimensional vector space multiplied by the scalar 2.7.

added and multiplied yielding new vector spaces. This is algebraic structure *among* vector spaces.

4.2   When is an Array a Matrix?

Aside 4.2.1. One day I was vehemently excoriated by my employer – an accomplished "applied mathematician" – for casually referring to an array of numbers as a "matrix." His objection was that an array of numbers – a table of numbers arranged in rows and columns, such as a rectangular region of a spreadsheet, or a commuter train schedule – is *not* a matrix because we were not discussing "standard" matrix operations involving the array in question. In other words, he was complaining that I used language inappropriate for the context. Fair enough, provided one ignores the possibilities for operations on tables even if they are not tables of numbers.

A matrix is an array of numbers, and matrices are added and multiplied by performing (sometimes a large number of) additions and multiplication operations on the numbers within them and arranging the resulting values in certain ways.

Bottom line? *Different operations are available for the same expressions depending on context. In object-oriented computer programming this kind of mutability of meaning is called "polymorphism."*

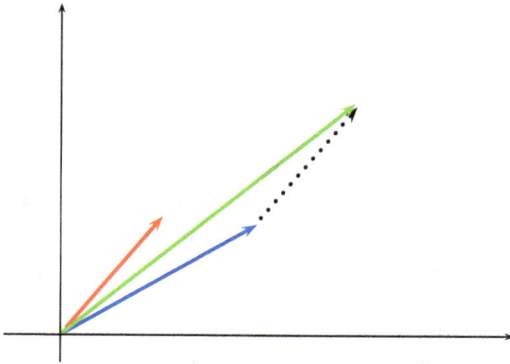

Fig. 4.3 Vector colored red added to vector colored blue yields vector colored green. The construction of the green vector depends on the dotted vector which is a copy of the red vector tacked onto the end of the blue vector.

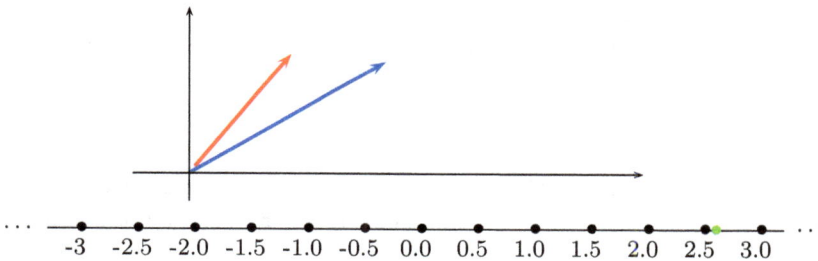

Fig. 4.4 The scalar product in green of a vector colored red and a vector colored blue.

4.3 List Algebra

Definition 4.1. Perhaps the simplest operation on expressions is to **list** them. There is no universal mathematical notation for a list of, say, the symbols a, b, x, 2 – *in that order* – but it is very common to arrange them in a row, separate them by commas, and surround the row by brackets of one kind or another. For example, ordinary parentheses are often used, so the list of those symbols could be represented by the expression $(a, b, x, 2)$. In this book I skip the commas and separate the items in a list by spaces, and I use square brackets, as in $[\,a\ b\ x\ 2\,]$. (This happens to be convention in MATLAB as well.) For the extreme case of an **empty list** the notation is [ ].

4.3.1 *Abstract Row List*

The mathematician would like to have a general expression for a list with a fixed but arbitrary number, say n, of items of a given set, say X. The trick is to define a notation for an abstract list of n locations, and label the locations with items selected from X.

Definition 4.2. Let $[1.n]$ denote the locations in one row of n columns. This is the **abstract row list** of length n. In particular, $[1.0]$ denotes one row of no locations, so the one and only map $[1.0] \rightarrow X$ is the empty list [ ] of items of set X. The map $1 \xrightarrow{\mu_j} [n]$ for $1 \leq j \leq n$ selects the jth location of $[1.n]$.

For $n > 0$ the diagram $[1.n] \xrightarrow{a} X$ represents the general list of n items of set X. Another expression for the same list is $[a_1, \ldots, a_n]$, where a_j is the label at position j: a_j is the label selected by the composite map $a \circ \mu_j$:

$$1 \xrightarrow{\mu_j} [1.n] \xrightarrow{a} X. \tag{4.1}$$

Remark 4.3. An abstract row list is the pure idea of left-to-right succession of a finite number of locations. It is most concretely realized by assigning expressions for items of a set to the locations, and writing that succession upon the Surface.

4.3.2 *Set of Row Lists*

Definition 4.4. For a given set X the **row lists** of length n for $n \geq 0$ form a new set by the Exponential Axiom, namely $X^{[1.n]}$. By the Union Axiom these sets may be joined together into one giant set of *all* lists formed from items of X, and it may be denoted by

$$X^{[1.*]} := X^{[1.0]} \cup X^{[1.1]} \cup X^{[1.2]} \cdots .$$

Note that for $m \neq n$ the sets $X^{[1.m]}$ and $X^{[1.n]}$ have no items in common, so that the $\cup$ operations above are *disjoint* unions of sets. As usual the $*$ is a "wildcard," in this case standing for the successive indexes $0, 1, 2, \cdots$.

4.3.3 *Inclusion of Row Lists*

Since the set of all lists of items from X is a union, each set in the union comes equipped automatically with an inclusion map

Definition 4.5.

$$X^{[1.n]} \overset{\mu_n}{\lhook\joinrel\longrightarrow} X^{[1.*]}, \quad n > 0$$

where $\mu_n(a) := a$ but "a" has a tiny difference in meaning on the two sides of this defining equation. On the left "a" means a list, a, considered as selected from the set $X^{[1.n]}$, whereas on the right it means the same list, but now regarded as selected from the set $X^{[1.*]}$.

4.3.4 *Projection of Row Lists*

Definition 4.6. The set of lists of length n of items of set X is automatically endowed with one **projection map** for each selected location $1 \overset{\mu_j}{\longrightarrow} [n]$, $n > 0$, namely

$$X^{[1.n]} \overset{\pi_j}{\longrightarrow} X$$

defined by $\pi_j [\, a_1 \dots a_n \,] := a_j$. In other words, $\pi_j(a) = (a \circ \mu_j)(1)$.

The set $X^{[1.n]}$ together with the projections π_j is exactly the set of lists whose existence is guaranteed by the List Axiom (2.4.4):

$$X^{[1.n]} = \overbrace{X \times \cdot \times X}^{n}.$$

Aside 4.3.1. A really compulsive pedant would want to decorate π_j with n as in π_j^n to distinguish notationally between the projection maps to the item at location j for lists of two different lengths greater than j. But I am not that way, most of the time.

4.3.5 *Row List Algebra*

Roughly speaking, an "algebra" is a mathematical structure for calculating with expressions that represent elements of one or more sets.

Definition 4.7. In the context of lists of items of set X the basic algebraic operation is **concatenation.**[1] Given two lists, say $[\, a \; x \; 23 \; Z \; b \,]$ and

---

[1]Of obscure origin, the word "concatenate" means *to chain (or link) together* – it may be related to a net or a helmet [Partridge (1966)].

[$qr8\ e\ p1\ zx3$], their **concatenate** is formed by arranging their items with the items of the second list following from left to right the items of the first list:

$$[\,a\ x\ 23\ Z\ b\ qr8\ e\ p1\ zx3\,].$$

The **binary concatenation operator** $*$

$$X^{[1..m]} \times X^{[1..n]} \overset{*}{\longrightarrow} X^{[1..m+n]}$$

is defined for two lists – f of length m and g of length n – by the formula

$$(f * g)(i) := \begin{cases} f(i) & \text{if } 1 \le i \le m \\ g(i) & \text{if } m+1 \le i \le m+n. \end{cases}$$

Theorem 4.8. *Concatenation is an associative binary operator:*

$$(f * g) * h = f * (h * h).$$

Proof. On the one hand,

$$((f * g) * h)(i) = \begin{cases} (f * g)(i) & \text{if } 1 \le i \le m+n \\ h(i) & \text{if } m+n+1 \le i \le m+n+p \end{cases}$$

$$= \begin{cases} \begin{cases} f(i) & \text{if } 1 \le i \le m \\ g(i) & \text{if } m+1 \le i \le m+n \end{cases} \\ h(i) & \text{if } m+n+1 \le i \le m+n+p \end{cases}$$

$$= \begin{cases} f(i) & \text{if } 1 \le i \le m \\ g(i) & \text{if } m+1 \le i \le m+n \\ h(i) & \text{if } m+n+1 \le i \le m+n+p \end{cases}$$

and on the other hand,

$$(f * (g * h))(i) = \begin{cases} f(i) & \text{if } 1 \le i \le m \\ (g * h)(i) & \text{if } m+1 \le i \le m+n+p \end{cases}$$

$$= \begin{cases} f(i) & \text{if } 1 \le i \le m \\ \begin{cases} g(i) & \text{if } m+1 \le i \le m+n \\ h(i) & \text{if } m+n+1 \le i \le m+n+p \end{cases} \end{cases}$$

$$= \begin{cases} f(i) & \text{if } 1 \le i \le m \\ g(i) & \text{if } m+1 \le i \le m+n \\ h(i) & \text{if } m+n+1 \le i \le m+n+p \end{cases} \qquad \square$$

Theorem 4.9. *The empty list is a left and right unit for the concatenation operation:*

$$f * [\,] = f = [\,] * f.$$

Proof.

$$(f * [\,])(i) = \begin{cases} f(i) & \text{if } 1 \le i \le n \\ \{\}(i) & \text{if } n+1 \le i \le n \quad \text{which is impossible} \end{cases}$$

$$([\,] * f)(i) = \begin{cases} \{\}(i) & \text{if } 1 \le i \le 0 \quad \text{which is impossible} \\ f(i) & \text{if } 1 \le i \le n \end{cases}$$

$\square$

Bottom line? The set of lists of items of a given set has an associative binary operator for which the empty list is both a left and right unit.

This kind of structure is extremely common in mathematics – the real numbers form such a structure both under addition, with unit 0, and under multiplication, with unit 1. When a structure is ubiquitous mathematicians single out a name for it.

Definition 4.10. A set with an associative binary operator for which there is a left and right unit is called a **monoid**.

4.3.6 *Monoid Constructed from a Set*

The category $\mathbb{M}on$ of monoids as a special kind of diagram was introduced in Sec. 2.4.10, and Theorem 2.58 introduced the forgetful functor $\mathbb{M}on \to \mathbb{S}gr$, which when composed with the forgetful functor $\mathbb{S}gr \to \mathbb{S}et$ yields a forgetful functor $\mathbb{M}on \to \mathbb{S}et$. There exists a functor going the other way which has a familiar "universal property."[2]

Theorem 4.11.

(1) There exists a functor $\mathbb{S}et \xrightarrow{\ Mon\ } \mathbb{M}on$ *defined on a set X by*

$$\mathrm{Mon}(X) := X^{[1..*]} \times X^{[1..*]} \xrightarrow{\ *\ } X^{[1..*]} \xleftarrow{\ [\,]\ } 1$$

(so $\mathrm{Set}(\mathrm{Mon}(X)) = X^{[1..*]}$*) and on a map* $X \xrightarrow{f} Y$ *by* $\mathrm{Mon}(f)[\,x_1 \cdots x_m\,] := [\,f(x_1) \cdots f(x_m)\,] \, ;$

---

[2]A technical definition with many examples of "universal property" is offered at http://en.wikipedia.org/wiki/Universal_property.

(2) for every set X there exists a diagram

$$A \xrightarrow{\eta_X} \text{Set}(\text{Mon}(X)) : x \mapsto [x] \; ;$$

(3) for every monoid $M := A \times A \xrightarrow{*} A \xleftarrow{e} 1$ (so $\text{Set}(M) = A$) and diagram

$$X \xrightarrow{f} A$$

there exists a unique diagram

Remark 4.12. This result about mere lists of elements of a set expresses a *universal property* about the concatenation monoid of a set that is exactly analogous to corresponding properties encountered in Theorem 2.89 about the free magma generated by a set, and in Theorem 2.90 about the free category generated by a directed set. The generators (X) are included in the (underlying set $\text{Mon}(X)$) of the free algebra, and *any* map of the generators into *any* (underlying set of a) monoid automatically defines a monoid map from the free monoid to the given monoid which restricts to the inclusion on the generators. A way to think about this is that mapping generators to a monoid possibly introduces relationships – equations – between the generators due to the binary operation of the monoid. Such equations do not hold in the set of lists, the set of lists is *free* of any equations (except for the trivial ones involving concatenation with the empty list). In other words, adjoint functors lurk deeply.

Remark 4.13. For icing on the cake, observe that a category with exactly one object corresponds exactly to a monoid. For, if $\mathbb{M}$ is a category with one object then the Associative and Identity Laws to which the morphisms of $\mathbb{M}$ conform are exactly the Associative and Identity Laws of the monoid whose elements are the morphisms: $\text{Mor}(M)$ is the underlying set of monoid corresponding to $\mathbb{M}$. This observation concludes with noting that a functor between one-object categories corresponds exactly to a monoid map as in Definition 2.55.

4.3.7 *Column List Algebra & Natural Transformation*

Aside 4.3.2. I have been taking for granted that a list is written as a *row* of expressions. Hence the title of the preceding section, "Row List Algebra." I daresay, however, that most people think of a list, say a shopping list, as a column of expressions.

Definition 4.14. For a set X the **column lists** of length $n, n \geq 0$ form a new set, namely $X^{[n.1]}$. These sets may be joined together into one giant set of *all* column lists formed from items of X, and denoted by

$$X^{[*.1]} := X^{[0.1]} \cup X^{[1.1]} \cup X^{[2.1]} \cdots .$$

Aside 4.3.3. The notation $[n.1]$ stands for an abstract column list analogous to the notation $[1.n]$ for an abstract row list. It is neat that we can use horizontal rows of symbols, e.g., $[,], 1, ., n$, to distinguish between horizontal and vertical lists of items on the Surface . We use *one-dimensional ordering* of symbols to represent two-dimensional distinctions. A *correspondence* mediated by the human mind between algebra operations on the Surface and geometrical or physical experience on the Surface, or even "out there" in physical space, undergirds much of mathematics – and physics.

Every row list concept has an exactly analogous columnar counterpart: we have maps (still denoted μ_j) to select locations in abstract column lists; projection maps (still denoted π_i); the set $X^{[*.1]}$ of all column lists; and the monoid structure given by *vertical concatenation*. Any row list can be converted into a column list by a quarter-turn clockwise rotation, so that the left end of the row becomes the top of the column, as in Fig. 4.5.

Definition 4.15. Let $\top$ denote the **rotation operator** defined by

$$[\, a_1 \, a_2 \cdots a_n \,]^\top = \begin{bmatrix} a_1 \\ a_2 \\ \vdots \\ a_n \end{bmatrix} .$$

Just as there exists a functor $\mathbb{S}\text{et} \xrightarrow{\text{Mon}} \mathbb{M}\text{on}$ that assigns to a set X the free monoid $\text{Mon}(X)$ of row lists $X^{[1.*]}$ generated by the elements of X, there exists a functor $\mathbb{S}\text{et} \xrightarrow{\text{Mon}^\top} \mathbb{M}\text{on}$ that assigns to X the free monoid $\text{Mon}^\top(X)$ of column lists $X^{[*.1]}$.

$$\begin{bmatrix} a \\ b \\ c \\ x \\ y \\ z \end{bmatrix}$$

$$[\, a \quad b \quad c \quad x \quad y \quad z \,]$$

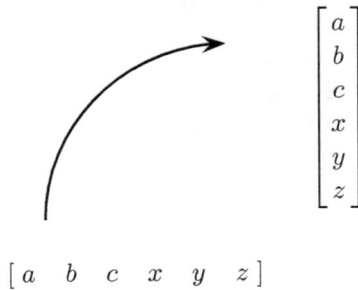

Fig. 4.5   Rotating a row list to become a column list.

Theorem 4.16. *For each set X there exists a map $X^{[1,*]} \xrightarrow{\top_X^*} X^{[*,1]}$ such that for any map $X \xrightarrow{f} Y$ in $\mathbb{S}$et there exists a diagram*

$$
\begin{array}{ccc}
X^{[1.*]} & \xrightarrow{\ Mon(f)\ } & Y^{[1.*]} \\
\top_X^* \downarrow & & \downarrow \top_Y^* \\
X^{[*.1]} & \xrightarrow{\ Mon^{\top}(f)\ } & Y^{[*.1]}
\end{array}
\qquad Mon\ .
$$
(4.2)

Proof.   For $n \geq 0$

$$
[\, x_1 \cdots x_n \,] \longmapsto [\, f(x_1) \cdots f(x_n) \,]
$$

$$
\begin{bmatrix} x_1 \\ \vdots \\ x_n \end{bmatrix} \longmapsto \begin{bmatrix} f(x_1) \\ \vdots \\ f(x_n) \end{bmatrix}
$$
     □

The situation in which there are two functors – Mon and $Mon^{\top}$ in this case – with the same domain category and the same codomain category, and such that on the same object – X in this case – and for a morphism – $X \xrightarrow{f} Y$ in this case – give rise to a square diagram as in Eq. (4.2) is ubiquitous in mathematics. What it means is that two different constructions on the same input are related in a natural way when the input is varied.

Definition 4.17. For two functors

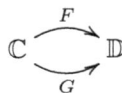

$$
\mathbb{C} \underset{G}{\overset{F}{\rightleftarrows}} \mathbb{D}
$$

a **natural transformation** is a diagram

$$\mathbb{C} \underset{G}{\overset{F}{\rightrightarrows}} \Downarrow \alpha \ \mathbb{D}$$

such that for every $\mathbf{1} \xrightarrow{X} \mathbb{C}$ there exists a diagram

$$
\begin{array}{c}
F(X) \\
\downarrow \alpha_X \\
G(X)
\end{array}
\qquad \mathbb{D}
$$

which represents a relationship between the two constructions F and G on the same input X such that for every variation of X to some object Y in $\mathbb{C}$ given by a map $X \xrightarrow{f} Y$ there exists a diagram

$$
\begin{array}{ccc}
F(X) & \xrightarrow{F(f)} & F(Y) \\
\alpha_X \downarrow & & \downarrow \alpha_Y \\
G(X) & \xrightarrow[G(f)]{} & G(Y)
\end{array}
\qquad \mathbb{D} \ .
$$

Theorem (4.16) is succinctly expressed by the diagram

$$\mathbb{S}\mathrm{et} \underset{\mathrm{Mon}^{\top}}{\overset{\mathrm{Mon}}{\rightrightarrows}} \Downarrow \top^* \ \mathbb{M}on \ .$$

Remark 4.18. The concepts of *natural transformation*, *functor*, and *category* were introduced in [Eilenberg and MacLane (1945)] and unified a diversity of mathematical phenomena. They regarded their work as "a continuation of the Klein Erlanger Programm, in the sense that a geometric space with its group of transformations is generalized to a category with its algebra of mappings."

Aside 4.3.4. The fact that natural transformations between two functors themselves are the morphisms of a category whose objects are those functors stretches the imagination towards "higher category theory," a subject with enormous ramifications and potential application to physics. That is all beyond the scope of my book, but see, for example [Baez and Stay (2009)] and [Coecke and Éric Oliver Paquette (2009)].

4.3.8   *Lists of Lists*

4.3.8.1   *Rows of Columns and Columns of Rows*

Aside 4.3.5. Starting with any set X I have defined the set $X^{[1.*]}$ of all rows of items of X, and it has associative binary concatenation operator with a two-sided unit, the empty row list. Likewise, for any set Y we have the corresponding concatenation monoid $Y^{[*.1]}$ of all columns of items of Y. In particular, if $Y = X^{[1.*]}$ then $Y^{[*.1]} = (X^{[1.*]})^{[*.1]}$. This is the set of all column lists of row lists from X. For example, if X is the set of all characters that can be typed on a standard computer keyboard, then a page of text corresponds to a column of rows of typed characters – with varying row length, to be sure. In this sense a book is a vertical concatenate of columns of rows.

$(X^{[1.*]})^{[*.1]}$ contains the set of all columns of rows of the same length m, namely $(X^{[1.m]})^{[*.1]}$, which in turn contains the set $(X^{[1.m]})^{[n.1]}$ of all columns of height n of rows of length m. O

Or, one has the set of all rows of length m of columns of height n: $(X^{[m.1]})^{[1.n]}$. Tacking on an exponent of the form $[1.p]$ or $[q.1]$ can be repeated endlessly, but I quickly run out of names after maybe three exponents. That is, we have rows, and columns, then maybe "piles," but then what? In this book rows and columns will do fine, because of certain bijections to be revealed soon enough.

4.3.8.2   *Flattening of rows of rows*

Abbreviate by $X^* := X^{[1.*]}$, and let $*$ also denote the free monoid functor $\mathbb{S}et \xrightarrow{\text{Mon}} \mathbb{M}on$. The empty list selected by $1 \xrightarrow{[]} X^*$ and the concatenation operation $X^* \times X^* \xrightarrow{j} X^*$ ("j" for juxtaposition) define the concatenation monoid with underlying set X^*.

For any map $X \xrightarrow{f} Y$ of sets there is a map $X^* \xrightarrow{f^*} Y^*$ from set of lists of items of X to the set of lists of items of Y defined by listing the value of f for each item. That is, $f^*([x_1 \cdots x_m]) := [f(x_1) \cdots f(x_m)]$ for given $[x_1 \cdots x_m]$.

For a set Y and any $1 \xrightarrow{a} Y$ there exists row list $[a]$ consisting of just a, hence there exists an injection $Y \rightarrowtail Y^*$. In particular, if $Y = X^*$ there exists an injection $X^* \xrightarrow{\mu} (X^*)^*$, so that a row list

$$[x_1 \cdots x_m] \xmapsto{\mu} [[x_1 \cdots x_m]] .$$

There also exists a map $(X^*)^* \xrightarrow{\pi} X^*$ which "collapses" or "flattens" a list of lists by obliterating the groupings into sublists, namely

$$[[x_{11} \cdots x_{1p_1}][x_{21} \cdots x_{2p_2}] \cdots [x_{m1} \cdots x_{mp_m}]]$$
$$\overset{\pi}{\mapsto} [x_{11} \cdots x_{1p_1} x_{21} \cdots x_{2p_2} \cdots x_{m1} \cdots x_{mp_m}] , \qquad (4.3)$$

where $p_1, \ldots p_m$ are the lengths of the lists in the given list of lists. This flattening of a list of lists to a list is a natural operation – it is completely determined by notation and not by any artificial choices – and this naturality is reflected by the observation that varying X along a map $X \overset{f}{\to} Y$ in $\mathbb{S}et$ preserves flattening in the sense that the same flattened list results from first flattening a list of lists in X and then applying $(f^*)^*$, or by first applying $(f^*)^*$ and then flattening a list of lists in Y. Indeed, something more general is going on here. Instead of considering the set $(X^*)^*$ of lists of lists of items of a set X, consider the set M^* of lists of items of an arbitrary monoid M.

Theorem 4.19. *For every monoid* M

(1) there exists a set map

$$M^* \overset{\varepsilon_M}{\longrightarrow} M : [m_1 \cdots m_k] \overset{\varepsilon_M}{\mapsto} m_1 * \cdots * m_k ;$$

(2) for every set X *and monoid map*

$$X^* \overset{g}{\to} M$$

there exists a unique diagram

$$
\begin{array}{ccc}
X^* & & \\
\ \ \downarrow f^* \searrow^{g} & & \\
& M & \\
M^* \ \nearrow_{\varepsilon_M} & &
\end{array}
\qquad \mathbb{M}on \qquad
\begin{array}{c}
X \\
\downarrow f \\
M
\end{array}
\qquad \mathbb{S}et .
$$

Proof. Given the definition of ε_M in (1), which is a generalization of "flattening," and a monoid map $X^* \overset{g}{\to} M$ as in (2), consider that f is completely determined by the requirements in the conclusion of (2), since on a generator $a \in X$ the triangle diagram imposes the condition that $f(a) = \varepsilon_M([f(a)]) = \varepsilon_M(f^*([a]) = (\varepsilon_M \circ f^*)([a]) = g([a])$. $\qquad \square$

Aside 4.3.6. One must be struck by the uncanny analogy between these diagrams and the diagrams in the Evaluation Axiom for set maps. Well, I am being a tad disingenuous. These diagrams, and their counterparts with arrows going in the opposite direction arising from the construction of a magma from a set, or a category from a directed graph, are all manifestations of the nearly metaphysical notion of adjoint functor.

4.4   Table Algebra

Definition 4.20. Directly generalizing the idea of abstract row or column list, the **abstract table** with m rows and n columns is denoted by $[m.n]$.

For $1 \leq i \leq m$ and $1 \leq j \leq n$ the map $\mathbf{1} \xrightarrow{i\mu_j} [m.n]$ selects the location at the ith row and j^{th} column of the abstract table $[m.n]$. A **table** is a map $[m.n] \xrightarrow{A} X$, and the item at location $_i\mu_j$ is selected by the composition

$$\mathbf{1} \xrightarrow{i\mu_j} [m.n] \xrightarrow{A} X, \qquad _iA_j := A \circ {}_i\mu_j.$$

The diagram of a table A on the Surface is of the form

$$\begin{bmatrix} _1A_1 & \cdots & _1A_n \\ \vdots & & \vdots \\ _mA_1 & \cdots & _mA_n \end{bmatrix}.$$

4.4.1   *The Empty and Unit Tables*

For $m, n \geq 0$ a table with m rows and no columns or a table with no rows and n columns is empty, and so it makes sense to set all of the following notations equal to one another:

$$[\,] = [m.0] = [0.n]\,.$$

Any of these is called the **empty table**. The abstract table $[1.1]$ with 1 row and 1 column is just an individual location. A map $[1.1] \xrightarrow{A} X$ is therefore just a location with an item of X assigned to it. But that is no different from selecting an element of X by a map $\mathbf{1} \to X$, nor from writing an item of X, so it also makes sense to abuse notations and write the equations

$$X = X^{\mathbf{1}} = X^{[1.1]}\,.$$

4.4.2   *The Set of All Tables*

Definition 4.21. The set of all m row n column tables with entries in X is $X^{[m.n]}$ for $m, n \geq 0$. The set of *all possible* tables with entries in X is

$$X^{[*.*]} := X^{[0.0]} \cup X^{[1.1]} \cup X^{[1.2]} \cup X^{[1.3]} \cup \cdots$$
$$\cup X^{[2.1]} \cup X^{[2.2]} \cup X^{[2.3]} \cup \cdots$$
$$\cup X^{[3.1]} \cup X^{[3.2]} \cup X^{[3.3]} \cup \cdots$$

$$\vdots$$

The **inclusion map** of the m row n column tables in the set of all possible tables is

$$X^{[m.n]} \overset{m\mu n}{\subset\!\!\!-\!\!\!-\!\!\!\longrightarrow} X^{[*.*]}.$$

For $1 \leq i \leq m$ and $1 \leq j \leq n$ the **projection map** of a table upon its entry at row i column j is the map

$$X^{[m.n]} \xrightarrow{i\pi j} X$$

defined by

$$i\pi j \left(\begin{bmatrix} {}_1A_1 & \cdots & {}_1A_n \\ \vdots & & \vdots \\ {}_mA_1 & \cdots & {}_mA_n \end{bmatrix} \right) := {}_iA_j \ .$$

4.4.3 *Juxtaposition of Tables is a Table*

Definition 4.22. Two tables with the same number of rows – but not necessarily the same number of columns – may be place side-by-side to form a new table. There exists a **horizontal juxtaposition map**

$$X^{[m.n]} \times X^{[m.p]} \xrightarrow{\text{®}} X^{[m.n+p]}.$$

Likewise, two tables with the same number of columns may be placed one above the other to form a new table, so there exists **vertical juxtaposition map**

$$X^{[m.n]} \times X^{[p.n]} \xrightarrow{\text{©}} X^{[m+p.n]}.$$

Theorem 4.23.

(1) The null table is a "multiplicative units" for both Horizontal and vertical concatenation:

$$A \, \text{®} \, [m.0] = [m.0] \, \text{®} \, A = A$$
$$B \, \text{©} \, [0.n] = [0.n] \, \text{©} \, B = B \ ;$$

(2) if

$$[m.n] \xrightarrow{A} X \qquad [m.q] \xrightarrow{B} X$$
$$[p.n] \xrightarrow{C} X \qquad [p.q] \xrightarrow{D} X$$

then

$$(A \, \text{®} \, B) \, \text{©} \, (C \, \text{®} \, D) = (A \, \text{©} \, C) \, \text{®} \, (B \, \text{©} \, D \ . \tag{4.4}$$

Remark 4.24. Tables may be organized into a category $\mathbb{T}ab$ in two directions by considering the row lists and column lists as objects, and tables as morphisms. From left to right the domain and codomain of a table are its left-most column and its right-most column; from top to bottom the domain and codomain of a table are its top-most row and its bottom-most row. The identity of a row list in the vertical direction is the list consisting of itself; likewise, the identity of a column list in the horizontal direction is the list consisting of itself. Horizontal composition of two tables such that the right-most column of one equals the left-most column of the other is the table formed by "gluing" the two tables together along their common column. Thus, if the left one of two tables has m rows and n columns, and the right one also has m rows but p columns, the composition has $n + p - 1$ columns. Similarly for vertical composition. The two compositions are related by the Exchange Law as in Eq. (4.4). The crucial difference between the two interpretations of that Exchange Law is that the one in Theorem 4.23 requires no composability condition as in this new double-fold category.

4.4.4 *Outer Product of Two Lists is a Table*

Aside 4.4.1. I find it fairly interesting that high school mathematics students persistently struggle to use the Distributive Law in algebra calculations.[3] Any number of times I have seen a student use the rule $a \times (b + c) = a \times b + c$ instead of $a \times (b + c) = a \times b + a \times c$. One can talk about the colloquial meaning of the word "distribution," and one can draw pretty pictures of sub-divided rectangles and talk about "the whole is equal to the sum of the parts," yet students will continue to make the "distributivity error."

Nevertheless, the idea of multiplying two lists to yield a table – like multiplying two positive numbers to yield an area – is another illustration of how algebra is the geometry of notation. Perhaps student practice with this notion will help educators convey the Distributive Law.

---

[3] Never mind that by using legalistic language in an algebra class there is an opportunity for the multiple meanings of the word "law" to subliminally undermine attention to the entirely benign use of the word in reference to calculation rules.

For example, a calculation can be arranged neatly:

$$(a + b + c + d) \times (x + y + z)$$
$$= \quad a \times (x + y + z)$$
$$+ \; b \times (x + y + z)$$
$$+ \; c \times (x + y + z)$$
$$+ \; d \times (x + y + z)$$
$$= \quad a \times x + a \times y + a \times z$$
$$+ \; b \times x + b \times y + b \times z$$
$$+ \; c \times x + c \times y + c \times z$$
$$+ \; d \times x + d \times y + d \times z$$

Students need to observe that such "multiplying out" leads to a *visual pattern* in which every item from the first sum is paired with and multiplied by every item from the second sum. Abstracting from this pattern of products and sums leads to the idea that any binary operation, say $\otimes$, gives rise to the same pattern:

$$\begin{bmatrix} a \\ b \\ c \\ d \end{bmatrix} \otimes [\, x \;\; y \; z\,] = \begin{bmatrix} a \otimes x & a \otimes y & a \otimes z \\ b \otimes x & b \otimes y & b \otimes z \\ c \otimes x & c \otimes y & c \otimes z \\ d \otimes x & d \otimes y & d \otimes z \end{bmatrix}.$$

Definition 4.25. Let $m, n > 0$, let X, Y be any two sets, and let $X \times Y \xrightarrow{\otimes} Z$ be a binary operation. Define a map

$$X^{[m.1]} \times Y^{[1.n]} \xrightarrow{\otimes} Z^{[m.n]}$$

for $[m.1] \xrightarrow{a} X$ and $[1.n] \xrightarrow{b} Y$ by the $m \cdot n$ equations $_i(a \otimes b)_j := a_i \otimes b_j$, where $1 \leq i \leq m, 1 \leq j \leq n$. The binary operation $\otimes$ is called the **outer product** associated with the given binary operator $\otimes$.

4.5 Vector Algebra

So far, operations on lists and tables depend only on location of expressions upon the Surface. This kind of algebra reflects the geometry of notation. However, In a context where the items positioned in a list – or arrayed in a table – are *numbers*, a new world of calculation opportunities comes into being. Not only that, the new opportunities are intimately related to geometry of the line, the plane, and space. In short, lists of numbers are representations of *vectors* and tables of numbers – *matrices* – are representations of *vector operators*. Vectors are often physical quantities that require lists of numbers for their mathematical representation. For example, *positions* of particles and *forces* that influence their motion are vectors. Vector operators map vector spaces to vector spaces, and represent physical relationships such as that between the stress vector and strain vector in a solid body. The study of vector spaces and vector operators between them is called *vector algebra.*

In its algebra of row lists $\mathbb{R}^{[1.n]}$ has only an "external" role because its items enter into calculations only in relation to other sets of lists $\mathbb{R}^{[1.p]}$, namely via concatenation

$$\mathbb{R}^{[1.n]} \times \mathbb{R}^{[1.p]} \xrightarrow{\quad \textcircled{R} \quad} \mathbb{R}^{[1.n+p]} .$$

But now, because $\mathbb{R}^{[1.n]}$ is built from an algebraic structure $\mathbb{R}$ to begin with, it has an "internal life" of its own as a vector space

$$\mathbb{R}^{[1.n]} \times \mathbb{R}^{[1.n]} \xrightarrow{\quad + \quad} \mathbb{R}^{[1.n]} \quad [[x_1 \cdots x_m][y_1 \cdots y_m]] \xmapsto{+} [x_1 + y_1 \cdots x_m + y_m]$$

$$\mathbb{R} \times \mathbb{R}^{[1.n]} \xrightarrow{\quad \cdot \quad} \mathbb{R}^{[1.n]} \qquad [r[x_1 \cdots x_m]] \mapsto [rx_1 \cdots rx_m] .$$

These internal and external aspects of the vector spaces of row lists are related.

Theorem 4.26. *Scalar multiplication $\cdot$ distributes over concatenation $\textcircled{R}$, and there exists an Exchange Law for row concatenation and vector addition. That is, for vectors $\overrightarrow{x}, \overrightarrow{y} \in \mathbb{R}^{[1.n]}, \overrightarrow{v}, \overrightarrow{w} \in \mathbb{R}^{[1.p]}$ and $r \in \mathbb{R}$,*

$$r \cdot (\overrightarrow{x} \, \textcircled{R} \, \overrightarrow{y}) = r \cdot \overrightarrow{x} \, \textcircled{R} \, r \cdot \overrightarrow{y} \tag{4.5}$$

$$(\overrightarrow{x} + \overrightarrow{y}) \, \textcircled{R} \, (\overrightarrow{v} + \overrightarrow{w}) = (\overrightarrow{x} \, \textcircled{R} \, \overrightarrow{v}) + (\overrightarrow{y} \, \textcircled{R} \, \overrightarrow{w}) . \tag{4.6}$$

4.5.1 *Category of Vector Spaces & Vector Operators*

Definition 4.27. For vector spaces $\mathbb{V}$ and $\mathbb{W}$ over $\mathbb{R}$ with underlying sets V and W respectively, a diagram $\mathbb{V} \xrightarrow{\overrightarrow{A}} \mathbb{W}$ is a **vector operator** if $V \xrightarrow{A} W$

in $\mathbb{S}et$ and for vectors $\overrightarrow{x}, \overrightarrow{y} \in V$ and $r \in \mathbb{R}$,

$$A(r \cdot \overrightarrow{x}) = r \cdot A(\overrightarrow{x}) \tag{4.7}$$

$$A(\overrightarrow{x} + \overrightarrow{y}) = A(\overrightarrow{x}) + A(\overrightarrow{y}) \tag{4.8}$$

Theorem 4.28.

*(1) The identity morphism $V \xrightarrow{1_V} V$ in $\mathbb{S}et$ determines the **identity vector operator** $1_V := \overrightarrow{1_V} : V \longrightarrow V$;*

(2) if $V \xrightarrow{\overrightarrow{A}} W$ and $W \xrightarrow{\overrightarrow{B}} Z$ are vector operators then the composite $V \xrightarrow{\overrightarrow{B} \circ \overrightarrow{A}} Z$ defined by $\overrightarrow{B} \circ \overrightarrow{A} := \overrightarrow{B \circ A}$ is a vector operator.

There exists a category $\mathbb{V}tr$ whose objects are the vector spaces over $\mathbb{R}$ and whose morphisms are the vector operators. There exists a forgetful functor from the category of vector spaces to the category of sets, namely by retaining only the underlying set. There also exists a forgetful functor to the category of commutative groups, namely by forgetting the scalar multiplication and retaining only the additive group of vectors.

Proof. $(\overrightarrow{B} \circ \overrightarrow{A})(r \cdot \overrightarrow{v}) = \overrightarrow{B}(\overrightarrow{A}(r \cdot \overrightarrow{v})) = \overrightarrow{B}(r \cdot \overrightarrow{A}(\overrightarrow{v})) = r \cdot \overrightarrow{B}(\overrightarrow{A}(\overrightarrow{v})) = r \cdot (\overrightarrow{B} \circ \overrightarrow{A})(\overrightarrow{v})$, and $(\overrightarrow{B} \circ \overrightarrow{A})(\overrightarrow{v} + \overrightarrow{v}') = \overrightarrow{B}(\overrightarrow{A}(\overrightarrow{v} + \overrightarrow{v}')) = \overrightarrow{B}(\overrightarrow{A}(\overrightarrow{v}) + \overrightarrow{A}(\overrightarrow{v}')) = \overrightarrow{B}(\overrightarrow{A}(\overrightarrow{v})) + \overrightarrow{B}(\overrightarrow{A}(\overrightarrow{v}')) = (\overrightarrow{B} \circ \overrightarrow{A})(\overrightarrow{v}) + (\overrightarrow{B} \circ \overrightarrow{A})(\overrightarrow{v}')$. $\square$

There are infinitely many vector spaces over $\mathbb{R}$, if only because $1 \xrightarrow{\mathbb{R}^{[1.n]}} \mathbb{V}tr$ for $n > 0$. Throughout this book vector spaces are over $\mathbb{R}$.

4.5.2 *Vector Space Isomorphism*

Definition 4.29. A vector operator $V \xrightarrow{\overrightarrow{A}} W$ is a **vector space isomorphism** if there is a vector operator $V \xleftarrow{\overrightarrow{B}} W$ – called an **inverse operator** to $\overrightarrow{A}$ – such that both $\overrightarrow{B} \circ \overrightarrow{A} = 1_V$ and $\overrightarrow{A} \circ \overrightarrow{B} = 1_W$:

$$\tag{4.9}$$

Theorem 4.30. *If a vector operator has an inverse then it has exactly one inverse.*

Proof.   If $\overrightarrow{C}$ is also an inverse to $\overrightarrow{A}$ – so that $\overrightarrow{C} \circ \overrightarrow{A} = 1_V$ and $\overrightarrow{A} \circ \overrightarrow{C} = 1_W$ – then $\overrightarrow{C} = \overrightarrow{C} \circ 1_W = \overrightarrow{C} \circ (\overrightarrow{A} \circ \overrightarrow{B}) = (\overrightarrow{C} \circ \overrightarrow{A}) \circ \overrightarrow{B} = (\overrightarrow{B} \circ \overrightarrow{A}) \circ \overrightarrow{B} = \overrightarrow{B} \circ (\overrightarrow{A} \circ \overrightarrow{B}) = \overrightarrow{B} \circ 1_W = \overrightarrow{B}$ proves that if $\overrightarrow{A}$ has an inverse then it has only one, and it can be referred to as *the* inverse of $\overrightarrow{A}$.   □

Definition 4.31. The notation

$$\overrightarrow{A} : V \overset{\longleftarrow}{\longrightarrow} W : \overrightarrow{B} \tag{4.10}$$

stands for "$\overrightarrow{A}$ is an isomorphism from V to W with inverse $\overrightarrow{B}$."

Theorem 4.32. *Vector space isomorphism is an equivalence relation among vector spaces.*

Proof.   It is required to demonstrate that the relation of vector space isomorphism is reflexive, symmetric, and transitive.

The unit vector operator $1_V : V \to V$ is a vector space isomorphism because it is its own inverse:

$$1_V : V \overset{\longleftarrow}{\longrightarrow} V : 1_V \; ; \tag{4.11}$$

by the notation symmetry of the triangle diagrams (4.9),

$$\overrightarrow{B} : W \overset{\longleftarrow}{\longrightarrow} V : \overrightarrow{A} \; . \tag{4.12}$$

This proves vector space isomorphism is a symmetric relation. Suppose also

$$\overrightarrow{C} : W \overset{\longleftarrow}{\longrightarrow} Z : \overrightarrow{D} \; . \tag{4.13}$$

then

$$(\overrightarrow{B} \circ \overrightarrow{D}) \circ (\overrightarrow{C} \circ \overrightarrow{A}) = \overrightarrow{B} \circ (\overrightarrow{D} \circ (\overrightarrow{C} \circ \overrightarrow{A})) \tag{4.14}$$

$$= \overrightarrow{B} \circ ((\overrightarrow{D} \circ \overrightarrow{C}) \circ \overrightarrow{A})) \tag{4.15}$$

$$= \overrightarrow{B} \circ (1_W \circ \overrightarrow{A}) \tag{4.16}$$

$$= \overrightarrow{B} \circ \overrightarrow{A} \tag{4.17}$$

$$= 1_V \tag{4.18}$$

and by symmetry of notation likewise $(\overrightarrow{C} \circ \overrightarrow{A}) \circ (\overrightarrow{B} \circ \overrightarrow{D}) = 1_Z$. Hence,

$$\overrightarrow{C} \circ \overrightarrow{A} : V \overset{\longleftarrow}{\longrightarrow} Z : \overrightarrow{B} \circ \overrightarrow{D} \; . \tag{4.19}$$

This proves vector space isomorphism is a transitive relation.   □

Vector space isomorphism is important because: if two vector spaces are isomorphic then any equation whatsoever that holds in one of them – no matter how convoluted the vector or scalar operations involved in the left and right hand expressions of the equation – can be transferred exactly to a corresponding equation holding in the other one. Briefly – except for labels of the vectors – in the context of vector calculations the two spaces are identical.

Definition 4.33. a vector space isomorphism

$$\overrightarrow{A} : \mathbb{V} \rightleftharpoons \mathbb{R}^{[1.n]} : \overrightarrow{B} \tag{4.20}$$

is a **coordinate system** for $\mathbb{V}$. The metaphor that considers $\overrightarrow{A}$ to be an assignment is most appropriate: for any vector $\overrightarrow{v}$ in $\mathbb{V}$, $\overrightarrow{A}$ assigns the *coordinate vector* $\overrightarrow{A}(\overrightarrow{v})$ to $\overrightarrow{v}$. The projection $\pi_i^n(\overrightarrow{A}(\overrightarrow{v}))$ is a number and is called the ith-**coordinate of** $\overrightarrow{v}$ **relative to** $\overrightarrow{A}$.

In $\mathbb{R}^{[1.n]}$ there is a distinguished set of especially simple vectors characterized by having exactly one non-zero projection of value 1. These are denoted by $\overrightarrow{e}_n^j$ for $1 \leq j \leq n$ and defined by

$$\overrightarrow{e}_n^1 := [1\,0\,0\,\ldots\,0\,0]$$
$$\overrightarrow{e}_n^2 := [0\,1\,0\,\ldots\,0\,0]$$

$$\vdots \qquad \vdots$$

$$\overrightarrow{e}^j := [0\,\ldots\,1\,\ldots\,0]$$

$$\vdots \qquad \vdots$$

$$\overrightarrow{e}_n^n := [0\,0\,0\,\ldots\,0\,1]$$

Putting it another way,

$$\pi_i^n(\overrightarrow{e}_n^j) := \begin{cases} 1 & \text{if} \quad i = j \text{ and } 1 \leq i \leq n \text{ and } 1 \leq j \leq n \\ 0 & \text{otherwise} \end{cases}$$

In the important case $n = 3$ another notation for these special vectors is given by

$$\overrightarrow{i} := \overrightarrow{e}_3^1$$
$$\overrightarrow{j} := \overrightarrow{e}_3^2$$
$$\overrightarrow{k} := \overrightarrow{e}_3^3.$$

Definition 4.34. A list $[\,\vec{b}_1\ \ldots\ \vec{b}_n\,]$ of vectors in a vector space $\mathbb{V}$ is a **basis** for $\mathbb{V}$ if for any vector $\vec{v} \in \mathbb{V}$ there is exactly one equation with $\vec{v}$ on the left-hand side and a sum of scalar multiples of the vectors $\vec{b}_1\ \ldots\ \vec{b}_n$ on the right-hand side:

$$\vec{v} = r_1\,\vec{b}_1 + \cdots + r_n\,\vec{b}_n$$

for $r_1, \ldots, r_n \in \mathbb{R}$. In particular,

$$0 = 0\,\vec{b}_1 + \cdots + 0\,\vec{b}_n$$

uniquely represents the 0 vector.

Theorem 4.35. *If $[\,\vec{b}_1\ \ldots\ \vec{b}_n\,]$ is a basis then there is no equation expressing any of its items as a sum of scalar multiples of the other items.*

Proof. If $\vec{b}_j = x_1\vec{b}_1 + \ldots + 0\,\vec{b}_j + \ldots + x_n\,\vec{b}_n$ then $0 = -0\,\vec{b}_j = x_1\vec{b}_1 + \cdots - 1\vec{b}_j + \cdots + x_n\,\vec{b}_n$, which implies the contradiction that $0 = -1$, so such representation is impossible. $\qquad\square$

Definition 4.36. Given a coordinate system (4.20) for $\mathbb{V}$ the list of vectors $[\,\vec{B}(\vec{e}_n^{\,1})\ \ldots\ \vec{B}(\vec{e}_n^{\,j})\ \ldots\ \vec{B}(\vec{e}_n^{\,n})\,]$ is the *basis for $\mathbb{V}$ relative to* (4.20) and its items are called *basis vectors*.

Theorem 4.37. *If $\vec{0} = x_1\vec{e}_n^{\,1} + \cdots + x_n\vec{e}_n^{\,n}$ then $x_1 = \cdots = x_n = 0$.*

Proof. Given the hypothesis,

$$\begin{aligned}
[0\ \ldots\ 0] = \vec{0} &= x_1\vec{e}_n^{\,1} + \cdots + x_n\vec{e}_n^{\,n} \\
&= [\,x_1\,0\ \ldots\ 0\,] + \cdots + [\,0\,0\ \ldots\ x_n\,] \\
&= [\,x_1\ x_2\ \ldots\ x_n\,].
\end{aligned}$$

$\qquad\square$

Theorem 4.38. *The list $[\,\vec{e}_n^{\,1}\ \ldots\ \vec{e}_n^{\,n}\,]$ is a basis for $\mathbb{R}^{[1.n]}$.*

Proof. If $\vec{v} = [\,v_1\ \ldots\ v_n\,]$ then $\vec{v} = v_1\vec{e}_n^{\,1} + \cdots v_n\vec{e}_n^{\,n}$. This proves existence. If also $\vec{v} = a_1\vec{e}_n^{\,1} + \cdots a_n\vec{e}_n^{\,n}$ then $\vec{0} = (v_1\vec{e}_n^{\,1} + \cdots v_n\vec{e}_n^{\,n}) - (a_1\vec{e}_n^{\,1} + \cdots a_n\vec{e}_n^{\,n}) = (v_1 - a_1)\vec{e}_n^{\,1} + \cdots (v_n - a_n)\vec{e}_n^{\,n}$. By the Lemma, $v_1 = a_1, \ldots, v_n = a_n$. This proves uniqueness. $\qquad\square$

Definition 4.39. Call $[\,\vec{e}_n^{\,1}\ \ldots\ \vec{e}_n^{\,n}\,]$ the *standard basis* for $\mathbb{R}[1.n]$.

Theorem 4.40. *If $[\,\vec{b}_1\ \ldots\ \vec{b}_m\,]$ and $[\,\vec{c}_1\ \ldots\ \vec{c}_n\,]$ are bases for $\mathbb{V}$ then $m=n$.*

Proof. Without loss of generality, assume $m < n$. Then $m + 1 \leq n$ and since $[\vec{b}_1 \ldots \vec{b}_m]$ is a basis, $\vec{c}_{m+1}$ may be expressed as a sum of scalar multiples of the b basis vectors. But since $[\vec{c}_1 \ldots \vec{c}_n]$ is a basis, each of the b basis vectors may be expressed as a sum of scalar multiples of the c basis vectors. Combining these expressions, $\vec{c}_{m+1}$ is expressed as a scalar multiple of the other c basis vectors, in contradiction to (4.35). $\square$

Definition 4.41. If a vector space $\mathbb{V}$ has a basis of length n then n is called the *dimension* of $\mathbb{V}$ and $\mathbb{V}$ is called *n-dimensional*.

Theorem 4.42. *Every n-dimensional vector space is isomorphic to $\mathbb{R}^{[1,n]}$.*

Proof. Choose a basis $[\vec{b}_1 \ldots \vec{b}_m]$ for $\mathbb{V}$ and set it into correspondence with the standard basis for $\mathbb{R}^{[1,n]}$. $\square$

Theorem 4.43. *Any two n-dimensional vector spaces are isomorphic.*

Proof. Vector space isomorphism is a transitive relation. $\square$

4.5.3 Inner Product

Definition 4.44. The binary operation on vectors

$$\mathbb{R}^{[1,n]} \times \mathbb{R}^{[1,n]} \xrightarrow{\langle \_|\_ \rangle} \mathbb{R} \tag{4.21}$$

defined by the equation

$$\langle [\, a_1 \ldots a_n \,] | [\, b_1 \ldots b_n \,] \rangle := a_1 b_1 + \cdots a_n b_n \tag{4.22}$$

is called the **inner product**.

Theorem 4.45. *For the inner product in $\mathbb{R}^{[1,n]}$ and $\vec{v}, \vec{w} \in \mathbb{R}^{[1,n]}$, $r \in \mathbb{R}$ the following equations exist*

$$\langle r\vec{v} | \vec{w} \rangle = r\langle \vec{v} | \vec{w} \rangle \tag{4.23}$$

$$\langle \vec{v} + \vec{w} | \vec{z} \rangle = \langle \vec{v} | \vec{z} \rangle + \langle \vec{w} | \vec{z} \rangle \tag{4.24}$$

$$\langle \vec{v} | r\vec{w} \rangle = r\langle \vec{v} | \vec{w} \rangle \tag{4.25}$$

$$\langle \vec{v} | \vec{w} + \vec{z} \rangle = \langle \vec{v} | \vec{w} \rangle + \langle \vec{v} | \vec{z} \rangle \tag{4.26}$$

$$\langle \vec{v} | \vec{w} \rangle = \langle \vec{w} | \vec{v} \rangle \tag{4.27}$$

$$\vec{v} = 0 \quad if \quad \langle \vec{v} | \vec{w} \rangle = 0 \quad for\ all \quad \vec{w} \tag{4.28}$$

$$\langle \vec{0} | \vec{0} \rangle = 0 \quad and\ if \quad \vec{v} \neq \vec{0} \quad then \quad \langle \vec{v} | \vec{v} \rangle > 0. \tag{4.29}$$

Theorem 4.46.

$$\langle \vec{e}^{\,i}, \vec{e}^{\,j} \rangle = \begin{cases} 1 & \text{if } i = j \\ 0 & \text{otherwise} \end{cases}$$

In other words, the inner product is a vector operator with respect to each of its two arguments if the other argument is held fixed:

$$\mathbb{R}^{[n]} \xrightarrow{\langle \vec{v} | \_\rangle} \mathbb{R}$$

$$\mathbb{R}^{[n]} \xrightarrow{\langle \_ | \vec{w} \rangle} \mathbb{R}$$

are both vector operators for any choice of vectors $\vec{v}, \vec{w}$.

4.5.4 *Vector Operator Algebra*

Recall that $\mathrm{Mor}(\mathbb{V}, \mathbb{W})$ is the set of all vector operators from $\mathbb{V}$ to $\mathbb{W}$ in the category $\mathbb{V}tr$.

Theorem 4.47.

(1) $\mathrm{Mor}(\mathbb{V}, \mathbb{W})$ *is a vector space with addition and scalar multiplication for vector operators $\vec{A}, \vec{B}$ and scalar r defined by*

$$\mathbb{V} \xrightarrow{\vec{A} + \vec{B}} \mathbb{W} \qquad (\vec{A} + \vec{B})(\vec{v}) = \vec{A}(\vec{v}) + \vec{B}(\vec{v})$$

$$\mathbb{V} \xrightarrow{r\vec{A}} \mathbb{W} \qquad (r\vec{A})(\vec{v}) = r(\vec{A}(\vec{v})) \; ;$$

(2) $\mathrm{Mor}(\mathbb{V}, \mathbb{V})$ *is a ring of operators with binary multiplication operation given by composition in $\mathbb{V}tr$ and binary addition operation as in (1), so in particular*

$$\vec{A} \circ \left(\vec{B} + \vec{C} \right) = \vec{A} \circ \vec{B} + \vec{A} \circ \vec{B} \; .$$

Remark 4.48. The high school mathematics educator will observe the analogies of these algebraic facts with Distributive and Associative Laws.

The underlying field $\mathbb{R}$ of the vector spaces in $\mathbb{V}tr$ is also a vector space over itself, and

Theorem 4.49. *There exists a vector space isomorphism*

$$\vec{A} : \mathbb{V} \overset{\longleftarrow}{\longrightarrow} \mathrm{Mor}(\mathbb{R}, \mathbb{V}) : \vec{B} \tag{4.30}$$

defined by the equations $\vec{A}(\vec{v})(r) := r\vec{v}$ and $\vec{B}(f) := f(1)$ for all $r \in \mathbb{R}, \vec{v} \in \mathbb{V}$, and $\mathbb{R} \xrightarrow{f} \mathbb{V}$ in $\mathbb{V}tr$.

Remark 4.50. This isomorphism is defined without reference to a coordinate system, and holds for any vector space $\mathbb{V}$ regardless of whether it is finite dimensional or not. Technically, there exists a natural isomorphism

$$\mathbb{V}tr \underset{\text{Mor}(\mathbb{R},\_)}{\overset{1_{\mathbb{V}tr}}{\Longrightarrow}} \mathbb{V}tr \ .$$

4.5.5 Dual Vector Space

For any vector space $\mathbb{V}$, Theorem (4.49) introduced a natural isomorphism

$$\overset{\Rightarrow}{A} : \mathbb{V} \xrightleftharpoons{\hspace{1cm}} \text{Mor}(\mathbb{R}, \mathbb{V}) : \overset{\Rightarrow}{B} \ . \tag{4.31}$$

Interchanging $\mathbb{R}$ and $\mathbb{V}$ leads to a new concept:

Definition 4.51. For any vector space $\mathbb{V}$ its **dual vector space** is $\text{Mor}(\mathbb{V}, \mathbb{R})$. In other words, the dualdual!space of a vector space is the set of real-valued vector operators defined on the vector space. The dual space is so important that a brief notation for it is introduced and defined by

$$\mathbb{V}^* := \text{Mor}(\mathbb{V}, \mathbb{R}) \ .$$

Theorem 4.52. *Let* $[\ \pi_1 \ \dots \ \pi_n\ ]$ *be the list of projection operators* $\pi^j :$ $\mathbb{R}^{[1.n]} \to \mathbb{R}$. *Then* $[\ \pi_1 \ \dots \ \pi_n\ ]$ *is a basis for* $(\mathbb{R}^{[1.n]})^*$. *Therefore,* $(\mathbb{R}^{[1.n]})^*$ *is an n-dimensional vector space.*

Proof. For any $\vec{v} \in \mathbb{V}$ the calculation

$$\begin{aligned}
f(\vec{v}) &= f([\ v_1 \ \dots \ v_n\ ]) \\
&= v_1 f(\vec{e}{}^1_n) + \cdots + v_n f(\vec{e}{}^n_n) \\
&= \pi_1(\vec{v}) f(\vec{e}{}^1_n) + \cdots + \pi_n(\vec{v}) f(\vec{e}{}^n_n) \\
&= (f(\vec{e}{}^1_n)\pi_1 + \cdots + f(\vec{e}{}^n_n)\pi_n)\,(\vec{v})
\end{aligned}$$

shows that there exists an equation $f = f(\vec{e}{}^1_n)\pi_1 + \cdots + f(\vec{e}{}^n_n)\pi_n$. As for uniqueness of such an equation representing f as a sum of scalar multiples of projection operators, suppose $f = a_1 \pi_1 + \cdots + a_n \pi_n$, hence $f(\vec{e}_n^1) \pi_1 + \cdots + f(\vec{e}_n^n) \pi_n = a_1 \pi_1 + \cdots + a_n \pi_n$. Then for $1 \le j \le n$

$$
\begin{aligned}
a_j &= (a_1 \pi_1 + \cdots + a_n \pi_n) \vec{e}_n^j \\
&= \left(f(\vec{e}_n^1) \pi_1 + \cdots + f(\vec{e}_n^n) \pi_n \right) \vec{e}_n^j \\
&= f(\vec{e}_n^j),
\end{aligned}
$$

so there is exactly one such equation. $\qquad\square$

Theorem 4.53. *If $[\, \vec{b}_1 \ \ldots \ \vec{b}_n \,]$ is a basis for $\mathbb{V}$ then $[\, \vec{b}_1^{\,*} \ \ldots \ \vec{b}_n^{\,*} \,]$ is a basis for $\mathbb{V}^*$, where $\mathbb{V} \xrightarrow{\ \vec{b}_j^{\,*}\ } \mathbb{R}$ is defined by $\vec{b}_j^{\,*}((\vec{v})) := v_j$ for $1 \le j \le n$ if $\vec{v} = v_1 \vec{b}_1 + \cdots + v_n \vec{b}_n$.*

Proof.

$$
\begin{aligned}
f(\vec{v}) &= f(v_1 \vec{b}_1 + \cdots + v_n \vec{b}_n) \\
&= v_1 f(\vec{b}_1) + \cdots + v_n f(\vec{b}_n) \\
&= f(b_1) \vec{b}_1^{\,*}(\vec{v}) + \cdots + f(b_n) \vec{b}_n^{\,*}(\vec{v}) \\
&= (f(b_1) \vec{b}_1^{\,*} + \cdots + f(b_n) \vec{b}_n^{\,*})(\vec{v})
\end{aligned}
$$

proves existence of the equation

$$
f = f(b_1) \vec{b}_1^{\,*} + \cdots + f(b_n) \vec{b}_n^{\,*} \, ,
$$

and

$$
\begin{aligned}
0 &= (f(b_1) - a_1) \vec{b}_1^{\,*} + \cdots + (f(b_n) - a_n) \vec{b}_n^{\,*} \\
0 &= (f(b_1) - a_1) \vec{b}_1^{\,*}(b_j) + \cdots + (f(b_n) - a_n) \vec{b}_n^{\,*}(b_j) \\
0 &= f(b_j) - a_j
\end{aligned}
$$

proves uniqueness. $\qquad\square$

Definition 4.54. The basis $[\, \vec{b}_1^{\,*} \ \ldots \ \vec{b}_n^{\,*} \,]$ is called the **dual basis** of $[\, \vec{b}_1 \ \ldots \ \vec{b}_n \,]$.

Therefore, $[\, \pi_1 \ \ldots \ \pi_n \,] = [\, \vec{e}_1^{\,*} \ \ldots \ \vec{e}_n^{\,*} \,]$ is the dual of the standard basis $[\, \vec{e}_1 \ \ldots \ \vec{e}_n \,]$ for $\mathbb{R}^{[1.n]}$.

4.5.6 *Double Dual Vector Space*

Theorem 4.55. *For any vector space* $\mathbb{V}$ *the map* $\mathbb{V} \xrightarrow{\vec{D}} \mathbb{V}^{**} := (\mathbb{V}^*)^*$ *defined for* $\vec{v}$ *in* $\mathbb{V}$ *and* f *in* $\mathbb{V}^*$ *by*

$$(\vec{D}(\vec{v}))(f) = f(\vec{v})$$

is a vector operator.

Proof. For any $f \in \mathbb{V}^*$

$$\vec{D}(\vec{v} + \vec{w}) = \vec{D}(\vec{v}) + \vec{D}(\vec{w})$$

$$\Leftrightarrow \vec{D}(\vec{v} + \vec{w})(f) = \vec{D}(\vec{v})(f) + \vec{D}(\vec{w})(f) \text{ for any } f \in \mathbb{V}^*$$

$$\Leftrightarrow f(\vec{v} + \vec{w}) = f(\vec{v}) + f(\vec{w}) \text{ for any } f \in \mathbb{V}^*,$$

but the last assertion holds since f is a vector operator, so the first equation holds. Likewise,

$$(\vec{D}(r\vec{v}) = r\,\vec{D}(\vec{v})$$

$$\Leftrightarrow (\vec{D}(r\vec{v})(f) = r\,\vec{D}(\vec{v})(f) \text{ for any } f \in \mathbb{V}^*$$

$$\Leftrightarrow f(r\vec{v}) = rf(\vec{v}) \text{ for any } f \in \mathbb{V}^*. \qquad \square$$

Remark 4.56. The vector operator $\mathbb{V} \xrightarrow{\vec{D}} \mathbb{V}^{**}$ is defined "naturally" – without an "arbitrary choice" of a coordinate system. Indeed, there exists a natural transformation

$$\mathbb{V}tr \underset{(\_)^{**}}{\overset{1_{\mathbb{V}tr}}{\Longrightarrow}} \mathbb{V}tr \ .$$

4.5.7 *The Unique Extension of a Vector Operator*

Theorem 4.57. *If* $[\ \vec{b}_1 \ \dots \ \vec{b}_n\ ]$ *is a basis for* $\mathbb{V}$ *and* $\mathbb{V} \xrightarrow{\vec{A}} \mathbb{W}$ *is a linear operator, then* $\vec{A}$ *is uniquely determined by its values* $[\ \vec{A}(\vec{b}_1) \ \dots \ \vec{A}(\vec{b}_n)\ ]$. *That is, there is an equation for* $\vec{A}(\vec{v})$ *entirely in terms of the values* $[\ \vec{A}(\vec{b}_1) \ \dots \ \vec{A}(\vec{b}_n)\ ]$, *and if* $\mathbb{V} \xrightarrow{\vec{B}} \mathbb{W}$ *is a linear operator such that* $[\ \vec{A}(\vec{b}_1) \ \dots \ \vec{A}(\vec{b}_n)\ ] = [\ \vec{B}(\vec{b}_1) \ \dots \ \vec{B}(\vec{b}_n)\ ]$, *then* $\vec{A} = \vec{B}$.

Proof. If $\vec{v} \in \mathbb{V}$ then there exists an equation $\vec{v} = v_1 \vec{b}_1 + \cdots + v_n \vec{b}_n$ hence $\vec{\vec{A}}(\vec{v}) = v_1 \vec{\vec{A}}(\vec{b}_1) + \cdots + v_n \vec{\vec{A}}(\vec{b}_n)$ since $\vec{\vec{A}}$ is a linear operator. This proves $\vec{\vec{A}}$ is determined by its values on the given basis. Then

$$\vec{\vec{A}}(\vec{v}) = v_1 \vec{\vec{A}}(\vec{b}_1) + \cdots + v_n \vec{\vec{A}}(\vec{b}_n)$$
$$= v_1 \vec{\vec{B}}(\vec{b}_1) + \cdots + v_n \vec{\vec{B}}(\vec{b}_n)$$
$$= \vec{\vec{B}}(\vec{v})$$

proves it is uniquely determined. $\square$

The vectors of a basis "generate" the whole vector space through operations of addition and scalar multiplication. Any linear operator in $\mathbb{V}tr$ restricts to a map in $\mathbb{S}et$ defined on the generators. Conversely, if a map is defined on the generators then it automatically extends uniquely to a linear operator on the whole vector space:

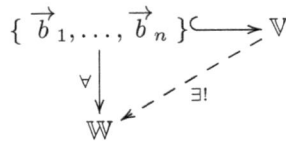

Theorem 4.58. *For finite-dimensional vector spaces there is a bijection between maps on the generators and the vector space of linear operators.*

Remark 4.59. Again, the notion of adjoint functor lurks deeply.

Theorem 4.60. *If $\mathbb{V}$ is finite-dimensional, then the natural vector operator $\vec{\vec{D}}$ is a vector space isomorphism.*

Proof. Let $[\,\vec{b}_1 \ \ldots \ \vec{b}_n\,]$, $[\,\vec{b}_1^* \ \ldots \ \vec{b}_n^*\,]$ and $[\,\vec{b}_1^{**} \ \ldots \ \vec{b}_n^{**}\,]$ be the standard, dual, and dual dual bases of $\mathbb{V}$, $\mathbb{V}^*$ and $\mathbb{V}^{**}$. Define $\mathbb{V}^{**} \xrightarrow{\vec{\vec{E}}} \mathbb{V}$ on generators by $\vec{\vec{E}}(\vec{b}_j^{**}) := \vec{b}_j$. Then the calculation

$$\vec{\vec{D}}(b_j)(\vec{b}_k^*) = b_k^*(\vec{b}_j)$$
$$= \delta_{kj}$$
$$= \delta_{jk}$$
$$= \vec{b}_j^{**}(b_k^*)$$

proves that $\vec{\vec{D}}(b_j) = \vec{b}_j^{**}$, so $\vec{\vec{E}}(\vec{\vec{D}}(\vec{b}_j)) = \vec{\vec{E}}(\vec{b}_j^{**}) = b_j$. In the other direction, $\vec{\vec{D}}(\vec{\vec{E}}(b_j^{**})) = \vec{\vec{D}}(\vec{b}_j) = \vec{b}_j^{**}$. Together these equations prove that $\vec{\vec{E}}$ is inverse to $\vec{\vec{D}}$. $\square$

4.5.8   *The Vector Space of Matrices*

Definition 4.61. Given m, n non-negative integers the **abstract matrix** $[m.n]$ consists of the items $1 \xrightarrow{i\mu_j} [m.n]$ for $1 \le i \le m$ and $1 \le j \le n$. A **matrix with entries in set** X is a map $[m.n] \xrightarrow{A} X$. The i, j **entry of** A is the composite map $_iA_j := A \circ {}_i\mu_j$:

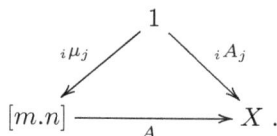

$$
\begin{array}{ccc}
 & 1 & \\
{}_i\mu_j \swarrow & & \searrow {}_iA_j \\
[m.n] & \xrightarrow{\quad A \quad} & X \ .
\end{array}
$$

Theorem 4.62. *The set* $\mathbb{R}^{[m.n]}$ *of matrices with real number entries is a vector space with addition for* $A, B : [m.n] \to \mathbb{R}$ *defined by*

$$
{}_i(A + B)_j := {}_iA_j +_i B_j
$$

and scalar multiplication defined by

$$
{}_i(rA)_j := r({}_iA_j) \ .
$$

4.5.9   *The Matrix of a Vector Operator*

Theorem 4.63. *In* $\mathbb{V}tr$ *there exists a vector space isomorphism*

$$
M : \mathrm{Mor}(\mathbb{R}^{[1.m]}, \mathbb{R}^{[1.n]}) \xrightleftharpoons{\qquad} \mathbb{R}^{[m.n]} : L \tag{4.32}
$$

defined for $\overrightarrow{A} \in \mathrm{Mor}(\mathbb{R}^{[1.m]}\mathbb{R}^{[1.n]})$ *by* $_i(M(\overrightarrow{A}))_j := \pi_j^n(\overrightarrow{A}(\overrightarrow{e}{}^i_m))$ *and for* $A \in \mathbb{R}^{[m.n]}$ *by* $\pi_j^n(L(A)(\overrightarrow{e}{}^i_m)) := {}_iA_j$.

Proof. The verifications are trivial calculations from the definitions:

$$
\begin{aligned}
{}_i(M \circ L)(A)_j &= {}_iM(L(A))_j \\
&= \pi_j^n(L(A)(\overrightarrow{e}{}^i_m)) \\
&= {}_iA_j,
\end{aligned}
$$

so $(M \circ L)(A) = A$, and

$$
\begin{aligned}
\pi_j^n((L \circ M)(\overrightarrow{A})(\overrightarrow{e}{}^i_m)) &= \pi_j^n(L(M(\overrightarrow{A}))(\overrightarrow{e}{}^i_m) \\
&= {}_iM(\overrightarrow{A})_j \\
&= \pi_j^n \overrightarrow{A}(\overrightarrow{e}{}^i_m),
\end{aligned}
$$

so $(L \circ M)(\overrightarrow{A}) = \overrightarrow{A}$. $\qquad\square$

Theorem 4.64. *For an m-dimensional vector space $\mathbb{V}$ and an n-dimensional vector space $\mathbb{W}$ there is a vector space isomorphism*

$$\vec{K} : \mathrm{Mor}(\mathbb{V}, \mathbb{W}) \rightleftharpoons \mathrm{Mor}(\mathbb{R}^{[1.m]}, \mathbb{R}^{[1.n]}) : \vec{L}. \tag{4.33}$$

Proof. Choose coordinate systems for $\mathbb{V}$ and $\mathbb{W}$ and vector operators as in the diagram

$$
\begin{array}{ccc}
\mathbb{V} & \xrightarrow{\vec{T}} & \mathbb{W} \\
\vec{B} \big\updownarrow \vec{A} & & \vec{D} \big\updownarrow \vec{C} \\
\mathbb{R}^{[1.m]} & \xrightarrow[\vec{M}]{} & \mathbb{R}^{[1.n]}.
\end{array}
$$

Define vector operators by $\vec{K}(\vec{T}) := \vec{C} \circ \vec{T} \circ \vec{B}$ and by $\vec{L}(\vec{M}) := \vec{D} \circ \vec{M} \circ \vec{A}$. Then by associativity of map composition, $(\vec{K} \circ \vec{L})(\vec{M}) = \vec{M}$ and likewise $(\vec{L} \circ \vec{K})(\vec{T}) = \vec{T}$. □

Theorem 4.65. *If $\mathbb{V}, \mathbb{W}$ are finite-dimensional vector spaces then there exist non-negative integers m, n such that there is a vector space isomorphism between $\mathrm{Mor}(\mathbb{V}, \mathbb{W})$ and the matrices $\mathbb{R}^{[m.n]}$.*

4.5.10 *Operator Composition & Matrix Multiplication*

Theorem 4.66. *If end-to-end vector operators are given by*

$$\mathbb{R}^{[1.m]} \xrightarrow{\vec{T}} \mathbb{R}^{[1.n]} \xrightarrow{\vec{U}} \mathbb{R}^{[1.p]}$$

then the entries of the matrix of the composition $\vec{U} \circ \vec{T}$ are given by the equation

$$_k(\vec{U} \circ \vec{T})_i = {_kU_1} \cdot {_1T_i} + \cdots + {_kU_n} \cdot {_nT_i}, \quad 1 \le i \le m, 1 \le k \le p.$$

Proof. It follows from $_jT_i = \pi_j^n(\vec{T}(\vec{e}_m^{\,i})), \quad 1 \le j \le n$ that

$$\vec{T}(\vec{e}_m^{\,i}) = {_1T_i}\,\vec{e}_n^{\,1} + \cdots + {_nT_i}\,\vec{e}_n^{\,n}.$$

Therefore,

$$
\begin{aligned}
{}_k(\vec{\vec{U}} \circ \vec{\vec{T}})_i &= \pi_k^p(\vec{\vec{U}} \circ \vec{\vec{T}})(\vec{e}_m^i) \\
&= \pi_k^p(\vec{\vec{U}}(\vec{\vec{T}})(\vec{e}_m^i)) \\
&= \pi_k^p(\vec{\vec{U}}({}_1T_i \vec{e}_n^1 + \cdots + {}_nT_i \vec{e}_n^n)) \\
&= \pi_k^p(\vec{\vec{U}}({}_1T_i \vec{e}_n^1) + \cdots + \vec{\vec{U}}({}_nT_i \vec{e}_n^n)) \\
&= \pi_k^p(\vec{\vec{U}}({}_1T_i \vec{e}_n^1)) + \cdots + \pi_k^p(\vec{\vec{U}}({}_nT_i \vec{e}_n^n)) \\
&= {}_1T_i \pi_k^p(\vec{\vec{U}}(\vec{e}_n^1)) + \cdots + {}_nT_i \pi_k^p(\vec{\vec{U}}(\vec{e}_n^n)) \\
&= {}_kU_1 \cdot {}_1T_i + \cdots + {}_kU_n \cdot {}_nT_i, \quad 1 \le i \le m, 1 \le k \le p
\end{aligned}
$$

since entries of the matrix corresponding to $\vec{\vec{U}}$ are given by ${}_kU_j = \pi_k^p(\vec{\vec{U}}(\vec{e}_n^j))$. $\qquad\square$

This theorem prompts the

Definition 4.67. Matrices $T \in \mathbb{R}^{[m.n]}$ and $U \in \mathbb{R}^{[n.p]}$ are called *multiplicable* and their *product* is defined by the entries

$$
{}_k(U * T)_i = {}_kU_1 \cdot {}_1T_i + \cdots + {}_kU_n \cdot {}_nT_i, \quad 1 \le i \le m, 1 \le k \le p.
$$

In summary, finite dimensional vector spaces are isomorphic to vector spaces of rows of real numbers, vector operators correspond to matrices of real numbers, and composition of vector operators corresponds to multiplication of matrices.

The vector space isomorphism between finite-dimensional vector operators and matrices means that what may appear at first sight to be an infinite amount of information – after all, a vector operator may be applied to infinitely many vectors – is actually summarized compactly in a finite amount of information: the matrix of the operator with respect to given coordinate systems.

4.5.11 *More on Vector Operators*

Theorem 4.68. *For any two vector operators*

$$
\vec{\vec{A}} : \mathbb{V} \xrightleftharpoons{\hspace{1cm}} \mathbb{V} : \vec{\vec{B}}
$$

the following are equivalent:

$$\langle \vec{A}(\vec{x})|\vec{y}\rangle = \langle \vec{x}|\vec{A}(\vec{y})\rangle \quad \text{for all} \quad \vec{x}, \vec{y} \in \mathbb{V} \tag{4.34}$$

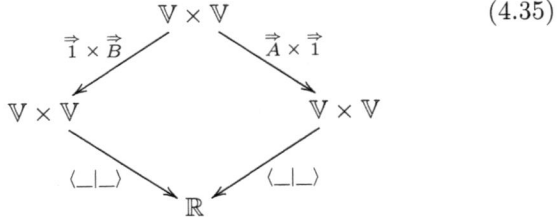

$$\tag{4.35}$$

Theorem 4.69. *([Lang (1987)] p.181) For any vector operator $\vec{A}: \mathbb{V} \to \mathbb{V}$ there exists a unique vector operator $\vec{B}: \mathbb{V} \to \mathbb{V}$ such that $\langle \vec{A}(\vec{x})|\vec{y}\rangle = \langle \vec{x}|\vec{A}(\vec{y})\rangle$ for all $\vec{x}, \vec{y} \in \mathbb{V}$.*

Definition 4.70. Denote that unique vector operator by $\vec{A}^{\top}$ and call it the *transpose* of $\vec{A}$. Say $\vec{A}$ is *symmetric* if $\vec{A}^{\top} = \vec{A}$.

Theorem 4.71. *([Curtis (1974)] p.274) If $\vec{A}: \mathbb{V} \to \mathbb{V}$ is symmetric and $[A]$ is the matrix representation of $\vec{A}$ relative to an orthonormal basis of $\mathbb{V}$, then $[A]$ is a symmetric matrix, and vice versa.*

Theorem 4.72. *If $\vec{A}: \mathbb{V} \to \mathbb{V}$ is a symmetric vector operator then there exists an orthonormal basis of $\mathbb{V}$ consisting of eigenvectors of $\vec{A}$.*

Theorem 4.73. *If $[A]$ is a symmetric matrix then $[A]$ has n eigenvalues and eigenvectors of distinct eigenvalues are orthonormal vectors.*

Theorem 4.74. *If $[A]$ is a symmetric matrix then there exists an invertible matrix $[P]$ such that $[P]^{-1}[A][P]$ is a diagonal matrix whose entries are the eigenvalues of $[A]$.*

Definition 4.75. Vectors $\vec{x}, \vec{y}$ are **parallel** and I write $\vec{x} \parallel \vec{y}$ if $\vec{x} = r\vec{y}$ for some $1 \xrightarrow{r} \mathbb{R}$.

Theorem 4.76. $\parallel$ *is an equivalence relation.*

Theorem 4.77. *If $\mathbb{V} \xrightarrow{\vec{A}} \mathbb{V}$ is an operator such that $|\vec{n}| = 1$ implies $\vec{A}(\vec{n})$ is parallel to $\vec{n}$, then there exists a number $1 \xrightarrow{p} \mathbb{R}$ such that $\vec{A} = p \cdot 1_{\mathbb{V}}$.*

Proof. If $\vec{A}(\vec{n}) \| \vec{n}$ for $|\vec{n}| = 1$ then $\vec{A}(\vec{n}) = r\vec{n}$ for some $\mathbf{1} \xrightarrow{r} \mathbb{R}$. Then

$$\frac{\vec{A}(\vec{n})}{|\vec{A}(\vec{n})|} = \frac{r\vec{n}}{|r\vec{n}|} = \frac{\vec{n}}{|\vec{n}|} = \vec{n} \ ,$$

so

$$\vec{A}(\vec{n}) = |\vec{A}(\vec{n})|\vec{n} \quad \text{for any unit vector } \vec{n} \ .$$

Let $\vec{m}, \vec{n}$ be unit vectors, and so

$$\vec{m} = \frac{\vec{A}(\vec{m})}{|\vec{A}(\vec{m})|}$$

$$\vec{n} = \frac{\vec{A}(\vec{n})}{|\vec{A}(\vec{n})|} \ .$$

Note that

if $\vec{m}, \vec{n}$ are non-parallel, then $x\vec{m} = y\vec{n}$ implies $x = y = 0$,

for, if $x\vec{m} = y\vec{n}$ with $x \neq 0$ then $\vec{m} = \frac{y}{x}\vec{n}$, which contradicts the assumption that $\vec{m}, \vec{n}$ are non-parallel.

Suppose $\vec{m}$ and $\vec{n}$ are non-parallel unit vectors. Prove that $|\vec{A}(\vec{m})| = |\vec{A}(\vec{n})|$ as follows.

$$\frac{|\vec{A}(\vec{m})|\vec{m} + |\vec{A}(\vec{n})|\vec{n}}{|\vec{A}(\vec{m} + \vec{n})|} = \frac{\vec{A}(\vec{m}) + \vec{A}(\vec{n})}{|\vec{A}(\vec{m} + \vec{n})|}$$

$$= \frac{\vec{A}\left(\frac{\vec{m}+\vec{n}}{|\vec{m}+\vec{n}|}\right)}{\left|\vec{A}(\frac{\vec{m}+\vec{n}}{|\vec{m}+\vec{n}|})\right|}$$

$$= \frac{\vec{m} + \vec{n}}{|\vec{m} + \vec{n}|}.$$

Therefore,

$$|\vec{A}(\vec{m})|\vec{m} + |\vec{A}(\vec{n})|\vec{n} = \frac{|\vec{A}(\vec{m} + \vec{n})|}{|\vec{m} + \vec{n}|}(\vec{m} + \vec{n}),$$

so that

$$\left(|\vec{A}(\vec{m})| - \frac{\vec{A}(\vec{m} + \vec{n})}{|\vec{m} + \vec{n}|}\right)\vec{m} = \left(-|\vec{A}(\vec{n})| + \frac{\vec{A}(\vec{m} + \vec{n})}{|\vec{m} + \vec{n}|}\right)\vec{n}.$$

Consequently

$$|\vec{\vec{A}}(\overrightarrow{m})| = \frac{|\vec{\vec{A}}(\overrightarrow{m} + \overrightarrow{n})|}{|\overrightarrow{m} + \overrightarrow{n}|} = |\vec{\vec{A}}(\overrightarrow{n})|.$$

Thus, the definition $p := |\vec{\vec{A}}(\overrightarrow{m})|$ is independent of the choice of unit vector $\overrightarrow{m}$. $\qquad\square$

PART 3
Particle Mechanics

This part of the book interconnects Newtonian, Lagrangian, and Hamiltonian equations for a cloud of particles flying around in empty space. Famous arguments by George Stokes, Paul Langevin, and Albert Einstein extend the picture to include a cloud of particles moving around in a liquid. For a certain group of muscle contraction researchers, this is the context "where the rubber meets the road," since the molecules responsible for motion of living beings are within cells packed with all sorts of other molecules that get in the way.

Chapter 5

Particle Universe

A particle universe is nothing but massive particles zooming around in space. To say "particle" implies a point-like item, that is to say, an item with a size of magnitude 0 (in every direction). Nevertheless, a "massive" particle has a single positive numerical magnitude called its mass. The mass of a particle completely characterizes its response to force. That each massive particle moves in a "flat 3-dimensional space" implies that there exists at least one coordinate system such that the position of the particle is completely specified by a triple of numbers, and that at any time the distance between two particles may be calculated using Pythagoras' Formula: square root of the sum of squares of coordinate differences.

Any changes in particle motion are due to forces. In other words, the "ontology" of a particle universe is only interesting if in addition to the particles there are forces that influence their motion.

In particular, a "uniform force field" is a model of gravity. If there is an obstacle to motion called the "ground" a particle is on the ground if its third coordinated is 0, and above the ground if its third coordinate is greater than 0. A particle initially above the ground falls to the ground because gravity forces it down.

The very idea of motion implies change in position over a period of time. Time is a "flat 1-dimensional space" extending from a beginning that is to time as the ground is to space, except that particles fall upward through time and downward through space.

The state of a particle at a specified time is its position and velocity at that time, both of which may change as time goes on. Hence, the state of a particle is a trajectory in a 6-dimensional space, with 3 coordinates for position and another 3 for velocity. All the particles of a particle universe together have a state in a high-dimensional space with six independent coordinates for each individual particle.

The momentum of a particle in a given direction is proportional to its mass and to the velocity in that direction. The constant force of gravity imparts upon each above-ground particle a proportional change in its momentum. That the change in momentum is proportional to the force upon it is Newton's Second Law of motion. Whatever the state of a particle universe at the start of time, the evolution of the state is completely determined by Newton's Second Law.

More precisely, Newton's Second Law declares that the gravitational force on a particle equals the rate of change of its momentum. In other words, force equals mass times acceleration. Therefore, an above-ground particle accelerates towards the ground. In particle free fall, the higher it starts, the more velocity it has upon hitting the ground. Newton's Second Law is in fact equivalent to conservation of total energy – the sum of potential energy and kinetic energy ([Arons (1965)] Ch.18).

However, in contrast to free fall, interaction between particles – collisions – can instantaneously change their momenta. Nevertheless, momentum is conserved in a collision.

Moving a particle in a particular direction means changing its position, and that requires effort only if there is some resistance to moving it. If it has no mass then there is no resistance to moving it. If it has mass then some effort is required to move it. The effort required to move a particle is called the **force** and for a given distance through which the particle moves, greater effort means greater force; for a given force the greater distance the particle moves, the greater the effort expended. Hence, the **work** performed to move the particle is defined to be the applied force multiplied by the distance moved. If multiple forces are applied to the particle their resultant force does work on it equal to the sum of the works done by each individually. The work done by the resultant force on a particle equals the change in its kinetic energy.

If the gravitational force is a constant, the potential energy of a particle is proportional to its mass and its height above the ground.

Newton's First Law of motion is that a particle subject to no force continues in its motion without change in momentum. A particle with unchanging momentum is said to be in equilibrium. Hence, between collisions a particle is in equilibrium.

When a system is in such a state that after any slight temporary disturbance of external conditions it returns rapidly or slowly to the initial state, this state is said to be one of equilibrium. *A state of equilibrium is a state of rest* [Lewis and Randall (1923)].

A state of "rest" carries implicitly a relationship to a coordinate system with respect to which is measured the opposite of "rest," which is "motion." This in turn contains an assumption about what are the relevant magnitudes comprising the coordinates of the state.

5.1 Conservation of Energy & Newton's Second Law

In this context x is the position of a particle of mass m, then its velocity is $\dot{x}$ and its acceleration is $\ddot{x}$. By definition, the total energy

$$E = E(x, \dot{x}) := K + V$$

is the *sum* of energy of motion, the kinetic energy

$$K = K(\dot{x}) := \frac{1}{2}m\dot{x}^2$$

and energy of position, the potential energy $V = V(x)$. **Conservation of energy** means the energy is constant, which expressed formally is that

$$\frac{dE}{dt} = 0 .$$

Theorem 5.1. *If the vector force field is the negative gradient* $-\dfrac{dV}{dx}$ *of scalar potential field V then conservation of energy is logically equivalent to Newton's Second Law.*

Proof. ($\Rightarrow$) Calculate

$$0 = \frac{dE}{dt}$$

$$= \frac{d}{dt}(K + V)$$

$$= \frac{dK}{d\dot{x}}\frac{d\dot{x}}{dt} + \frac{dV}{dx}\frac{dx}{dt}$$

$$= m\dot{x}\ddot{x} + \frac{dV}{dx}\dot{x} .$$

Dividing both sides by $\dot{x}$ yields the forward implication.

($\Leftarrow$) Calculate:

$$-(V_1 - V_0) = -\int_{V_0}^{V_1} dV \qquad \text{Fundamental Rule}$$

$$= \int_{x_0}^{x_1} \left(-\frac{dV}{dx}\right) dx \qquad \text{Change-of-Variable Rule}$$

$$= \int_{t_0}^{t_1} \left(-\frac{dV}{dx}\right) \frac{dx}{dt} dt \qquad \text{Change-of-Variable Rule}$$

$$= \int_{t_0}^{t_1} m\ddot{x}\frac{dx}{dt} dt \qquad \text{by hypothesis}$$

$$= \int_{t_0}^{t_1} m\ddot{x}\dot{x} dt \qquad \text{dot notation for time derivative}$$

$$= m \int_{t_0}^{t_1} \frac{d\dot{x}}{dt}\dot{x} dt \qquad \text{integration commutes with}$$
$$\text{scalar multiplication}$$

$$= m \int_{t_0}^{t_1} \dot{x} d\dot{x} \qquad \text{cancellation of } dt$$

$$= m \left[\frac{1}{2}\dot{x}^2\right]_{t_0}^{t_1} \qquad \text{Fundamental Rule}$$

$$= \frac{1}{2}m\dot{x_1}^2 - \frac{1}{2}m\dot{x_0}^2 \qquad \text{definition of } [\cdots]_{t_0}^{t_1}.$$

Consequently after re-arrangement,

$$E_1 = \frac{1}{2}m\dot{x_1}^2 + V_1 = \frac{1}{2}m\dot{x_0}^2 + V_0 = E_0 \, ,$$

so the energy does not change during the interval from t_0 to t_1, which is the reverse implication. $\qquad\qquad\square$

5.2   Lagrange's Equations & Newton's Second Law

Aside 5.2.1. Chapter 19 of Richard P. Feynman's *Lectures on Physics, Volume II* is intended for "entertainment" and is "almost verbatim" the record of a special lecture on The Principle of Least Action [Feynman *et al.* (1964)]. The difference $L = L(q, \dot{q}) := K(\dot{q}) - V(q)$, which –unlike the total energy $K + V$ – is the kinetic energy *minus* the potential energy, is generally called the **Lagrangian** of a mechanical system in motion. The variable q represents the position (vector) of the system and $\dot{q}$ represents its velocity (vector). He discusses the **action** A of a particle in motion, which is the integral of the Lagrangian over the time of a motion, as in

$$A = A(q) := \int_{t_0}^{t_1} L(q, \dot{q}) dt \ .$$

In his enthusiastic way he demonstrates that if there exists a motion q during t_0 to t_1 such that A has an extreme value (either minimum or maximum) relative to variations of q (with endpoints fixed), then q satisfies Newton's Second Law. I prove something a little interesting, but much simpler.

Definition 5.2. For a mechanical system with position $q = \overrightarrow{q} = (q_1, \ldots q_n)$, velocity $\dot{q} = \overrightarrow{\dot{q}} = (\dot{q}_1, \ldots \dot{q}_n)$, and Lagrangian $L = L(\overrightarrow{q}, \overrightarrow{\dot{q}})$,

$$\frac{d}{dt} \frac{\partial L}{\partial \dot{q}_j} = \frac{\partial L}{\partial q_j} \quad j = 1, \ldots, n \tag{5.1}$$

are called **Lagrange's Equations.** If the system is 1-dimensional and $L = L(x, \dot{x})$ these equations reduce to

$$\frac{d}{dt} \frac{dL}{d\dot{x}} = \frac{dL}{dx} \ .$$

Theorem 5.3. *The 1-Dimensional Lagrange's Equation is logically equivalent to Newton's Second Law.*

Proof. Kinetic energy is independent of position, and potential energy is independent of velocity. Hence

$$0 = \frac{d}{dt} \frac{d(K - V)}{d\dot{x}} - \frac{d(K - V)}{dx}$$

$$= \frac{d}{dt} \frac{dK}{d\dot{x}} - \frac{d}{dt} \frac{dV}{d\dot{x}} - \frac{dK}{dx} + \frac{dV}{dx}$$

$$= \frac{d}{dt} \frac{dK}{d\dot{x}} - \frac{d}{dt} 0 - 0 + \frac{dV}{dx}$$

$$= m\ddot{x} + \frac{dV}{dx} \ .$$

$\square$

5.3   The Invariance of Lagrange's Equations

Theorem 5.4. *Assume there exists equations for change of coordinates* $\vec{x} = x(\vec{q})$ *as in*

$$x_1 = x_1(q_1, \ldots, q_n)$$

$$\cdots\cdots\cdots$$

$$x_m = x_m(q_1, \ldots, q_n).$$

Then

(1)

$$\frac{d\dot{x}_i}{dq_k} = \frac{d}{dt}\frac{dx_i}{dq_k} \; ; \tag{5.2}$$

(2)

$$\frac{d\dot{x}_i}{d\dot{q}_j} = \frac{dx_i}{dq_j} \; ; \tag{5.3}$$

(3)

$$\frac{d}{dt}\left(\frac{dK}{d\dot{x}_i}\frac{dx_i}{dq_j}\right) = \frac{dK}{d\dot{x}_i}\frac{d\dot{x}_i}{dq_j} + \left(\frac{d}{dt}\frac{dK}{d\dot{x}_i}\right)\frac{d\dot{x}_i}{d\dot{q}_j} \; . \tag{5.4}$$

Proof.

(1)

$$\frac{d\dot{x}_i}{dq_k} = \frac{d}{dq_k}\sum_j \frac{dx_i}{q_j}\dot{q}_j$$

$$= \sum_j \frac{d}{dq_k}\left(\frac{dx_i}{dq_j}\dot{q}_j\right)$$

$$= \sum_j \left(\frac{dx_i}{dq_j}\frac{d\dot{q}_j}{dq_k} + \frac{d^2 x_i}{dq_k\,dq_j}\dot{q}_j\right)$$

$$= \sum_j \left(\frac{dx_i}{dq_j}\cdot 0 + \frac{d^2 x_i}{dq_k\,dq_i}\dot{q}_j\right)$$

$$= \frac{d}{dt}\frac{dx_i}{dq_k}.$$

(2)

$$\frac{d\dot{x}_i}{d\dot{q}_j} = \frac{d}{d\dot{q}_j} \sum_k \frac{dx_i}{dq_k} \dot{q}_k$$

$$= \sum_k \frac{d}{d\dot{q}_j} \left(\frac{dx_i}{dq_k} \dot{q}_k \right)$$

$$= \sum_k \left(\frac{d}{d\dot{q}_j} \frac{dx_i}{dq_k} \dot{q}_k + \frac{dx_i}{dq_k} \frac{d\dot{q}_k}{d\dot{q}_j} \right)$$

$$= \sum_k \left(0 \cdot \dot{q}_k + \frac{dx_i}{dq_k} \cdot \delta_j^k \right)$$

$$= \frac{dx_i}{dq_j} .$$

(3)

$$\frac{d}{dt} \left(\frac{dK}{d\dot{x}_i} \frac{dx_i}{dq_j} \right) = \frac{dK}{d\dot{x}_i} \frac{d}{dt} \frac{dx_i}{dq_j} + \left(\frac{d}{dt} \frac{dK}{d\dot{x}_i} \right) \frac{dx_i}{dq_j}$$

$$= \frac{dK}{d\dot{x}_i} \frac{d\dot{x}_i}{dq_j} + \left(\frac{d}{dt} \frac{dK}{d\dot{x}_i} \right) \frac{d\dot{x}_i}{d\dot{q}_j} .$$

$\square$

Theorem 5.5. *If Newton's Second Law is true then a change of coordinates* $\vec{x} = x(\vec{q})$ *as in*

$$x_1 = x_1(q_1, \dots, q_n)$$

$$\dots\dots\dots$$

$$x_m = x_m(q_1, \dots, q_n) .$$

does not change the form of Langrange's Equations.

Proof. Assume the potential energy is given by

$$V = V(x_1, \dots, x_m) = V(x_1(q_1, \dots q_n), \dots, x_m(q_1, \dots q_n))$$

Then

$$\frac{dV}{dq_j} = \sum_i \frac{dV}{dx_i} \frac{dx_i}{dq_j} \quad \text{and} \quad \frac{dV}{d\dot{q}_j} = 0 .$$

Calculate

$$-\frac{dV}{dq_j} = \sum_i \left(-\frac{dV}{dx_i}\right) \frac{dx_i}{dq_j} \qquad \text{Chain Rule} \qquad (5.5)$$

$$= \sum_i (m_i \ddot{x}_i) \frac{dx_i}{dq_j} \qquad \text{Newton's Second Law} \qquad (5.6)$$

$$= \sum_i \left(\frac{d}{dt}\frac{dK}{d\dot{x}_i}\right) \frac{dx_i}{dq_j} \qquad \text{definition of } K \qquad (5.7)$$

$$= \sum_i \left[\frac{d}{dt}\left(\frac{dK}{d\dot{x}_i}\frac{dx_i}{dq_j}\right) - \frac{dK}{d\dot{x}_i}\frac{d\dot{x}_i}{dq_j}\right] \qquad \text{Eq. (5.2), Product Rule} \qquad (5.8)$$

$$= \frac{d}{dt}\sum_i \frac{dK}{d\dot{x}_i}\frac{d\dot{x}_i}{d\dot{q}_j} - \sum_i \frac{dK}{d\dot{x}_i}\frac{d\dot{x}_i}{dq_j} \qquad \text{Eq. (5.3)} \qquad (5.9)$$

$$= \frac{d}{dt}\frac{dK}{d\dot{q}_j} - \frac{dK}{dq_j} \qquad \text{Chain Rule} \qquad (5.10)$$

Consequently,

$$-\frac{dV}{dq_j} = \frac{d}{dt}\frac{dK}{d\dot{q}_j} - \frac{dK}{dq_j} \quad \text{so}$$

$$\frac{d}{dt}\frac{dK}{d\dot{q}_j} - \frac{d(K-V)}{dq_j} = 0 \quad \text{and therefore} \quad \frac{d}{dt}\frac{d(K-V)}{d\dot{q}_j} - \frac{d(K-V)}{dq_j} = 0.$$

By definition of the Lagrangian,

$$\frac{d}{dt}\frac{dL}{d\dot{q}_j} = \frac{dL}{dq_j},$$

which is exactly the same form as Lagrange's Equations (5.1) in terms of x. $\qquad \square$

Remark 5.6. The basic idea that an equation expressing a Law of Physics has the same form regardless of the coordinate system is called *covariance*. Covariance is discussed in greater detail than appropriate for this book in ([Goldstein (1980)] p. 277) and in [Marsden and Hughes (1983)].

5.4   Hamilton's Principle

Here are formal details of the "entertainment" provided by Richard P. Feynman introduced in Sec. 5.2.

$$
\begin{array}{ccc}
[\vec{x}] & \dfrac{[\vec{x}]}{[T]} & [E] \\
\mathbb{R}^n \times \mathbb{R}^n & \xrightarrow{\quad L \quad} & \mathbb{R} \\
\vec{q} & \dot{\vec{q}} & L = L(\vec{q}, \dot{\vec{q}})
\end{array}
\tag{5.11}
$$

The setup (5.11) represents the Lagrangian of a mechanical system with n coordinates $\vec{q}$ having physical dimensions $[\vec{x}]$. The Lagrangian also depends on the variables $\dot{\vec{q}}$ representing the time rates of change of the coordinates.

To represent the idea of an *imaginary variation* of a path in the n-dimensional coordinate space of $\vec{q}$, the basic setup is given in Eq. (5.12),

$$
\begin{array}{ccc}
[T][iT] & [X]^n & \\
\mathbb{I} \times \mathbb{I} & \xrightarrow{\quad\quad} \mathbb{R}^n & \underset{b}{\overset{a}{\rightleftarrows}} 1 \\
t \quad e & \vec{q} &
\end{array}
\tag{5.12}
$$

wherein $\mathbb{I} = [0,1] \subset \mathbb{R}$ denotes the closed unit interval. This map $\vec{q}$ carries the closed unit square $\mathbb{I} \times \mathbb{I}$ into $\mathbb{R}^n$.

Definition 5.7. For a setup

$$
\begin{array}{ccc}
[T] & [X]^n & \\
\mathbb{I} & \xrightarrow{\quad \underline{q} \quad} \mathbb{R}^n & \underset{b}{\overset{a}{\rightleftarrows}} 1 \\
t & \underline{q} &
\end{array}
\tag{5.13}
$$

say q **varies** $\underline{q}$ **with fixed endpoints** a **and** b

$$q(t,0) = \underline{q}(t) \qquad\qquad \text{for } t \in \mathbb{I} \tag{5.14}$$

$$q(0,e) = a \qquad\qquad \text{for } e \in \mathbb{I} \tag{5.15}$$

$$q(1,e) = b \qquad\qquad \text{for } e \in \mathbb{I} \tag{5.16}$$

There are two time coordinates in the setup Eq. (5.12). Variable t represents the time of the actual system trajectory represented by $\underline{q}$. Variable e is an "imaginary" time during which the actual system trajectory is varied.

Thus, at imaginary time e the map $q(\_, e)$ is a particular variation of $\underline{q}$, but starting at a and ending at b.

By the substitution

$$q = q(t, e)$$

in the Lagrangian

$$L = L(q, \dot{q}),$$

the Lagrangian is defined for the imaginary variations of $\underline{q}$,

$$L = L(q(t, e), \dot{q}(t, e)),$$

wherein it is assumed that

$$\dot{q} = \frac{dq}{dt}.$$

Given these setups and equations for imaginary variation of $\underline{q}$ by q, a new variable $J = J(e)$ called the *action* along the imaginary path at e is defined by the setup

$$
\begin{array}{lll}
[iT] & [E][T] & \qquad (5.17)\\[4pt]
\mathbb{I} \longrightarrow \mathbb{R} & \\[4pt]
e & J_q &
\end{array}
$$

and the equation

$$J_q = J_q(e) := \int_0^1 L(q(t, e), \dot{q}(t, e))dt.$$

Note for sure that the physical dimension of action is *energy* × *time*, **[NRG][TME]**. Also, please bear in mind that – by definition – the Lagrangian is the kinetic energy minus potential energy. The following theorem declares that the actual motion $\underline{q} = \underline{q}(t)$ conforms to Lagrange's Equations precisely when the action is at an extreme value for *any* variation q of $\underline{q}$.

Now, an object thrown up in a gravitational field does rise faster first and then slow down. That is because there is also the potential energy, and we must have the least *difference* of kinetic and potential energy on the average. Because the potential energy rises as we go up in space, we

will get a lower difference if we can get as soon as possible up to where there is a high potential energy. Then we can take that potential away from the kinetic energy and get a lower average. So it is better to take a path which goes up and gets a lot of negative stuff from the potential energy. On the other hand, you can't go up too fast, or too far, because you will then have too much kinetic energy involved – you have to go very fast to get way up and come down again in the fixed amount of time available. So you don't want to go too far up, but you want to go up some. So it turns out that the solution is some kind of balance between trying to get more potential energy with the least amount of extra kinetic energy – trying to get the difference, kinetic minus the potential as small as possible [Feynman *et al.* (1964)].

Theorem 5.8. *The following are equivalent:*

$$\text{Lagrange's Equations} \quad \left(\frac{d}{dt} \frac{d}{d\dot{q}_j} - \frac{d}{dq_j} \right) L(\underline{q}, \underline{\dot{q}}) = 0 \tag{5.18}$$

For any variation q of $\underline{q}$ there exists an equation $\dfrac{dJ_q}{de}(0) = 0$ (5.19)

Proof. (5.18 $\Rightarrow$ 5.19) Suppose q satisfies Langrange's Equation, and that q is a variation of $\underline{q}$. Then

$$\frac{dJ_q}{de} = \frac{d}{de} \int_0^1 L(q(t,e), \dot{q}(t,e)) dt \qquad \text{definition of } J_q$$

$$= \int_0^1 \frac{d}{de} L(q(t,e), \dot{q}(t,e)) dt \qquad \text{Differentiation Under the Integral}$$

$$= \int_0^1 \left(\frac{dL}{dq} \frac{dq}{de} + \frac{dL}{d\dot{q}} \frac{d\dot{q}}{de} \right) dt \qquad \text{Chain Rule} \tag{5.20}$$

$$= \int_0^1 \left(\frac{dL}{dq} \frac{dq}{de} + \frac{dL}{d\dot{q}} \frac{d}{dt} \frac{dq}{de} \right) dt \qquad \text{Eq. (5.2)} \tag{5.21}$$

$$= \int_0^1 \left[\frac{dL}{dq}\frac{dq}{de} + \frac{d}{dt}\left(\frac{dL}{d\dot{q}}\frac{dq}{de} \right) - \left(\frac{d}{dt}\frac{dL}{d\dot{q}} \right)\frac{dq}{de} \right] dt \quad \text{Product Rule} \quad (5.22)$$

$$= \int_0^1 \frac{d}{dt}\left(\frac{dL}{d\dot{q}}\frac{dq}{de} \right) dt + \int_0^1 \left[\left(\frac{d}{dq} - \frac{d}{dt}\frac{d}{d\dot{q}} \right) L \cdot \frac{dq}{de} \right] dt$$

Commutative & Distributive Laws.   (5.23)

The climactic moment in the calculation above is the use of the trick

$$\frac{d}{dt}\left(\frac{dL}{d\dot{q}}\frac{dq}{de} \right) = \frac{dL}{d\dot{q}}\left(\frac{d}{dt}\frac{dq}{de} \right) + \left(\frac{d}{dt}\frac{dL}{d\dot{q}} \right)\frac{dq}{de} \quad \text{Product Rule} \quad (5.24)$$

upon that middle term $\dfrac{d}{dt}\left(\dfrac{dL}{d\dot{q}}\dfrac{dq}{de} \right)$ in (5.21). Now there are two integrals
to evaluate in (5.22).

The first integral is 0 according to the following calculation:

$$\int_0^1 \frac{d}{dt}\left(\frac{dL}{d\dot{q}}\frac{dq}{de} \right) dt = \left[\frac{dL}{d\dot{q}}\frac{dq}{de} \right]_0^1$$

$$\text{Fundamental Rule}$$

$$= \frac{dL}{d\dot{q}}(q(1,e),\dot{q}(1,e))\frac{dq}{de}(1,e) - \frac{dL}{d\dot{q}}(q(0,e),\dot{q}(0,e))\frac{dq}{de}(0,e)$$

$$\text{definition of } [\cdots]_0^1$$

$$= \frac{dL}{d\dot{q}}(q(1,e),\dot{q}(1,e)) \cdot 0 - \frac{dL}{d\dot{q}}(q(0,e),\dot{q}(0,e)) \cdot 0$$

$$\text{Constant Rule \& definition of variation}$$

$$= 0.$$

The second integral, evaluated at 0, yields

$$\frac{dJ_q}{de}(0) = \left(\int_0^1 \left[\left(\frac{dL}{dq} - \frac{d}{dt}\frac{dL}{d\dot{q}} \right) \cdot \frac{dq}{de} \right] dt \right)(0)$$

$$\text{evaluation of the second integral at } 0$$

$$= \int_0^1 \left(\frac{dL}{dq}(q(t,0),\dot{q}(t,0)) - \frac{d}{dt}\frac{dL}{d\dot{q}}(q(t,0),\dot{q}(t,0)) \right)\frac{dq}{de}(t,0)dt$$

$$\text{substitution of } 0 \text{ for } e \text{ in the integrand}$$

$$= \int_0^1 \left(\frac{dL}{dq}(\underline{q}(t), \underline{\dot{q}}(t)) - \frac{d}{dt}\frac{dL}{d\dot{q}}(\underline{q}(t), \underline{\dot{q}}(t)) \right) \frac{dq}{de}(t, 0) dt$$

hypothesis that $\underline{q}$ satisfies Lagrange's Equations

$$= \int_0^1 0 \cdot \frac{dq}{de}(t, 0) dt$$

$$= 0 ,$$

as was to be shown.

(5.18 ⇐ 5.19) Assume $\frac{dJ_q}{de}(0) = 0$ for any variation q of $\underline{q}$. In particular, suppose $\eta : \mathbb{I} \to \mathbb{R}$ is a function

$$\boxed{\begin{array}{cc} [T] & [X] \\ \mathbb{I} \xrightarrow{} & \mathbb{R} \xleftarrow{0} 1 \\ t & \eta \end{array}}$$

such that $\eta(0) = \eta(1) = 0$ and that the variation is given by $q(t, e) := \underline{q}(t) + e \cdot \eta(t)$. Then

$$0 = \left(\frac{d}{de} J_{\underline{q}+e\eta} \right) \qquad\qquad \text{hypothesis} \qquad (5.25)$$

$$= \left(\frac{d}{de} \int_0^1 L(\underline{q}(t) + e \cdot \eta(t), \underline{\dot{q}}(t) + e \cdot \dot{\eta}(t)) dt \right)(0)$$

substitution in the definition of J_q \qquad (5.26)

$$= \left(\int_0^1 \left(\frac{L}{dq} - \frac{d}{dt}\frac{dL}{d\dot{q}} \right) \frac{d}{de}(\underline{q}(t) + e \cdot \eta(t)) \right)(0)$$

as in (5.20)–(5.22) and the climax \qquad (5.27)

$$= \left(\int_0^1 \left(\frac{L}{dq} - \frac{d}{dt}\frac{dL}{d\dot{q}} \right) \eta(t) \right)(0). \qquad (5.28)$$

Since this is true regardless of the choice of η, it follows that

$$0 = \left(\frac{dL}{dq} - \frac{d}{dt}\frac{dL}{d\dot{q}} \right)(\underline{q}(t) + 0 \cdot \eta(t), \underline{\dot{q}}(t) + 0 \cdot \dot{\eta}(t)) .$$

Therefore,

$$\left(\frac{d}{dt}\frac{d}{d\dot{q}} - \frac{d}{dq} \right) L = 0 .$$

$\square$

5.5   Hamilton's Equations

Theorem 5.9. *Assume that the setup*

$$\boxed{\begin{array}{ccccc} \mathbb{R} \longleftarrow & \mathbb{R}^n \times \mathbb{R}^n & \rightleftharpoons & \mathbb{R}^n \times \mathbb{R}^n & \longrightarrow \mathbb{R} \\ \mathcal{L} & q \quad \dot{q} & & q \quad p & \mathcal{H} \end{array}} \qquad (5.29)$$

satisfies

$$\begin{aligned} p &= p(q, \dot{q}(q,p)) \\ \dot{q} &= \dot{q}(q, p(q,\dot{q})) \\ q &= q(t) \\ \dot{q} &= \frac{dq}{dt} \,. \end{aligned}$$

Then the following are true:
[I]

$$If \qquad \mathcal{L}(q,\dot{q}) = \langle \dot{q}, p(q,\dot{q}) \rangle - \mathcal{H}(q, p(q,\dot{q})) \qquad (5.30)$$

$$and \qquad \frac{d\mathcal{L}}{d\dot{q}} = p \qquad (5.31)$$

$$and \qquad \frac{d}{dt}\frac{d\mathcal{L}}{d\dot{q}} = \frac{d\mathcal{L}}{dq}, \qquad (5.32)$$

$$then \qquad \frac{d\mathcal{H}}{dq} = -\frac{dp}{dt}\,. \qquad (5.33)$$

[II]

$$If \qquad \mathcal{H}(q,p) = \langle p, \dot{q}(q,p) \rangle - \mathcal{L}(q, \dot{q}(q,p)) \qquad (5.34)$$

$$and \qquad \frac{d\mathcal{H}}{dp} = \dot{q} \qquad (5.35)$$

$$and \qquad \frac{d\mathcal{H}}{dq} = -\frac{dp}{dt}, \qquad (5.36)$$

$$then \qquad \frac{d}{dt}\frac{d\mathcal{L}}{d\dot{q}} = \frac{d\mathcal{L}}{dq}\,. \qquad (5.37)$$

Proof of **[I]**.

Proof.   Assumed Eqs. (5.30)–(5.31) imply Eq. (5.34) by Theorem 3.59. This implies – in coordinate form, using the definition of scalar product,

Scalar Product Rule, and Chain Rule – that

$$\frac{d\mathcal{H}}{dq_j} = \sum_i p_i \frac{d\dot{q}_i}{dq_j} + \sum_i \frac{dp_i}{dq_j}\dot{q}_i - \sum_i \frac{d\mathcal{L}}{dq_i}\frac{dq_i}{dq_j} - \sum_i \frac{d\mathcal{L}}{d\dot{q}_i}\frac{d\dot{q}_i}{dq_j}. \tag{5.38}$$

By Eq. (5.31) the first and last sums cancel. The second sum vanishes since p, q are independent variables. The third sum is

$$= -\sum_i \frac{d\mathcal{L}}{dq_i}\delta_j^i \qquad \text{since the } q_i, q_j \text{ are independent} \tag{5.39}$$

$$= \frac{d\mathcal{L}}{dq_j} \qquad \text{except for } i = j \tag{5.40}$$

$$= -\frac{d}{dt}\frac{d\mathcal{L}}{d\dot{q}} \qquad \text{Eq. (5.32)} \tag{5.41}$$

$$= -\frac{dp_j}{dt} \qquad \text{by (5.31).} \tag{5.42}$$

Together with (5.38) this proves Eq. (5.33). □

Proof of [**II**].

Proof. Assumed Eqs. (5.34)–(5.35) imply (5.30)–(5.31) by Theorem 3.59. Therefore,

$$\frac{d\mathcal{L}}{dq_j} = \frac{d}{dq_j}\langle \dot{q}, p(q,\dot{q})\rangle - \frac{d\mathcal{H}}{dq_j}(q, p(q,\dot{q})) \qquad \text{in coordinates} \tag{5.43}$$

$$= \sum_i \frac{d\dot{q}_i}{dq_j}p_i + \sum_i \dot{q}_i\frac{dp_i}{dq_j} \qquad \text{Scalar Product Rule} \tag{5.44}$$

$$- \sum_i \frac{d\mathcal{H}}{dq_i}\frac{dq_i}{dq_j} - \sum_i \frac{d\mathcal{H}}{dp_i}\frac{dp_i}{dq_j} \qquad \text{Chain Rule} \tag{5.45}$$

$$= \sum_i \frac{d\dot{q}_i}{dq_j}p_i + -\sum_i \frac{d\mathcal{H}}{dq_i}\frac{dq_i}{dq_j} \qquad \text{by Eq. (5.35)} \tag{5.46}$$

$$= 0 + \sum_i \frac{dp_i}{dt}\frac{dq_i}{dq_j} \qquad q, \dot{q} \text{ are independent \& assumption Eq. (5.36)} \tag{5.47}$$

$$= \frac{dp_j}{dt} \qquad q_i, q_j \text{ are independent except when } i = j \tag{5.48}$$

$$= \frac{d}{dt}\frac{d\mathcal{L}}{d\dot{q}_j} \qquad \text{derivative of Eq. (5.31).} \tag{5.49}$$

This proves Eq. (5.37). □

Corollary. If $\mathcal{L}$ and $\mathcal{H}$ form a Legendre pair, then the following systems of equations are equivalent:

Lagrange's Equations

$$\frac{d}{dt}\frac{d\mathcal{L}}{d\dot{q}} = \frac{d\mathcal{L}}{dt}$$

Hamilton's Equations

$$\frac{d\mathcal{H}}{dp} = \frac{dq}{dt} \quad \text{and} \quad \frac{d\mathcal{H}}{dq} = -\frac{dp}{dt}.$$

5.6 A Theorem of George Stokes

Theorem 5.10. *(Sir George Gabriel Stokes (1819–1903)) If a spherical solid body of radius r is moving in a liquid of viscosity η* $[\mathbf{FRC}][\mathbf{TME}][\mathbf{ARA}]^{-1}$ *with velocity* $\dfrac{dx}{dt}$ $[\mathbf{DST}][\mathbf{TME}]^{-1}$*, the viscous drag force is given by*

$$F = -\zeta \frac{dx}{dt}, \qquad \text{where} \quad \zeta := 6\pi\eta r. \tag{5.50}$$

Remark 5.11. The **frictional drag coefficient** is

$$\zeta \quad [\mathbf{FRC}][\mathbf{TME}][\mathbf{DST}]^{-1}.$$

Like so many great formulas in physics, the simple Eq. (5.50) requires a delicate proof [Landau and Lifshitz (1987)] and is enormously useful, as will be seen.

Remark 5.12. Dimensional confirmation of formulas in physics is crucial for establishing their integrity, and always useful for boosting the intuition. Hence, we make an effort to display the measurement units for physics equations. The convention here is that square brackets around the symbol for a physical quantity represents its physical dimension, such as $[\mathbf{DST}]$ for distance or length, $[\mathbf{TME}]$ for time, $[\mathbf{MSS}]$ for mass, and $[\mathbf{TMP}]$ for temperature. Also, $[\mathbf{FRC}]$ is for force and $[\mathbf{NRG}]$ is for energy, so there exists an equation $[\mathbf{NRG}]=[\mathbf{FRC}][\mathbf{DST}]$. Also, $[\mathbf{AMT}]$ is the dimension symbol for amount of substance, so if N is a number of particles and k_A is Avogadro's Constant, then $N/k_A = [\mathbf{AMT}]$.

5.7   A Theorem on a Series of Impulsive Forces

Theorem 5.13. *[Butkov (1968)] If the motion of a particle of mass m moving along a line with location given by $x = x(t)$ is both impeded by a force $-\zeta v$ proportional to its velocity $v := \frac{dx}{dt}$ and impelled by a non-negative force $f = f(t)$ that is positive only during a short interval $[\tau, \tau + \Delta\tau]$, then*

$$v(t) = \begin{cases} 0 & \text{for } 0 \leq t \leq \tau \\ \dfrac{I_f}{m} \, e^{-\zeta(t-\tau)/m} & \text{for } \tau + \Delta\tau \leq t \, . \end{cases} \tag{5.51}$$

where

$$I_f := \int_{\tau}^{\tau+\Delta\tau} f(t)dt$$

with units of momentum

$$[\text{FRC}][\text{TME}] = [\text{MSS}][\text{DST}][\text{TME}]^{-2} \cdot [\text{TME}] = [\text{MSS}][\text{DST}][\text{TME}]^{-1}$$

*is called the **impulse** delivered to the particle by f.*

Proof.   By Newton's Second Law,

$$m\frac{dv}{dt} = -\zeta v + f \, . \tag{5.52}$$

For $t \geq \tau + \Delta\tau$ the equation $m\frac{dx}{dt} = -\zeta v$ holds, so $v(t) = Ae^{-\zeta t/m}$ if $A := v(\tau + \Delta\tau)$, where A – to be determined – reflects the immediate effect of the impulse upon the motion of the particle. For any t the equation $mdv = -\zeta vdt + fdt$ holds, so calculate the change Δp of momentum $p := mv$ of the particle by

$$\Delta p = \Delta mv = m\Delta v = m \int_{\tau}^{\tau+\Delta\tau} dv = -\zeta \int_{\tau}^{\tau+\Delta\tau} vdt + I_f \cong I_f$$

where the last approximation depends on the unspoken assumption that f is so great, and $\Delta\tau$ so minute, that the change in velocity during the interval under consideration although large, when integrated it is a negligible amount. Since $v(\tau) = 0$, calculate

$$I_f = m\Delta v = mAe^{-\zeta(\tau+\Delta\tau)/m} \tag{5.53}$$

hence $A = \dfrac{I_f}{m} e^{\zeta(\tau+\Delta\tau)/m} = \dfrac{I_f}{m} e^{\zeta\tau/m} e^{\zeta\Delta\tau/m} \cong \dfrac{I_f}{m} e^{\zeta\tau/m}$,   (5.54)

from which the conclusion follows, where $\frac{I_f}{m}$ is identifiably the immediate change in velocity of the particle, and the exponential diminishes it by amount $1/e$ in time ζ/m.   □

Remark 5.14. A series of impulsive forces sufficiently far apart in time will impart rapid jumps in velocity, hence changes in location, followed by resumption of rest. Think of repeatedly striking a nail with a hammer.

5.8   Langevin's Trick

Aside 5.8.1. I only know a little about probability theory. But that is enough to convince me that most classical probabilistic arguments may be re-cast in terms of more modern Kolmogorov probability spaces [Billingsley (1986)].

Given a finite set Ω of particles ω and known, definite trajectories $x(t)(\omega)$ and known, definite forces $f(t)(\omega)$ upon them, create a probability space (Ω, Pr) with $\mathrm{Pr}[\omega] := 1/|\Omega|$ so that any functions associated with particles are random variables defined on this space. In particular, the deterministic trajectory $x(t)(\omega)$ of particle ω is a random variable $\tilde{x}(t) : \Omega \to \mathbb{R}$ defined by $\tilde{x}(t)(\omega) := x(t)(\omega)$, and likewise the deterministic force $f(t)(\omega)$ is the random variable $\tilde{f}(t) : \Omega \to \mathbb{R}$ given by $\tilde{f}(t)(\omega) := f(t)(\omega)$. With this trick it is easy to express probabilistic assumptions about the particle behavior.

The theorem due to Langevin is that – under certain certain assumptions – the variance in particle position across all particles in the cloud at a given time is simply proportional to the elapsed time, where the constant of proportionality depends directly on the temperature of the ambient liquid, and is inversely proportional to its viscosity.

To simulate one particle in the cloud with a computer program means to construct a probability space $(\Omega_1, \mathrm{Pr}_1)$ – whose "experiments" ω are runs of the computer program each with counting probability $1/\Omega_1$ – and a random process $\tilde{x}(t) : \Omega_1 \to \mathbb{R}$ satisfying the equations $\tilde{(x)}(0) = 0$ and $\mathrm{Var}\,(\tilde{x}(t)) = 2Dt$, where D is the diffusion constant in the liquid. But a computer program must proceed in discrete steps, it cannot compute a continuous function $\tilde{x}(t)(\omega)$. So, sub-divide time into short intervals Δt. If there is a sequence of independent identically distributed random variables

$\tilde{z}_n : \Omega_1 \to \mathbb{R}$ such that $\text{Var}(\tilde{z}_n) = 2D\Delta t$, then by defining $\tilde{x}(n\Delta t) :=$ $\tilde{z}_1 + \cdots + \tilde{z}_n = \sum\limits_{i=1}^{n} \tilde{z}_i$, the calculation

$$\text{Var}(\tilde{x}(n\Delta t)) = \text{Var}\left(\sum_{i=1}^{n} \tilde{z}_i\right) = \sum_{i=1}^{n} \text{Var}(\tilde{z}_i)$$

$$= \sum_{i=1}^{n} 2D\Delta t = 2D \sum_{i=1}^{n} \Delta t = 2D(n\Delta t) \qquad (5.55)$$

validates the simulation. Let $\tilde{Z}(\mu, \sigma^2)$ be the random variable with the Gaussian probability density function $N(\mu, \sigma^2)$. If the sequence $\tilde{z}_n$ is defined to be independent random variables with mean 0 and variance $2D\Delta t$, that is, if $\tilde{z}_n = \tilde{Z}(0, 2D\Delta t)$, then $\text{Var}(\tilde{z}_n) = \text{Var}\left(\tilde{Z}(0, 2D\Delta t)\right) = \sqrt{2D\Delta t}\tilde{Z}(0, 1)$ fulfills the requirement for the simulation since it depends only on the availability of values of a unit Gaussian probability density function $N(0, 1)$.

In summary, the discrete random process $\tilde{x}(n\Delta t) := \sum\limits_{i=1}^{n} \sqrt{2D\Delta t}\tilde{Z}(0, 1)$ which may also be written as a recursion

$$\boxed{\tilde{x}(0) = 0, \qquad \tilde{x}((n+1))\Delta t) = \tilde{x}(n\Delta t) + \sqrt{2D\Delta t}\tilde{Z}(0, 1)}$$

determines a computer program for simulating a single particle moving in a liquid. In other words, each run ω of the program generates a particle trajectory $0 = \tilde{x}(0)(\omega), \tilde{x}(\Delta t)(\omega), \ldots, \tilde{x}(n\Delta t)(\omega)$, also called a "realization" of the process. Functions on runs may be calculated, such as

$$\langle \tilde{x}(n\Delta t) \rangle = \frac{1}{|\Omega_1|} \sum_{\omega \in \Omega_1} \tilde{x}(n\Delta t)(\omega), \text{ and} \qquad (5.56)$$

$$\langle \tilde{x}(n\Delta t)^2 \rangle = \frac{1}{|\Omega_1|} \sum_{\omega \in \Omega_1} \tilde{x}(n\Delta t)^2(\omega). \qquad (5.57)$$

5.9 An Argument due to Albert Einstein

Albert Einstein reasoned and calculated as follows about the scenario where a large number N of spherical solid bodies of radius r – think of *a cloud of particles* – moving in a liquid of viscosity η [Nelson (1967)][Pais (1982)].

The particles of the cloud are analogous to the molecules of an ideal gas, thus are in constant motion and exert collectively a pressure $p = p(x, y, z, t)$, occupy a volume V and are at thermal equilibrium with temperature T of the ambient liquid. Therefore, by definition there are N/k_A moles of particles, where k_A is Avogadro's Constant, and by the Ideal Gas Law (10.7),

$$pV = \frac{N}{k_A} RT = Nk_B T \tag{5.58}$$

where R is the Gas Constant and $k_B := \frac{R}{k_A}$ is Boltzmann's Constant.

Remark 5.15. Dimensionally pressure is measured in units of force per area, $[\mathbf{FRC}][\mathbf{ARA}]^{-1} = [\mathbf{FRC}][\mathbf{DST}]^{-2}$ and volume is in units $[\mathbf{DST}]^3$ so Eq. (5.58) equates "mechanical energy" $[\mathbf{NRG}] = [\mathbf{FRC}][\mathbf{DST}] = [\mathbf{FRC}][\mathbf{DST}]^{-2} \cdot [\mathbf{DST}]^3$ on the left to "thermal energy" $[\mathbf{NRG}] = [\mathbf{AMT}] \cdot [\mathbf{NRG}][\mathbf{AMT}]^{-1}[\mathbf{TMP}]^{-1} \cdot [\mathbf{TMP}]$ on the right.

Let $\rho = \dfrac{N}{V}$ $[\mathbf{AMT}][\mathbf{DST}]^{-3}$ denote the cloud's particle density, and by the Ideal Gas Law,

$$\frac{\partial p}{\partial x} = k_B T \frac{\partial \rho}{\rho x} . \tag{5.59}$$

Assume the cloud is randomly distributed, and let

$$f = f(x) \quad [\mathbf{FRC}][\mathbf{AMT}]{-}1$$

be a force applied uniformly in the y, z coordinates to each particle, so the total force on the cloud is $f \cdot N$.

At equilibrium the force is balanced by the pressure gradient,

$$f\rho = \frac{\partial p}{\partial x} \quad [\mathbf{FRC}][\mathbf{AMT}]^{-1} \cdot [\mathbf{AMT}][\mathbf{DST}]^{-3} = [\mathbf{FRC}][\mathbf{ARA}]^{-2}[\mathbf{DST}]^{-1} . \tag{5.60}$$

"The state of dynamic equilibrium that we have just considered can be conceived as a superposition of two processes proceeding in opposite directions." On one hand is the motion of the cloud due to the force f. On the other hand, "a process of diffusion, which is to be conceived as the result of the random motions of the particles due to thermal molecular motions" [Einstein (1989)].

Thus, on one hand, dividing Eq. (5.60) by ζ yields

$$\frac{1}{\zeta} f \rho = \frac{1}{\zeta} \frac{\partial p}{\partial x} \quad [\mathbf{AMT}][\mathbf{DST}]^{-2}[\mathbf{TME}]^{-1} \tag{5.61}$$

that is, the flux of the cloud due to the applied force.

On the other hand if D [DST]^2[TME]$^{-1}$ is the **diffusion coefficient**, then $D\frac{\partial \rho}{\partial x}$ is the flux of particles due to diffusion. At equilibrium these opposed fluxes balance, so

$$D\frac{\partial \rho}{\partial x} = \frac{1}{\zeta}f\rho = \frac{1}{\zeta}\frac{\partial p}{\partial x} = \frac{k_B T}{\zeta}\frac{\partial \rho}{\partial x}. \qquad (5.62)$$

Dividing by $\frac{\partial \rho}{\partial x}$ magically eliminates f from the scenario, leaving the final result

$$\boxed{D = \frac{k_B T}{\zeta}.} \qquad (5.63)$$

This is called a Fluctuation-Dissipation Theorem because on the left diffusion movement represented by D is the result of fluctuating forces upon particles of the cloud, and on the right the drag force represented by ζ resists their motion.

5.10 An Argument due to Paul Langevin

A Fundamental Theorem of the Kinetic Theory of Gases

$$\boxed{Equipartition \ of \ Energy}$$

states that in thermal equilibrium at temperature T the energy of a single particle of any body is the sum of equal quantities $\frac{k_B T}{2}$ for each of the independent ways that the particle can move [Present (1958)]. In particular, a single particle moving in a fluid has kinetic energy $\frac{k_B T}{2}$ in each of three mutually perpendicular directions of motion. If it is buffeted by impacts from other particles, as in Brownian Motion, its *average* kinetic energy in each perpendicular direction is still $\frac{k_B T}{2}$. Or, instead of tracking one particle over time to consider its average kinetic energy, one may contemplate at one moment of time a cloud of particles and consider their average kinetic energy. In greater detail, the kinetic energy in the direction of the x-axis is $\frac{1}{2}m\left(\frac{dx}{dt}\right)^2$, where m is the particle mass, and Equipartition of Energy in this case declares

$$\boxed{\left\langle \frac{1}{2}m\left(\frac{dx}{dt}\right)^2 \right\rangle = \frac{k_B T}{2}.} \qquad (5.64)$$

Note that the mass of the particle does not appear on the righthand side of this equation.

A particle ... large relative to the average distance between the molecules of the liquid, and moving with respect
to the latter at the speed $[\frac{dx}{dt}]$ experiences a viscous resistance equal to $[F$ as in Eq. (5.50)] according to Stokes'
Formula. In actual fact, this value is only a mean, and
by reason of the irregularity of the impacts of the surrounding molecules, the action of the fluid on the particle
oscillates around the preceding value ... If we consider a
large number of identical particles, and take the mean of
the equations written for each one of them, the average
value of the term [for Brownian Motion] is evidently null
by reason of the irregularity of the complementary forces
[Lemons and Gytheil (1997)].

Theorem 5.16. *In Infinitesimal Calculus the following statements are true:*

(1) the general solution of the ordinary differential equation

$$\frac{dz}{dt} = a - bz$$

is

$$z(t) = \frac{a}{b} + Ce^{-bt} \; ;$$

(2) if $y = y(x), y(0) = y_0$ and $u = u(x)$, then the solution to the ordinary differential equation

$$\frac{dy}{dx} = -\frac{du}{dx}y$$

is

$$y(x) = y_0 e^{-u(x)} \; .$$

Theorem 5.17. *In Probability Theory the following statements are true:*

(1) If $\tilde{f}$ and $\tilde{x}$ are uncorrelated random variables and the mean of $\tilde{f}$ is 0, then the mean of the product $\tilde{f}\tilde{x}$ is 0, $\left\langle \tilde{f}\tilde{x} \right\rangle = 0$;

(2) the covariance of a random variable with itself is its variance, $\mathrm{Cov}(\tilde{x}, \tilde{x}) = \mathrm{Var}(\tilde{x})$.

These reminders out of the way, consider how to build a mathematical model of a fluid at temperature T containing a suspended cloud of independent, non-interacting particles (they do not collide if they cross paths, they just obliviously pass through each other), each impeded in motion by viscous drag force ζ and impelled by impacts of fluid molecules. Suppose there are N of them and let $(\Omega, \Pr)$ be the finite probability space whose items are the particles ω, all equal in probability, so $\Pr[\omega] = \frac{1}{N}$. Each particle is subjected to a time-varying force, and this may be modeled by a random variable, that is, for each particle $\omega \in \Omega$ and time t there is a force $\tilde{f}_\omega(t)$,

$$\Omega \xrightarrow{\tilde{f}(t)} \mathbb{R} \qquad \textbf{[FRC]} . \qquad (5.65)$$

Likewise, the location of each particle is given by a random variable,

$$\Omega \xrightarrow{\tilde{x}(t)} \mathbb{R} \qquad \textbf{[DST]} . \qquad (5.66)$$

For the sake of simplifying the visual appearance of formulas it is convenient to define abbreviations, namely $\widetilde{v_\omega} := \dfrac{d\widetilde{x_\omega}}{dt}$ and $\widetilde{u_\omega} = \widetilde{x_\omega}^2$. The subscripts ω are included to be perfectly clear that for each particle these "random" variables are garden-variety real-valued functions of $t \in \mathbb{R}$. These functions are assumed to be differentiable. Therefore, for example, the mean of the derivatives is the derivative of the mean, in the sense that for any of these random variables, say $\tilde{a}$, we have

$$\left\langle \frac{d\tilde{a}}{dt}(t) \right\rangle = \sum_\omega \frac{d\widetilde{a_\omega}}{dt}(t) \cdot \Pr[\omega]$$

$$= \left(\frac{d}{dt} \sum_\omega \widetilde{a_\omega}(t) \right) \Pr[\omega]$$

$$= \frac{d}{dt} \left(\sum_\omega \widetilde{a_\omega}(t) \Pr[\omega] \right)$$

$$= \frac{d \langle \tilde{a} \rangle}{dt}(t) .$$

Theorem 5.18. *(Paul Langevin, in [Lemons and Gytheil (1997)]) Assume that the cloud of suspended particles satisfies the following conditions:*

$$m\frac{d^2\tilde{x}}{dt^2} = -\zeta\tilde{v} + \tilde{f} \qquad \text{Newton's Second Law} \qquad (5.67)$$

$$\tilde{x}(0) = 0 \qquad \text{all particles are initially at the origin} \quad (5.68)$$

$$\left\langle \tilde{f}(t) \right\rangle = 0 \qquad \text{average force is 0} \qquad (5.69)$$

$$Cov(\tilde{f}(t), \tilde{x}(t)) = 0 \qquad \text{forces and locations are uncorrelated} \quad (5.70)$$

$$D = \frac{k_B T}{\zeta} \qquad \text{Einstein's Fluctuation-Dissipation Theorem}$$

$$(5.71)$$

$$\left\langle m\frac{d\tilde{x}}{dt} \right\rangle = k_B T \qquad \text{Equipartition of Energy .} \qquad (5.72)$$

Then

$$\boxed{Var(\tilde{x}(t)) = 2Dt .} \qquad (5.73)$$

Proof. Multiplying Eq. (5.67) by $\tilde{x}$ and applying Theorem 5.16 yields

$$\frac{m}{2}\frac{d^2\tilde{u}}{dt^2} - m\tilde{v}^2 = -\frac{\zeta}{2}\frac{d\tilde{u}}{dt} + \tilde{f}\tilde{x} . \qquad (5.74)$$

Averaging Eq. (5.74),

$$\frac{m}{2}\frac{d^2\langle\tilde{u}\rangle}{dt^2} - \langle m\tilde{v}^2 \rangle = -\frac{\zeta}{2}\frac{d\langle\tilde{u}\rangle}{dt} + \left\langle \tilde{f}\tilde{x} \right\rangle . \qquad (5.75)$$

By Theorem 5.17 and Eqs. (5.69)–(5.70) the term involving $\left\langle \tilde{f}\tilde{x} \right\rangle$ may be omitted. By Eq. (5.71) the term involving $\tilde{v}$ may be replaced by $k_B T$. Setting $z := \dfrac{d\langle\tilde{u}\rangle}{dt}$ and dividing both sides by m leads to

$$\frac{dz}{dt} = \frac{k_B T}{m} - \frac{\zeta}{2m}z , \qquad (5.76)$$

of which the general solution by Theorem 5.16.3 is

$$z(t) = \frac{2k_B T}{\zeta} + C \cdot e^{-\zeta t/2m} . \qquad (5.77)$$

This diminishes by $\dfrac{C}{e}$ after time $\dfrac{2m}{\zeta}$. After some time $z(t) \cong \dfrac{2K_B T}{\zeta}$, hence after unwinding the abbreviations we have for infinitesimal time increments

$$d\langle\tilde{x}^2\rangle = \frac{2k_B T}{\zeta}dt \qquad (5.78)$$

and for finite time increments

$$\Delta \left\langle \tilde{x}^2 \right\rangle = \frac{2k_B T}{\zeta} \Delta t \ . \tag{5.79}$$

The conclusion follows from Eq. (5.68), by substituting $t = \Delta t$ in Eq. (5.79), and appealing to the definition of variance. $\qquad \square$

Remark 5.19. It is peculiar that the two force terms of Eq. (5.67) refer to two different intuitions about the ambient liquid of the particle cloud. The viscosity term involving ζ brings to mind the intuition of a continuous liquid so thick that it impedes particle motion. The Brownian Motion term $\tilde{f}$ relies on intuition of a dense but discrete collection of liquid molecules impinging upon the much larger particles of the cloud.

Remark 5.20. The Langevin Equation (5.67) is actually a "cloud" of ordinary differential equations indexed by particles, the elements of the sample space Ω. Thus, the Langevin Equation is *not* what in modern literature is called a "stochastic differential equation" [Arnolt (1974)].

PART 4
Timing Machinery

The intuition for states that spontaneously time out and emit signals that trigger other states into activity comes from physics, chemistry, and neurobiology.

Chapter 6

Introduction to Timing Machinery

Aside 6.0.1. My friend Florian Lengyel and I have worked together on discovering the mathematical and computer programming implications of the idea for "timing machinery" that I initiated in 1989. This part of the book lays out the current state of the idea, and Appendix C provides our MATLAB code.

> **Timing machinery is a model of concurrent timed computation, in which a machine state may spontaneously time out and emit a signal that may trigger activity elsewhere within the machine. We derive a master ordinary differential equation for the machine state by imposing Poisson- and Markov-like restrictions on the behavior of a stochastic timing machine. This equation and the machine it describes generalize the chemical master equation and Gillespie stochastic exact simulation algorithm, used widely in studies of chemical systems with many species, prokaryotic genetic circuits, genetic regulatory networks, and gene expression in single cells** [Cooper and Lengyel (2009)].

The mathematics required for drawing and interpreting stochastic timing machinery models is elementary. At the least, stochastic timing machine models and their mathematics could be pedagogically useful for advanced high school students, their teachers, and undergraduate students in mathematics, physics, and computer science:

(1) Stochastic timing machinery is a general purpose diagram-based parallel programming language for simulating many different physically

interesting and sometimes mathematically challenging situations;
(2) the immediate intuitive relationship between stochastic timing machines, difference equations and differential equations can be of immense pedagogical value;
(3) the stochastic timing machine interpreter is easily implemented in a few lines of code in any modern programming language such as MATLAB;
(4) the stochastic timing machine interpreter is intrinsically parallelizable.

Motivation for timing machinery derives from well-known ideas in Physics, Chemistry, and Neurobiology:

Physics In the special theory of relativity one imagines space filled with a latticework of 1 meter rods and clocks attached at the intersections. A clock once started is considered to time out every 1 unit of time. To synchronize the clocks one distinguished clock - the "origin" - emits a light signal in all directions and then starts. Each of the other clocks is supposed to have been preset forward by exactly the amount of time it takes light to travel from the origin to the clock, and the clock only starts when the light signal arrives. Since the speed of light is independent of direction of propagation, and of relative motion of the latticework, this method is an absolute standard for synchronization, and such a latticework is used to measure the four coordinates of any physical event. [Taylor and Wheeler (1966)]

Photons interact with the electrons of an atom. In the *stimulated absorption* process an incident photon is absorbed and stimulates the atom to undergo a transition from a low energy state to a higher energy state. An atom in a higher energy state may undergo the process of *spontaneous emission*, thus leaving the higher energy state and entering a lower energy state after a period of time, and emitting a photon. The period of time is a random variable called the *lifetime* of the higher energy state, and may vary in duration from nanoseconds to milliseconds [Eisberg and Resnick (1985)].

Chemistry Photons can also induce molecules to change conformation. For example, nanosecond to microsecond relaxation times leading to tertiary and quaternary structural changes have been observed in hemoglobin using the technique of time-resolved spectroscopy. "Light initiation techniques such as photoinduced electron transfer and photoisomerization can be used to study such photobiologically active molecules as photosynthetic reaction centers, visual pigments, bacteriorhodopsin, and phytochromes." [Chen *et al.* (1997)]

Neurobiology Neurons receive electro-chemical signals called *action potentials* (a.k.a. "spikes") at spatially distributed location on their dendrites, and if sufficiently many signals occur over a sufficiently short period of time, then the soma of the neuron initiates an action potential along its axon. Thus, a neuron performs a of signals that may pass a threshold for generating an output signal. Abstractly, the neuron may be triggered from a ground state into an excited state which in about a millisecond decays back to the ground state while emitting a new signal. This "Platonic" model of the mammalian neuron is based on research on motor neurons and the assumption that the dendritic tree is electrically passive and linearly adds the incoming signals. A great deal of early artificial neural network research used this model. It has also been known for some time that in fact neurons can behave as "resonators" and even as spontaneous generators of action potentials.

By single-cell oscillator I mean a neuron capable of self-sustained rhythmic firing independent of synaptic input. By single-cell resonation, I mean rhythmic firing occurring in response to electrical or ligand-dependent oscillatory input. Resonance implies that the intrinsic electroresponsive properties of the target neurons are organized to respond preferentially to input at specific frequencies [Llinas (1988)].

It has been argued that in the brain "spike sequences encode memory," and specifically that "invariant spatio-temporal patterns" of spike sequences are relevant to hippocampal information processing [Nadásdy (1998)]. Indeed, although "traditional connectionist networks assume channel-based rate coding," *neural timing nets* have been proposed that operate on "time structured spike trains" to explain auditory computations [Cariani (2001)].

6.1 Blending Time & State Machine

Timing machinery blends the idea of state machine with the idea that a state may spontaneously time out and emit a signal that in turn triggers other states into activity. This intuition leads to a syntax for a simple but expressive "graphical assembly language" for building models of concurrent processes.

Remark 6.1. For the expert, a *deterministic* timing machine is essentially equivalent to a system of impulsive ordinary differential equations , for example

$$\frac{du}{dt} = v$$

$$\frac{dv}{dt} = -u$$

$$\frac{dy}{dt} = g(y) + k(y)\delta(v, v_0)\delta(u, u_0) \ .$$

Remark 6.2. Some *stochastic* timing machinery models correspond to stochastic differential equations based on Wiener, Poisson, and other stochastic processes. The relationship between stochastic timing machinery and stochastic differential equations in the literature (e.g., [Arnolt (1974)][Misra *et al.* (1999)]), merits further study.

6.2   The Basic Oscillator

The basic oscillator Fig. 6.1 has a single state a that times out to itself every **3** units of time.

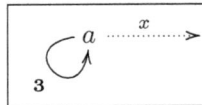

Fig. 6.1   The basic oscillator.

Each timeout results in the emission of a signal, x. The most important idea for elaboration of this first example of timing machine is replacement of the constant 3 by a (possibly random) variable, resulting in a *variable oscillator*. Thus, the waiting time for the next emission of x could be an with parameter λ, so that emission of x is a Poisson process.

6.3 Timing Machine Variable

The very notion of "variable," however, may be analyzed in terms of timing machinery. In the sense that a variable "holds" a value for reference, it has states as in Fig. 6.3.

Algorithm 6.3.

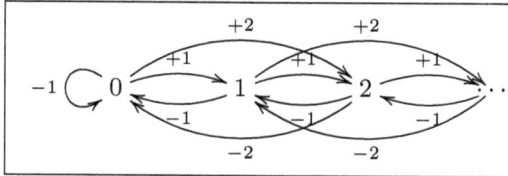

Fig. 6.2 In principle a variable taking values in the non-negative integers can be implemented with a (necessarily infinite) state machine - that is, a timing machine without timeout arrows. An input signal +1 triggers transition from any state to its rightmost neighbor, +2 to its second rightmost neighbor, and so on. Likewise, negative inputs trigger leftward transitions, except at 0.

Remark 6.4. To query the value of a variable implemented as in Fig. 6.3 the machine must respond to a query signal with an output signal carrying the identity of the active state of the machine. A way to achieve this is to add a state n' for each value state n such that arrival of a query signal, say ?, triggers n to n' which immediately times out back to n while emitting a signal labeled by the value, as in Fig. 6.4.

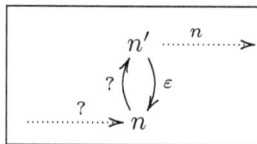

Fig. 6.3 Machinery for querying a variable.

Aside 6.3.1. It would be just as ridiculous to implement variables this way as it would be to write all computer programs using $0's$ and $1's$. My point in exhibiting this obviously over-the-top complex way to implement a variable as a timing machine is that timing machinery is in some sense "complete" because anything that can be done by a computer – for example, storing a value at a location that can be queried for its content – can be implemented by timing machinery. In other words, timing machinery is a minimal "universal machine language" for concurrent processes involving states and values. As with any computer language, there corresponds a virtual machine atop a tower of lower and lower level virtual machines, bottoming out at realization by a physical machine [Tanenbaum (1994)].

Remark 6.5. The concept of variable in timing machinery may be compared and contrasted to the concepts of variable in computer programming, mathematics, and physics. See Section 3.2.

6.4   The Robust Low-Pass Filter

Timing machines can process signals as well as generate them and store values. For a robust low-pass filter machine assume that a is the "resting state." Assume signal x is distributed to two states, a and b. When signal x arrives, a is triggered to b which times out in 2 units of time back to a, *unless* b is interrupted by arrival of another x, in which case b is triggered to itself and restarts its activation. In other words, the only way b can complete its activation and time out back to a is when input signals x arrive sufficiently slowly compared to 2. This machine performs this "low-pass filter" operation *robustly*, in the sense that the exact timing of input signals x is irrelevant, so long as on average they arrive slowly.

6.5   Frequency Multiplier & Differential Equation

Aside 6.5.1. I showed these timing machinery ideas to my friend Brian Stromquist while we were both employed as computer programmers at The Rockefeller University in 1989. Trained as an electronic engineer, his response a day later was to point out that timing machinery *with feedback* could easily simulate a *phased-lock loop* circuit. I realized that a little differential equation models the behavior of his timing machine. These observations cemented my conviction that timing machinery is a really good idea.

A timing machine is naturally partitioned into sub-machines that are connected only by signals. These are considered its *parts*. There are three parts to Brian Stromquist's frequency multiplier: a basic oscillator whose output is a decrement signal -1, a variable oscillator to receive that signal, and a period multiplier connected in a *feedback loop* that sends an increment signal $+1$ to the variable oscillator for every two of its outputs x. See Fig. 7.4.

The net effect of this circuit is to *multiply frequency*. To formalize this, there exists a differential equation to approximate the behavior of this continuous-time discrete-state system. Let v denote the period of the variable oscillator and let m denote the period of the basic oscillator. Thus, the frequency of the basic oscillator is $1/m$, and that is the rate at which v is decremented. So far,

$$\frac{dv}{dt} = -\frac{1}{m}.$$

On the other hand, v is incremented at half the rate of its output, so

$$\frac{dv}{dt} = -\frac{1}{m} + \frac{1}{2v}.$$

This equation is not easy to solve in closed form, but its equilibrium period is easily determined by solving for v in

$$-\frac{1}{m} + \frac{1}{2v} = 0$$

to get $v = m/2$. In other words, the variable oscillator and period multiplier combine *robustly* to multiply frequency by 2.

Aside 6.5.2. The frequency multiplier timing machine is a splendid example because it has multiple (three) concurrent parts, it interconnects states with a variable, it performs a useful function, and is not so simple to render it trivial, nor too complex to be analyzed completely.

6.6 Probabilistic Timing Machine

Remark 6.6. Molecular machinery such as muscle contraction is deeply influenced by the incessant impact of water molecules upon very large molecules. This behavior is intrinsically random and to simulate it with timing machinery requires the introduction of the "probabilistic state."

Definition 6.7. A state b is a **probabilistic state** if there are states $a_1, \ldots, a_n$ and numbers $p_1, \ldots, p_n$ such that $0 \le p_i \le 1$ with $p_1 + \cdots + p_n = 1$, and if a is active then it times out instantly to state a_i with probability p_i.

6.7   Chemical Reaction System Simulation

Aside 6.7.1. Computer simulation of chemical reactions got a big boost by Daniel T. Gillespie who invented an alternative to the conventional algorithm based on forward integration of differential equations that model chemical kinetics. In a way the story is strange. To begin with, the "molecular reality" of the world is a fact [Perrin (2005)][Newburgh *et al.* (2006)][Horne *et al.* (1973)][Lemons and Gythiel (1997)]. This means nature is discrete, but at the human scale much is achieved by conceiving the incredibly vast numbers of molecules as though they form continua, as described in my The Theory of Substances. Differential equations are the mathematical technology of choice for modeling continuous systems. In other words the discrete is modeled by the continuous. Then, to find out the behavior of the continuous equations, they are translated into discrete equations so that computers can simulate their behavior. This discrete-to-continuous-to-discrete series of approximations works fine, up to a point. The point is when – as for example in biology – one is down at the level of simulating relatively small numbers of molecules, say in the billions or less.

Gillespie basically says, wait, put aside the presumptions of continuous modeling, and actually model the discrete events of chemical reactions. The title of an early paper, "Exact Stochastic Simulation of Coupled Chemical Reactions" says it all [Gillespie (1977)]. Stochastic simulation has become a small but very successful industry, and in particular is applicable to Langevin Equations [Gillespie (2007)], which are a type of stochastic differential equation deployed in the muscle contraction modeling business ([C.P. Fall and Tyson (2002)] Ch.12). Indeed, stochastic simulation is employed generally in cell biology [Sun *et al.* (2008)].

The advantage of timing machinery over Gillespie stochastic simulation is that it is specifically designed to take advantage of parallel computers. The surprise is how this is done: instead of assigning a processor to each microscopic entity, as George Oster quickly suggested, the idea is to compute "all the next events" in parallel. This approach is, in a way, at right angles to the usual simulation of differential equations or stochastic equations by stepping along individual solution curves.

Aside 6.7.2. Every technology has limitations and timing machinery of course is no exception. "Signaling complexes [such as receptor complexes, adhesion complexes, and mRNA splicing complexes] typically consist of highly dynamic molecular ensembles that are challenging to study and to describe accurately. Conventional mechanical descriptions [parts lists and blueprints, finite state diagrams] misrepresent this reality and can be actively counterproductive by misdirecting us away from investigating critical issues" [Mayer *et al.* (2009)]. Fortunately for this book, those authors consider molecular motors – including the actin-myosin complex of muscle contraction – susceptible to "mechanical description."

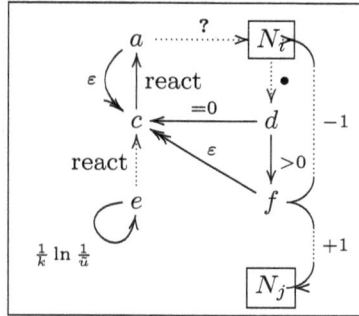

Fig. 6.4 Timing machine simulation of a chemical reaction that consumes species i and produces species j. The timeout value $\frac{1}{k}\ln\frac{1}{u}$ generates random wait times between successive reactions according to the exponential wait time of a Poisson process with parameter k. When state e times out the reaction signal triggers state c to state a which times out quickly (ϵ) and queries the variable N_i, which is the number of species i molecules that are available. If the reply ($\bullet$) is 0, meaning all of species i has been consumed, idle state c is activated. Otherwise, state f is activated, which times out (ε) back to c but decrements N_i and increments N_j.

6.8 Computer Simulation

> The hallmark of emergence is this sense of much coming from little.

> [Holland (1998)]

Aside 6.8.1. Computer simulations of complex systems tend to ignore selected details in the interest of obtaining results about salient features in a reasonable amount time. This is especially true when biological simulations of phenomena involving very, very large numbers of very, very tiny items must be performed. For example, there are billions of molecules in just one cell, and billions of cells in one organism. Great as computers are these days, no computer can simulate every last detail of even a single cell in a reasonable amount of time, say, a year. But scientists are not interested in such details, they are interested in patterns, the "laws of nature." What they want is a model, an abstraction that tears away the inessential and reveals the essence of some aspect of nature. No scientist works without the hope this can be done. The fact that it has been done many times, and so fruitfully, sustains him or her through dark periods of seemingly futile effort.

The beauty of differential equation models is that a small number of parameters in the equations can give rise to a great variety of behavior of the solutions to the equations. The difficulty with differential equations is that they usually cannot be solved. In fact, finding formulas for solutions to differential equations is a separate mathematical industry. The next best thing is computer simulation of solutions to differential equations. You have to bear in mind that by doing so the scientist is removed twice-fold from the phenomena under study. First, the differential equation has to to be set up – with parameters deemed appropriate for the phenomena at hand. So, that is already one step removed, one level of abstraction away from reality. Second, simulating the differential equation generally involves approximations most often in time, and frequently in space as well. In other words, time steps or space steps are chosen to skip across reality under the assumption that what is missed in the interstices is not of the essence in the phenomena under study.

The point of timing machinery simulations – which are proffered as an alternative to simulations using differential equations – is that timing machine diagrams are also condensed abstractions with carefully selected parameters, but the algorithm for simulating them is extremely simple and intrinsically adapted to take advantage of parallel computers. The same cannot be said about most algorithms for integrating differential equations.

Aside 6.8.2.

> **What is the physiologic basis of the force-velocity relationship? The force generated by a muscle depends on the total number of cross-bridges attached. Because it takes a finite amount of time for cross-bridges to attach, as filaments slide past one another faster and faster (i.e., as the muscle shortens with increasing velocity), force decreases**

due to the lower number of cross-bridges attached. Conversely, as the relative filament velocity decreases (i.e., as muscle velocity decreases), more cross bridges have time to attach and to generate force, and thus force increases [Holmes (2006)].

The basic unit of muscle contraction is the *sarcomere*. Half of a sarcomere contains 150 myosin molecules with tails twisted together in each thick filament [Lan and Sun (2005)][Reconditi *et al.* (2002)]. Computer simulations of as many as 30 [Duke (1999)] and even all 150 myosins [Lan and Sun (2005)] have been reported. Yet when I spoke with George Oster – a prominent muscle contraction researcher – and said I am writing a book about the mathematical science background for simulating 150 myosin molecules, he said, "You had better get 150 computers." I doubt that the reported simulations used so many computers, so what is going on here? My guess is that Professor Oster is absolutely right, of course, if one demands a great deal of detail from the simulation. Then again, as I implied above, detail is not necessarily the Holy Grail of simulation.

Chapter 7

Stochastic Timing Machinery

Stochastic timing machinery (STM) is a diagram-oriented parallel programming language. Examples in electrical engineering, chemistry, game theory, and physics lead to associated ordinary and stochastic differential and partial differential equations. There exists an abstract mathematical definition of an algorithm for interpreting timing machine diagrams, and compact MATLAB code for implementing simulations. In particular, a Brownian particle in an arbitrary force field is modeled by a stochastic timing machine associated with the Smoluchowski Equation. Simulation of a Brownian particle subjected to a spatially varying force field is a basic ingredient in modern models of the molecular motor responsible for muscle contraction. The mathematical theory and the technology for simulation is expected to be accessible to a high school AP-level physics, calculus and statistics student. The STM interpreter offers opportunities for efficient parallel simulations and can compute exact solutions of associated differential equations. The technique is similar in principle to the Gillespie algorithm in computational chemistry.

7.1 Introduction

Stochastic timing machinery (STM) is a diagram-oriented parallel programming language. According to the very well known idea of finite state machine, structureless states undergo transitions along labeled arrows in response to input signals – or automatically undergo transitions along arrows labeled with probabilities. This idea is augmented by adding timing structure to the states. In a stochastic timing machine a state may be inactive, or timing out for a certain duration of time at the end of which it becomes inactive and also emits signals to trigger other states. A differential equation may be associated with many kinds of stochastic timing machines and may be useful for qualitative analysis of machine behavior.

Stochastic timing machines are represented by diagrams involving a handful of drawing conventions. Circles for states, squares for variables, and four kinds of arrow can be combined into diagrams for an infinite variety of models. Because of this simple syntax it is easy to implement a computer program to interpret such diagrams, thus exhibiting their semantics. Since STM is intrinsically a method of describing concurrent processes, the interpreter program is directly suited for implementation on multi-core computers.

One type of timing machine already in the literature is the *timed automaton* [Dill and Alur (1990)], wherein the time-related structure is based on the metaphor of a "stopwatch", i.e., a gadget which can be reset to 0, which can be read at any time, and when halted reads out an amount of time elapsed. Such gadgets can be reset by state transitions, and read to qualify state transitions. The theory of *timed automata* was introduced because, "[a]lthough the decision to abstract away from quantitative time has had many advantages, it is ultimately counterproductive when reasoning about systems that must interact with physical processes." The crucial element of that theory is the addition of "stopwatches" to finite state machinery. That theory has been extended to also include a notion of *stochastic timed automaton* [Mateus *et al.* (2003)].

An important advantage of STM over conventional model-building technologies (for example, differential equations) is that the diagrams, mathematics, and computer code demand no more background than infinitesimal calculus, probability theory and basic programming experience at the level of an AP high school student. This implies that it should be of considerable pedagogical value in addition to its possible use by researchers for building and testing new models.

The purpose of this Part of the book is to introduce the syntax and semantics of STM, to give an abstract mathematical definition of it by analogy with well known algorithms, and to provide interesting examples. Section 7.1.1 defines the syntax and semantics of the stochastic timing machinery diagram-oriented language. Section 7.1.2 presents some examples with emphasis on a Brownian particle in a force field because models of that particular kind play an important role in simulations of molecular motors.

Appendix B provides a formal definition of STM by analogy with standard recursive definitions of algorithms for approximating solutions to different types of differential equations. Appendix C offers MATLAB code for implementing models using STM, and in particular specific models for a Brownian particle in a force field. The Figures section accumulates graphs produced by STM simulations of a Brownian particle bouncing around in response to assorted force fields. Appendix D discusses the theory of equilibrium and detailed balance in STM.

7.1.1 *Syntax for Drawing Models*

A stochastic timing machine is a diagram of labeled dots and arrows, so is a directed graph. The dots represent states of the machine if labeled by lower-case (Roman alphabet) letters, or variables if labeled by upper-case letters.

There are four types of arrows called timeout, signal, trigger, and stochastic arrows. Every arrow connects two dots, and an arrow may connect a dot to itself. The label on an arrow has a prefix to indicate its type: $tm : x$ labels a timeout arrow, and $sg : x$, $tg : x$, $pr : x$ label signal, trigger, and stochastic arrows, respectively.

A signal arrow must always start at a state but may end at either a state or a variable; a trigger arrow and a stochastic arrow must always connect from a state to a state.

A state is the start of at most one timeout arrow, but may be the start of any number of signal, trigger, or stochastic arrows. A state may be the end of any number of any type of arrow. Any arrow may be a loop at one state.

The stochastic arrows from one state to other states are labeled by positive numbers that sum to 1. These probabilities may be variables, so long as changing one is accompanied by adjustment to the others to maintain the sum at 1.

Figure 7.1 summarizes the building blocks of timing machine diagrams. Item (i) does double duty. If β is a state, say (b), then (i) represents (a) signaling to (b). If β is a variable, say X, then (a) signals X. Item (iv) is meant to suggest that if there are stochastic arrows from (b) to c_1, c_2 and c_3 labeled by probabilities $pr : 1$, $pr : 2$ and $pr : 3$, respectively, then the stochastic vector variable $[pr : 1, pr : 2, pr : 3]$ may be thought of as the barycentric coordinates of a point in the triangle with vertices c_1, c_2, c_3, so that a signal K to the stochastic vector variable repositions it inside the triangle. See "Semantics" below for greater insight.

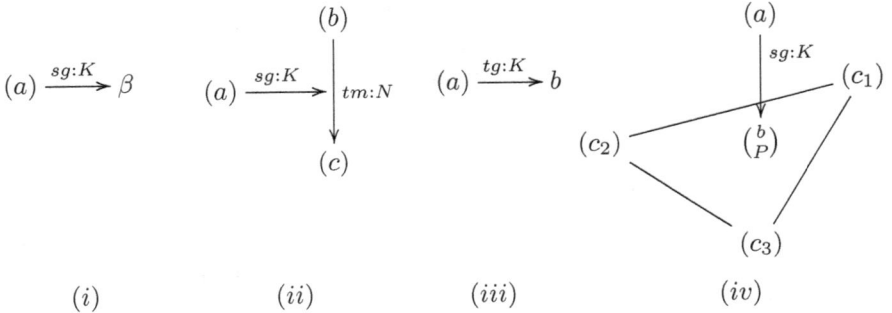

Fig. 7.1 Signal K is a variable. (i) State (a) signals K to state or variable β. (ii) State (a) signals the timeout variable from (b) to (c). (iii) State (a) may be triggered to (b). (iv) (a) signals the probability that (b) moves to c_1, c_2, or c_3.

7.1.2 *Semantics for Interpreting Models*

The root idea of timing machinery is that when an active state times out it sends signals that (a) trigger transitions between states, or (b) alter the values of variables.

The most basic scenario is a state with a timeout arrow to another state, $a \xrightarrow{tm:\Delta t} b$. At the moment t that (a) becomes activated it starts a timeout of duration Δt. During the interval $[t, t + \Delta t)$, (a) remains active. At the moment $t + \Delta t$ the timeout state (b) of (a) becomes active. The "baton" is passed from (a) to (b) at $t + \Delta t$. If (b) has a timeout arrow, then it stays active for its duration. There might be a long chain of successive timeout arrows, or cycles of them, or chains that end in cycles, or combinations of these chains and cycles, and so on.

Remark 7.1. A timing machine consisting exclusively of states and timeout arrows connecting them is the same as a "finite dynamical system" "or automaton" ([Lawvere and Schanuel (1997)] p. 137) except with the additional structure of time.

A more interesting scenario is exhibited in Fig. 7.2. A timing machine subdivides into parts connected by signals. Thus, a part of a timing machine consists of states connected exclusively by timeout and trigger arrows. In Fig. 7.2 there are two parts to the machine, namely ab and $cdef$. If (c) happens to be active when part ab sends the x signal, then (c) does not

$$
\begin{array}{ccc}
(a) \xrightarrow{\ sg:x\ } (c) \xrightarrow{\ tg:x\ } (e) \\
\Big\downarrow tm{:}\Delta t \qquad \Big\downarrow tm{:}\Delta t' \qquad \Big\downarrow tm{:}\Delta t'' \\
(b) \qquad\quad (d) \qquad\quad (f)
\end{array}
$$

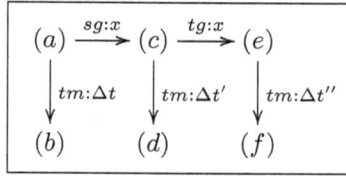

Fig. 7.2 Say (a) and (c) are active but (e) is not. If (a) times out it sends signal x to (c), and if this happens to occur while (c) is active then the match of that signal to the trigger arrow from (c) to (e) then the timeout of (c) is interrupted, so (c) is de-activated, and (e) is activated.

continue timing out to (d) but is triggered to (e) instead, so that (e) is (re-)activated to timeout towards (f). A state cannot be triggered if it is not active, but an active state may be re-activated to start over. There can only be a transition from an active state if there is a match between an incoming signal and some trigger arrow starting at the state. If multiple signals arrive simultaneously at an active state then exactly one match is selected uniformly at random to trigger the transition.

Remark 7.2. Each part of a timing machine sub-divides further into deterministic sub-parts consisting exclusively of timeout arrows. The deterministic parts are connected exclusively by trigger arrows. The adjective "deterministic" is appropriate since unless a signal arrives at an active state of a part, an active state times out eventually to its unique next active state, and so forth. Only if a signal arrives will this deterministic process be "derailed," so to speak, causing transition to activate a state of another deterministic sub-part, until the next signal, and so on.

A state with multiple stochastic arrows whose probabilities (necessarily) sum to 1 should be thought of as a state such that when activated it times out instantaneously to one of its targets with the corresponding probability, see Fig. 7.3.

The mental model for interpretation of timing machine diagrams is a kind of "chasing" around the diagram keeping in mind what states are active.

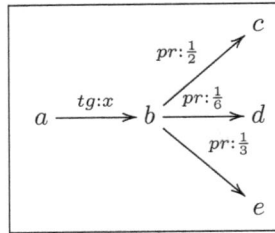

Fig. 7.3   If (a) is triggered to (b) then (b) undergoes transition to (c) with probability $\frac{1}{2}$, to (d) with probability $\frac{1}{6}$, or to (e) with probability $\frac{1}{3}$.

7.2   Examples

7.2.1   *The Frequency Doubler of Brian Stromquist*

Remark 7.3. In 1989 after absorbing a brief explanation of timing machinery the electronic engineer and computer programmer Brian Stromquist invented the first timing machine with feedback. He showed that frequency division in a feedback loop implies frequency doubling.

Figure 7.4 exhibits the "frequency doubler" stochastic timing machine. The state (a) times out with duration obtained by selecting a value for the

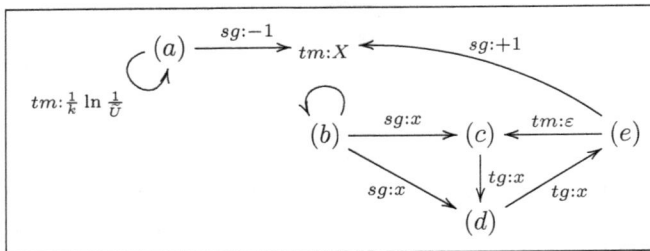

Fig. 7.4   Stochastic timing machine simulation of a *robust* frequency doubler. State (a) emits decrement signals to the variable X randomly according to the exponential wait time of a Poisson process with parameter k. Variable X is the timeout duration of basic oscillator (b), which upon timeout triggers either state (c) to (d) or (d) to (e). State (e) times out immediately back to (c), so the net effect is that (e) times out for every second x signal from (b). In other words, the sub-machine c, d, e divides the rate X of (b) by 2. However, when X is incremented due to timeout of (e), (b) slows down, while the decrement signals from (a) are speeding it up. These two opposing "forces" on X eventually balance out by having (b) time out twice as fast as (a) *on average*.

uniformly distributed random variable $\widetilde{U}$ and calculating $\widetilde{W} := \frac{1}{k} \ln \frac{1}{\widetilde{U}}$. The result is the exponentially distributed waiting time of a Poisson process [Billingsley (1986)]. An interesting non-linear stochastic differential equation can be associated with this machine by observing that the rate at which X decreases due to timeout of (a) is approximately $-\frac{1}{\widetilde{W}}$, and the rate at which X increases due to its own timeouts is $+\frac{1}{2X}$, so that

$$\frac{dX}{dt} = -\frac{1}{\widetilde{W}} + \frac{1}{2X}.$$

In the long run the decrements and increments balance out, so that at equilibrium

$$X = \frac{1}{2}\widetilde{W}.$$

In other words, the timeout duration of (b) is one-half the timeout duration of (a), that is, (b) doubles the frequency of (a). Actually, this is "on the average."

Remark 7.4. The use of feedback in this example to multiply frequency is reminiscent of the "phase-locked loop" control system.[1]

7.3   Zero-Order Chemical Reaction

Figure 7.5 is a stochastic timing machine that may be associated with the simple chemical reaction $i \rightarrow j$, where N_i and N_j are the numbers of molecules of the species i, j respectively. At exponentially distributed waiting times, a molecule of i is consumed while one of j is created. The point of the example is to show that the model "idles" when all of the i species is used up. This requires the query signal $?$ to the variable N_i and the result $\bullet$ tested by state (d). Only if the result is positive will (f) get a chance to time out and send the decrement and increment signals that represent consumption and production.

The associated system of stochastic differential equations is

$$\frac{dN_i}{dt} = -\frac{1}{\widetilde{W}}$$

$$\frac{dN_j}{dt} = +\frac{1}{\widetilde{W}}.$$

---

[1]http://en.wikipedia.org/wiki/Phase-Locked_Loop

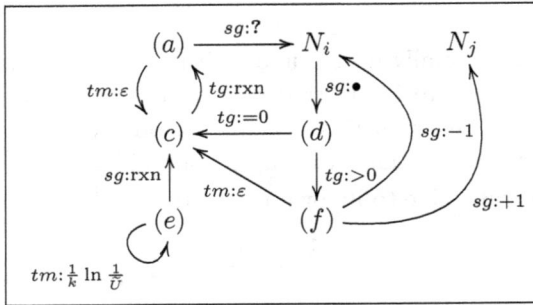

Fig. 7.5 Stochastic timing machine simulation of a chemical reaction that consumes species i and produces species j. The timeout value $\frac{1}{k} \ln \frac{1}{U}$ generates random wait times between successive reactions according to the exponential wait time of a Poisson process with parameter k. When state (e) times out the reaction signal triggers state (c) to state (a) which times out quickly (ϵ) and queries (?) the variable N_i, which is the number of species i molecules. If the reply ($\bullet$) is 0, meaning all of species i has been consumed, idle state (c) is re-activated. Otherwise, state (f) is activated, which times out (ε) back to (c) but decrements N_i and increments N_j.

This is a stochastic version of a "zero-order chemical reaction" and might be encountered in models of "heterogeneous reactions on surfaces" [Steinfeld *et al.* (1989)].

7.3.1 *Newton's Second Law*

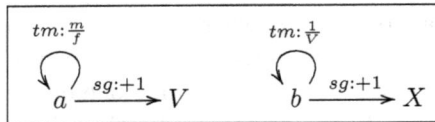

Fig. 7.6   Deterministic timing machine simulation of Newton's Second Law.

Figure 7.6 is a deterministic timing machine that models Newton's Second Law. The rate at which V increases is approximated by $\frac{f}{m}$ and the rate at which X increases likewise is approximated by V. Hence, the associated system of first-order ordinary differential equations is

$$\frac{dV}{dt} = \frac{f}{m}$$

$$\frac{dX}{dt} = V.$$

The equivalence between such a system of first-order differential equations and the second-order Newton's Second Law is a standard relationship and easy to establish [Hirsch and Smale (1974)]. Differentiate the second equation with respect to time and substitute from the first equation to yield Newton's Second Law in the form

$$\frac{dX^2}{dt^2} = \frac{f}{m}.$$

The implication is that STM can be used to simulate the behavior of higher-order differential equations.

7.3.2 *Gillespie Exact Stochastic Simulation*

The stochastic timing machine in Fig. 7.7 implements the Gillespie Exact Stochastic Simulation [Gillespie (1977)] described by a recursion in Appendix B, Algorithm (B.13).

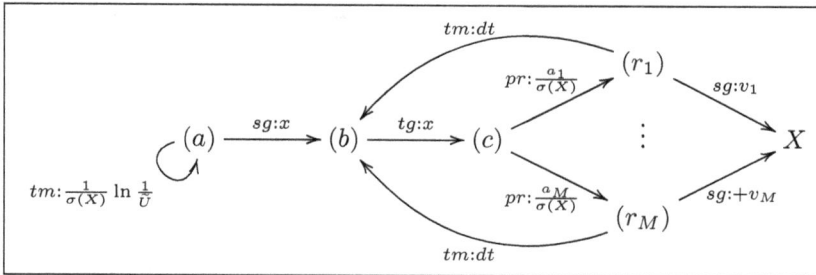

Fig. 7.7 The sum of propensities $\sigma(X) = \sum_{j=1}^{M} a_j(X)$. The average timeout of state (a) is given by the sum of the propensities of all the reactions. When (a) times out the signal x to (b) triggers transition immediately to stochastic state (c). The transition of (c) to (r_j) has probability $\frac{a_j}{\sigma(X)}$, and (r_j) times out back to (b) in time dt and changes the state X of the reaction by reaction vector v_j.

Remark 7.5. The Gillespie stochastic simulation algorithm (and its refined progeny [Gillespie (2007)]) is used widely in studies of chemical systems with many species, prokaryotic genetic circuits, genetic regulatory networks, and gene expression in single cells [McAdams and Arkin (1997)][McAdams and Arkin (1998)][Jong (2002)].

7.3.3 Brownian Particle in a Force Field

Until now the examples use ordinary differential equations to approximate timing machines. This example exhibits a stochastic timing machine associated with a partial differential equation for a probability distribution.

7.3.3.1 From Random Walk to Smoluchowski's Equation

Suppose a particle moves among $N + 1$ physical locations $0, \ldots, N$ spaced Δx apart, and that if it is at location 0 it can move only to location 1, if it is at k for $0 < k < N$ then it can move only to $k - 1$ or $k + 1$, and if at N only to $N - 1$ or 0. Suppose further that each move takes dt units of time, and furthermore that the probability of a rightward move equals the probability of a leftward move. If it is at 0 then half the time it stays put, and half the time it moves to 1; finally, assume that if it is at N then half the time it moves to $N - 1$, and half the time it starts over again back at 0. Note that this is not a symmetrical situation: the particle never moves directly from 0 to N but it may move from N to 0. A way to think of this is that a particle is placed initially at 0, half the time it starts over again, and half the time it has some chance of getting to N, and then starts over again at 0.

Call the history of a particle that starts at 0 a "run," so placing a particle at 0 results in an infinite sequence of runs – some possibly of great total duration. In any case, maintain a record of the number of runs and of how many times a particle visits each location, so that over a long sequence of runs those counts build up to certain values, say x_k. Then x_k divided by the number of runs is the frequentist probability that the particle visits location k.

Since the moves in either direction are equi-probable, there is insufficient reason to believe that any location would end up with a higher probability than any other location.

On the other hand, change the story, and suppose that moves to the left are more likely than moves to the right. Then there would be reason to believe that locations to the left would have higher probabilities than locations to the right. It would be as if there is a "force" – albeit random – pushing leftward against the particle as it moves.

Although this mental model is for one particle bouncing around, the model fits equally well with the idea of a large "cloud" of particles moving independently and without colliding but all according to the exact same rules. The resulting distribution of the particles in the cloud would have the same probability distribution as a single particle. Figure 7.8 is a stochastic timing machine model of this thought experiment. Note that Δx does not appear in the diagram. The columns correspond to the $N + 1$ locations. For the general situation, assume that the probability of moving right is $p_k := \frac{1}{2} - f_k \Delta x$ and that of moving left is $q_k = \frac{1}{2} + f_k \Delta x$, where $f_k \Delta x \leq \frac{1}{2}$ for $k = 0, \ldots, f_N$ are given "force" values – bearing in mind that force is the negative gradient of potential energy. Appendix D for further discussion.

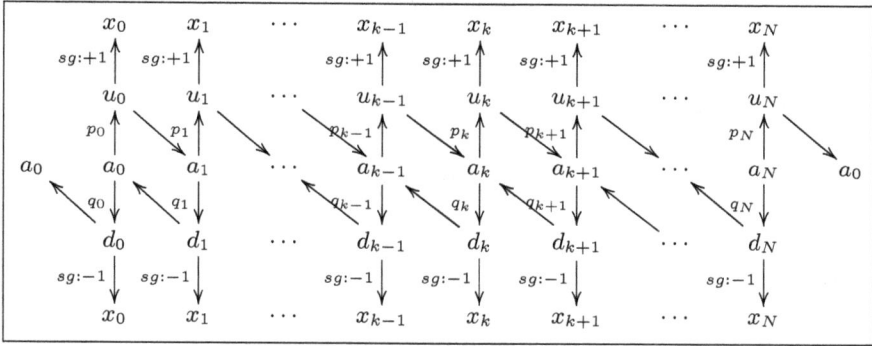

Fig. 7.8 Stochastic timing machine simulation of a Brownian particle in a conservative force field. $a_0, \ldots, a_N$ are stochastic states, meaning that upon activation a_k times out immediately to u_k with probability p_k and to d_k with the complementary probability q_k. All diagonal arrows are timeout arrows with understood label $tm : dt$. Therefore, arrival at u_k leads to a_{k+1} as the next state and simultaneously increments the variable x_k with probability $p_k = pr : p_k$, or leads to a_{k-1} and decrements x_k with probability $q_k = pr : q_k$. In other words, arrival of a Brownian particle at position k is represented by activation of the stochastic state a_k. If the particle gets to position N then with probability p_N it bounces back to position 0, and likewise if it gets to position 0 it re-starts at position 0 with probability q_0. The net effect is that the Brownian particle bounces from position to position, may start over again, but all the while accumulating counts in the variables x_k. After many such runs the values $x_0, \ldots, x_N$ represent the probability of the Brownian particle at positions $0, \ldots, k$. If $p_k = q_k$ the net result is simple diffusion. Otherwise, there is a bias in the resulting distribution $x_0, \ldots, x_N$ that represents the effect of a "force" upon the bouncing particle.

Remark 7.6. The timeout arrow from d_0 to a_0 corresponds to the idea that the bouncing particle is "absorbed" at location 0. If the arrow from d_0 terminated at a_1 instead then the bouncing particle would be "reflected" at location 0.

Let $P_n(k)$ denote the probability that in many runs the particle is at location k. In other words, in each experiment of many runs $x_k \to P_n(k)$.

The only ways that a particle can move to location k are that it is at $k-1$ and moves to the right with probability p_{k-1}, or that it is at $k+1$ and moves to the left with probability q_{k+1}. Therefore, there is a balance equation relating gains at k and losses at $k-1$ and $k+1$:

$$P_{n+1}(k) = \left(\frac{1}{2} - f_{k-1}\Delta x \right) P_n(k-1) + \left(\frac{1}{2} + f_{k+1}\Delta x \right) P_n(k+1). \quad (7.1)$$

This is a finite difference equation, and the associated differential equation is obtained by the replacements $n+1 \to t+dt$, $k \to x$, $\Delta x \to dx$, $k-1 \to x-dx$ and $k+1 \to x+dx$, where dt, dx are infinitesimals. Hence Eq. (7.1) becomes

$$P_{t+dt}(x) = \left(\frac{1}{2} - f(x-dx)dx \right) P_t(x-dx) + \left(\frac{1}{2} + f(x+dx)dx \right) P_t(x+dx). \quad (7.2)$$

Theorem 7.7.

$$\frac{dP_t(x)}{dt} = \frac{\partial^2 P_t(x)}{\partial x^2} + 4\frac{\partial}{\partial x}(f(x)P_t(x)) . \quad (7.3)$$

Proof. For reference, recall the Increment Rule from the infinitesimal calculus, namely for any smooth function g,

$$g(x+dx) \simeq g(x) + g'(x)dx$$
$$g(x-dx) \simeq g(x) - g'(x)dx.$$

To get a derivative with respect to time on the left of Eq. (7.2), subtract $P_t(x)$ and divide by dt on both sides, then calculate:

$$\frac{dP_t(x)}{dt} \simeq \frac{1}{dt}\left(\frac{1}{2}P_t(x+dx) - P_t(x) + \frac{1}{2}P_t(x-dx) \right)$$

$$+ \frac{dx}{dt}\left(f(x+dx)P_t(x+dx) - f(x-dx)P_t(x-dx) \right)$$

$$= \frac{dx^2}{2dt}\frac{P_t(x+dx) - 2P_t(x) + P_t(x-dx)}{dx^2}$$

$$+ \frac{dx}{dt}\left(f(x+dx)P_t(x+dx) - f(x-dx)P_t(x-dx) \right)$$

$$= \frac{dx^2}{2dt}\frac{\partial^2 P_t(x)}{\partial x^2} + \frac{dx}{dt}\left(f(x+dx)P_t(x+dx) - f(x-dx)P_t(x-dx) \right)$$

$$= \frac{dx^2}{2dt} \frac{\partial^2 P_t(x)}{\partial x^2} + \frac{dx}{dt} \left((f(x) + f'(x)dx)(P_t(x) + \frac{\partial P_t(x)}{\partial x} dx) \right.$$

$$\left. - (f(x) - f'(x)dx)(P_t(x) - \frac{\partial P_t(x)}{\partial x} dx) \right)$$

$$= \frac{dx^2}{2dt} \frac{\partial^2 P_t(x)}{\partial x^2} + 4 \frac{dx^2}{2dt} \left(f(x) \frac{\partial P_t(X)}{\partial x} + f'(x) P_t(x) \right)$$

$$= \frac{dx^2}{2dt} \left(\frac{\partial^2 P_t(x)}{\partial x^2} + 4 \frac{\partial}{\partial x} (f(x) P_t(x)) \right) .$$

Therefore, assuming $dx^2 = 2dt$ (i.e., that the diffusion constant $D := \frac{dx^2}{2dt} = 1$) the result is

$$\boxed{\frac{dP_t(x)}{dt} = \frac{\partial^2 P_t(x)}{\partial x^2} + 4 \frac{\partial}{\partial x} (f(x) P_t(x))}$$

which is the legendary equation of Marian Ritter von Smolan Smoluchowski (1872–1917) for a particle in a field of force [Kac (1954)] [Nelson (1967)]. □

Rewriting the Smoluchowski Equation as

$$\frac{dP_t(x)}{dt} = \frac{\partial}{\partial x} \left(\frac{\partial P_t(x)}{\partial x} + 4(f(x) P_t(x)) \right)$$

and identifying the expression $J_t(x) := \frac{\partial P_t(x)}{\partial x} + 4f(x)P_t(x)$ as "probability flux" immediately leads to noting that a sufficient condition for a cloud of particles to come to equilibrium at $t = \infty$, in the sense that the probability flux is 0, is that

$$\frac{\partial P_\infty(x)}{\partial x} + 4f(x)P_\infty(x) = 0 . \tag{7.4}$$

Equivalently,

$$\frac{\partial P_\infty(x)}{\partial x} = -4f(x)P_\infty(x) ,$$

which by Theorem 5.16(4) has the equilibrium solution

$$P_\infty(x) = P_\infty(0)e^{-4 \int_0^x f(x)dx} , \tag{7.5}$$

namely, the Boltzmann Distribution.

Remark 7.8. This is the formula implemented in Code (C.3.6).

Remark 7.9. Dimensional analysis helps clarify this expression. The position variable k discretely or x continuously has dimension L. Probability is dimensionless, hence for $f = f(x)$ to somehow represent force at location x it is necessary to recall that energy has the same dimensions as force times distance: "the work, or energy expended, to move a body is the force applied times the distance moved." Therefore, to render fx dimensionless let $F(x)$ be the magnitude of the force at location x, and let K be a basic unit of energy, so that if $f := \frac{F}{4K}$ then $fx = \frac{Fx}{4K}$ is the dimensionless quantity of energy measured in units of size K. Therefore, the Boltzmann Distribution takes the more familiar form

$$P_\infty(x) = P_\infty(0)e^{-\frac{E}{K}}, \tag{7.6}$$

where $E := Fx$.

Note furthermore that if $f(x) = 0$, that is to say, if there is no force on the particle, then the Smoluchowski Equation reduces to

$$\frac{dP_t(x)}{dt} = \frac{\partial^2 P_t(x)}{\partial x^2},$$

which is the Diffusion Equation associated with the difference equation

$$P_{n+1}(k) = \left(\frac{1}{2}\right) P_n(k-1) + \left(\frac{1}{2}\right) P_n(k+1), \tag{7.7}$$

in other words, the case when $p_k = q_k$.

The Smoluchowski Equation and closely related (Fokker-Planck) equations ([Van Kampen (2007)] [Zwanzig (2001)]) are the starting point in numerous mathematical and simulation studies of molecular motors [Astumian and Bier (1996)] [Jülicher *et al.* (1997)] [Keller and Bustamente (2000)] [Bustamente *et al.* (2001)] [Reimann (2001)] [C.P. Fall and Tyson (2002)] [Lan and Sun (2005)] [Woo and Moss (2005)] [Xing *et al.* (2005)] [Lindén (2008)] [Wang (2008a)] [Wang (2008b)].

7.3.3.2 *A Simpler, Equivalent Stochastic Timing Machine*

Although the timing machine in Fig. 7.8 adequately evokes the intuition of a particle bouncing among locations, the much simpler – but equivalent – diagram in Fig. 7.9 is more suitable for simulation. See Appendix C for MATLAB (C.3.1).

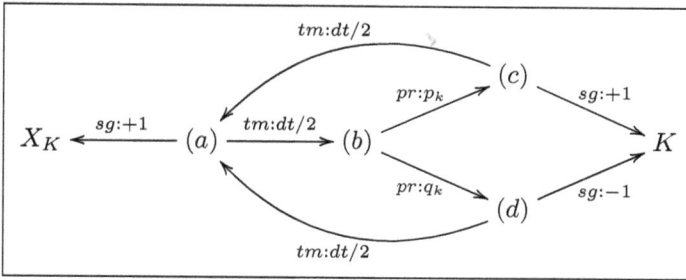

Fig. 7.9 In time increment $dt/2$ state (a) times out to (b) which is a stochastic state. Hence (b) transitions immediately to (c) – with probability p_k – or to (d) with complementary probability $q_k := 1 - p_k$. In time $dt/2$ either (c) times out back to (a) and increments the location index K, or (d) times out back to (a) and decrements the location index. In any case, whenever (a) times out the visit counter X_K at location K is incremented. Not shown are initialization steps such as setting K to location 1, and all visit counts to 0 except for $X_1 = 1$ signifying the start location of the particle. Also not shown are that when K gets to the right end its increment does not move the particle beyond the right end, and likewise when K gets to the left end, decrement does not move the particle beyond the left end.

PART 5

Theory of Substances

Classical thermodynamics is the most general mathematical science of the transport and transformation of substances – including immaterial but indestructible ones such as entropy and virtual but conserved ones such as energy. This chapter explains what that means by blending category theory with calculus to express the axioms for a Theory of Substances in a *Universe* of discourse. There is a general consensus in the muscle contraction research community that muscle contraction must be explained as transduction of energy bound in chemicals to energy released in momentum current – A.K.A. force.

Chapter 8

Algebraic Thermodynamics

8.1 Introduction

> One of the best ways to learn something is to reinvent it. Mathematics needs to be constantly reinvented to stay alive and prosper. Every new generation is reinventing mathematics. Category theorists are permanently reinventing mathematics.
>
> ———————————————
> André Joyal, September, 2010

I believe that everybody present here will be greatly interested if I say that there is a physics course-book which can be used to explain, in a way readily comprehensible to the average senior high school students, and in one single class hour (40 minutes), the concept of "entropy" that has been considered to be so abstract and complicated, the definition of "heat pump", and so on. This book expounds explicitly all three Newton's Laws of motion by using only one law; in addition, this book has solved the problems regarding the forms and transformation of energy by introducing the concept of energy carrier. You may ask: is there really such a book, and what kind of book is it? My answer is definitely in the affirmative. The course-book I have in mind is a German one entitled Der Karlsruher

Physikkurs (KPK), or Karlsruhe Physics Course (KPC) in English.[1]

Aside 8.1.1. In the United States, high school mathematics and science teachers seem unaware of the Karlsruhe Physics Course. This Part of my book offers a rigorous mathematical foundation for the Karlsruhe Physics Course. I reinvent thermodynamics with category theory. It is my hope that our teachers may find here a valuable resource for understanding more than they are responsible for teaching, and that their understanding is based on the modern ideas in the Karlsruhe Physics Course. The text by Hans U. Fuchs extends this view of thermodynamics for physics and engineering students with a vast fund of detailed examples and exercises.

[D]idactic tools have been built that make it not just simple, but rather natural and inevitable to use entropy as the thermal quantity with which to start the exposition. The outcome is a course that is both fundamental and geared toward applications in engineering and the sciences.   In continuum physics an intuitive and unified view of physical processes has evolved: That it is the flow and the balance of certain physical quantities such as mass, momentum, and entropy which govern all interactions. The fundamental laws of balance must be accompanied by proper constitutive relations for the fluxes and other variables. Together, these laws make it possible to describe continuous processes occurring in space and time. The image developed here lends itself to a presentation of introductory material simple enough for the beginner while providing the foundations upon which advanced courses may be built in a straightforward manner. Entropy is understood as the everyday concept of heat, a concept that can be turned into a physical quantity comparable to electrical charge or momentum. With the recognition that heat (entropy) can be created, the law of balance of heat, i.e., the most general form of the

---

[1]Wu, Guobin, 11th National Symposium on Physics Education, East China Normal University, China, "Thoughts on the Karlsruhe Physics Course,"www.physikdidaktik. uni-karlsruhe.de/publication/wu_kpk.pdf.

second law of thermodynamics, is at the fingertips of the student [Fuchs (1996)].

Personally, my primary motivation for The Theory of Substances is the fact that thermodynamics offers the theory of energy transduction required for analyzing the relationship between chemical energy and mechanical energy: the core of muscle contraction. This Part presents this Theory. My secondary motivation is that a question about thermodynamics of radiation addressed by Max Planck initiated quantum mechanics [Kuhn (1978)]. Here this is not discussed any further.

Aside 8.1.2. In Winter the tires of my car bulge more.

It is related that Albert Einstein as a child was intrigued by the behavior of a magnetic compass shown to him by his father. What startled him was the motion of the needle *without direct contact* [Ohanian (2008)]. But even direct contact of bodies – not colliding, not rubbing, but just touching – may result in noticeable changes in shape. Mechanics of rigid bodies is the scientific study of their motion, for the only change they may undergo is that of position, and their response to forces depends only on their mass. The science arising from study of changes in shape of bodies in *direct contact* is thermodynamics. The concept of *temperature* may be introduced by reference to changes in shape – specifically volume – of a standard body composed of a standard *material substance*. Material substances – chemical elements, compounds, and mixtures – form and transform in an endless variety of ways. Chemical thermodynamics is the scientific study of changes in temperature associated with chemical activity.

8.2 Chemical Element, Compound & Mixture

Let $\mathbb{W} \hookrightarrow \mathbb{Z}$ denote the set of positive integers $1, 2, 3, \ldots$, and let $\mathbb{F} \hookrightarrow \mathbb{Q}$ denote the set of non-negative rational numbers. If X is a set let $\Pi X_{\mathbb{W}}$ denote the set of finite formal products of elements of X subscripted by positive integers, as in

$$x^1_{n_1} \cdots x^p_{n_p} \qquad x^1, \ldots, x^p \in X, \quad n_1, \ldots, n_p \in \mathbb{W} \,.$$

Likewise, $\Sigma \mathbb{F} X$ denotes the set of finite formal sums of elements of X with non-negative rational coefficients ("fractions"), as in

$$a_1 x_1 + \cdots + a_m x_m \qquad x_1, \ldots, x_m \in X, \quad a_1, \ldots, a_m \in \mathbb{F} \,.$$

Aside 8.2.1. Always aware that unconventional notations may be dangerous – eyebrow raisers – I hasten to note that ΠX_{W} is intended merely to be a mnemonic for the symbolic representations of chemical compounds ordinarily used in chemistry. The notation $\Sigma \mathbb{F} X$ is a bit more problematic because there are standard ways to denote the vector space of finite formal sums with coefficients in a field, say $\mathbb{Q}$: the *direct sum* is denoted by $\bigoplus_X \mathbb{Q}$ or $\coprod_X \mathbb{Q}$ or $\sum_X \mathbb{Q}$. So, to be perfectly clear, $\Sigma \mathbb{Q} X = \bigoplus_X \mathbb{Q}$. Of course $\mathbb{F}$ is not a field (no negatives), so $\Sigma \mathbb{F} X$ is a specialized notation for this book.

The *cartesian product* of $\mathbb{Q}$ with itself X many times is by definition $\mathbb{Q}^X$, which is the set of all sequences of elements of $\mathbb{Q}$ indexed by the elements of X. To bring out a certain dual relationship between the direct sum and the cartesian product, these are given the notations $\coprod_X \mathbb{Q}$ and $\prod_X \mathbb{Q}$, and as vector spaces $\coprod_X \mathbb{Q} \hookrightarrow \prod_X \mathbb{Q}$ [Mac Lane (1967)].

Axiom 8.2.1. $\boxed{\textbf{Chemical Substance}}$

There exists a non-void set $\mathcal{A}$ of **chemical elements** and a non-void set $\mathcal{C} \hookrightarrow \Pi \mathcal{A}_{\mathrm{W}}$ of **chemical compounds** (also known as **species**). For example, if $H, O \in \mathcal{A}$ then possibly $H_2 O \in \mathcal{C}$. Every element is a compound. More formally, the injection of $\mathcal{A} \rightarrowtail \Pi \mathcal{A}_{\mathrm{W}}$ defined by $e \mapsto e_1$ factors through $\mathcal{C}$:

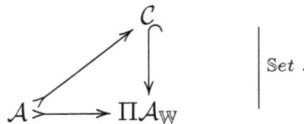

$$
\begin{array}{ccc}
 & \mathcal{C} & \\
 & \nearrow \quad \downarrow & \\
\mathcal{A} & \longrightarrow & \Pi \mathcal{A}_{\mathrm{W}}
\end{array}
\qquad \Big| \; \text{\small Set} \, .
$$

The elements of the set $\mathcal{M} := \Sigma \mathbb{F} \mathcal{C}$ are called **mixtures** and assigned unit of measurement [**AMT**]. Each species is a **pure mixture**, that is there exists an injection $\mathcal{C} \rightarrowtail \mathcal{M}$ defined by $X \mapsto \frac{1}{1} X$, and in particular for $e \in \mathcal{A}$ there exists the composition $e \mapsto e_1 \mapsto \frac{1}{1} e_1$.

Theorem 8.1. $\mathcal{M}$ *is closed under (formal) addition and multiplication by non-negative rational numbers. Moreover, the function* $\mathcal{M} \xrightarrow{\Phi} \mathcal{M}$ *defined to be the identity on* $\mathcal{A}$ *and extended first to* $\mathcal{C}$ *by*

$$
\Phi(e_{n_1}^1 \cdots e_{n_p}^p) := \frac{n_1}{1} e^1 + \cdots \frac{n_p}{1} e^p
$$

*and then extended to all of $\mathcal{M}$ by linearity, namely $\Phi(aX + bY) := a\Phi(X) + b\Phi(Y)$ for $X, Y \in \mathcal{C}$, is **idempotent**, that is, $\Phi \circ \Phi = \Phi$.*

Proof. Formal addition is associative by definition, and likewise multiplication by scalars is distributive. Idempotence follows directly from the inductive definition of Φ. $\qquad\square$

8.3 Universe

Definition 8.2. A **universe** is a diagram

$$\Sigma \in \mathbb{T}op \qquad\qquad \textbf{state space}$$
$$\mathbb{U} \hookrightarrow \mathbb{P}th(\Sigma) \in \mathbb{C}at \qquad\qquad \textbf{process category}$$
$$\mathcal{S} \in \mathbb{S}et \qquad\qquad \textbf{substances}$$
$$\mathcal{X} \in \mathbb{S}et \qquad\qquad \textbf{bodies}$$

In general a map $\Sigma \xrightarrow{Q} \mathbb{R}$ is called a **quantity**. A finite set

$$\mathcal{B} := \{\, B_1 \cdots B_n \,\}$$

of bodies is called a **system**. Assume that $\mathcal{X}$ is a finite set. Then $\mathcal{X} \backslash \mathcal{B}$ is a system, and may be referred to as the **environment** (or **surroundings**) of $\mathcal{B}$. The inclusion of categories $\mathbb{U} \hookrightarrow \mathbb{P}th(\Sigma)$ implies that every process is a path, but suggests there may be paths that are not processes. Indeed, axioms to be enunciated subsequently will definitely distinguish between those paths that are and those that are not processes. In any case, it is assumed that *every state $u \in \Sigma$ is an object of* $\mathbb{U}$.

Definition 8.3. The underlying set of $\mathbb{E} := \mathbb{R}^3$ is called **space**.

In classical physics physical space *per se* is empty but it has a rich mathematical structure. Pre-eminent is that space admits three-dimensional orthogonal coordinate systems, of which one is explicit in the representation $\mathbb{R}^3$.

Axiom 8.3.1. $\boxed{\textbf{Spatial Region}}$

For each body $B \in \mathcal{X}$ there is a diagram $B \in \mathbb{S}et$ such that for each diagram $p \in B$, there are 3 quantities

$$\Sigma \xrightarrow{x_p} \mathbb{R}$$

$$\Sigma \xrightarrow{y_p} \mathbb{R}$$

$$\Sigma \xrightarrow{z_p} \mathbb{R}$$

An item p in B is called a **point** of B. A body with exactly one point is called a **particle**. For a state u in Σ the values $x_p(u), y_p(u), z_p(u)$ are called the x, y, z **spatial coordinates** of p in state u. The set $B(u) \subset \mathbb{E}$ of all lists $[\,x_p(u)\ y_p(u)\ z_p(u)\,]$ for p a point of B is called the **spatial region** of B in state u. The map $B \to B(u)$ is a bijective correspondence: distinct points of B have distinct spatial coordinates for any state.

Definition 8.4. Let $B \in \mathcal{X}$ be a body. If for each $p \in B$ there is a quantity

$$\Sigma \xrightarrow{Y_p} \mathbb{R}$$

then Y is called a **spatially-varying quantity** in B. Since distinct points of B have distinct spatial coordinates, a spatially-varying quantity Y determines a map

$$B(u) \xrightarrow{Y(u)} \mathbb{R}$$

such that $Y(u)(\,x_p(u)\ y_p(u)\ z_p(u)\,) = Y_p(u)$.

For a non-zero spatial vector $\overrightarrow{v}$ if the standard part of the ratio of the inifintesimal change in $Y_p(u)$ in the direction of $\overrightarrow{v}$ to an infinitesimal change in distance in that direction is independent of the size of that infinitesimal, then that standard part is called the **spatial derivative of** $Y(u)$ **at** p **along** $\overrightarrow{v}$. Formally,

$$DY_p(u)(\overrightarrow{v}) := \text{st}\left(\frac{Y_{p+\varepsilon\overrightarrow{v}}(u) - Y_p(u)}{\varepsilon}\right)$$

if the right-hand side is defined for some positive infinitesimal $\varepsilon > 0$ and is independent of the choice of ε.

In particular, the spatial derivatives along the coordinate axes are defined by

$$\partial_x Y_p(u) = \frac{\partial Y(u)}{\partial x}(p) := DY_p(u)(1,0,0)$$

$$\partial_y Y_p(u) = \frac{\partial Y(u)}{\partial y}(p) := DY_p(u)(0,1,0)$$

$$\partial_z Y_p(u) = \frac{\partial Y(u)}{\partial z}(p) := DY_p(u)(0,0,1)\,.$$

If $DY_p(u) = 0$ for all $p \in B$ then $Y(u)$ is a constant map on B, in which case the definition $Y_B(u) := Y_p(u)$ is independent of the choice of $p \in B$. Otherwise put, if $Y(u)$ is not spatially varying in B then there exists a map

$$\Sigma \xrightarrow{Y_B} \mathbb{R} \ .$$

Axiom 8.3.2. $\boxed{\textbf{Basic Substance Quantities}}$

If X is a substance and A, B, C are bodies, then there are quantities

$$\Sigma \xrightarrow{X_A} \mathbb{R} \qquad \qquad \textbf{total amount}$$

$$\Sigma \xrightarrow{\dot{X}_A} \mathbb{R} \qquad \qquad \textbf{rate of change of amount}$$

$$\Sigma \xrightarrow{\varphi_A^{(X)}} \mathbb{R} \qquad \qquad \textbf{potential}$$

$$\Sigma \xrightarrow{X_B^A(C)} \mathbb{R} \qquad \qquad \textbf{transport rate}$$

$$\Sigma \xrightarrow{K_C^{(X)}(A,B)} \mathbb{R} \qquad \qquad \textbf{conductivity}$$

with units of measurement

$$\Sigma(\mathcal{A}) \xrightarrow{X_A} \mathbb{R} \qquad \textbf{[AMT]}$$

$$\Sigma(\mathcal{A}) \xrightarrow{\dot{X}_A} \mathbb{R} \qquad \textbf{[AMT][TME]}^{-1}$$

$$\Sigma(\mathcal{A}) \xrightarrow{X_B^A(C)} \mathbb{R} \qquad \textbf{[AMT][TME]}^{-1}$$

$$\Sigma(\mathcal{A}) \xrightarrow{\varphi_A^{(X)}} \mathbb{R} \qquad \textbf{[NRG][AMT]}^{-1} \ .$$

$X_A(u)$ is the total amount of substance X in body A at state u, and $\dot{X}_A(u)$ is the rate at which that total amount is changing. For each substance X there is a potential energy $\varphi_A^{(X)}(u)$ per amount of X in A at state u. The rate $X_B^A(C)(u)$ at which X is transported from A to B through C at state u is related to the conductivity $K_C^{(X)}(A,B)(u)$ by the next axiom.

The distinction between the points of a body and the spatial region occupied by a body is reflected in a distinction between two basic kinds of quantity. On one hand, there is the total amount of a quantity in the spatial region occupied by a body (for a specified state of the universe).

For example the volume, or the mass, or the charge, or the water in the body. If the spatial region of a body is considered in the imagination to be subdivided into several disjoint parts which together cover the entire region of the body, then the sum of the total amounts of the quantity in the parts equals the total amount in the entire body. This total is a characteristic of the whole body (for a specified state of the universe). In general a variable total amount $X_A = X_A(x)$ is called an **extensive variable**.

On the other hand, at a point of a body the ratio, of the total quantity in a small part of the spatial region of the body surrounding the point, to the total quantity of some other quantity in the small part, in the limit as the small part shrinks to nothingness, is a number that is characteristic of that point of the body (for a specified state of the universe). In general a variable limiting amount $Q(p)$ for $p \in B$ is called an **intensive variable**. In particular, the limit of the ratio of a quantity in a small part around a point to the volume of the small part is a **density variable**. For example the mass density or charge density at a point. Especially interesting is the limit of the ratio of energy to entropy at a point: that intensive variable will be identified as the *temperature*.

These definitions imply an intimate relationship between extensive and intensive variables and basic operations of the infinitesimal calculus. The total mass of a body, for example, is the integral of the mass density over the spatial region of the body. Inversely, the mass density is the derivative of the mass with respect to volume.

The substance "ontology" in Fig. 8.1 distinguishes material from immaterial substances in (the spatial region of) a body. For example, the spatial region of a body may contain water, which is a material substance, and it has a volume, which is an immaterial substance independent of whatever occupies the region of the body.

Definition 8.5. For any body $B \in \mathcal{X}$ of a universe the **material substance** in the body is denoted by $|B|$.

Remark 8.6. The sequence of equations and injections

$$\mathcal{M} := \Sigma \mathbb{F} \mathcal{C} \hookrightarrow \coprod_{\mathcal{C}} \mathbb{Q} \hookrightarrow \prod_{\mathcal{C}} \mathbb{Q} = \mathbb{Q}^{\mathcal{C}} \hookrightarrow \mathbb{R}^{\mathcal{C}}$$

merely highlights the fact that for a reactor B the chemical substance $\Sigma(B) \xrightarrow{|B|} \mathcal{M}$ is truly a vector of quantities in the sense of the composed map $\Sigma(B) \xrightarrow{|B|} \mathbb{R}^{\mathcal{C}}$.

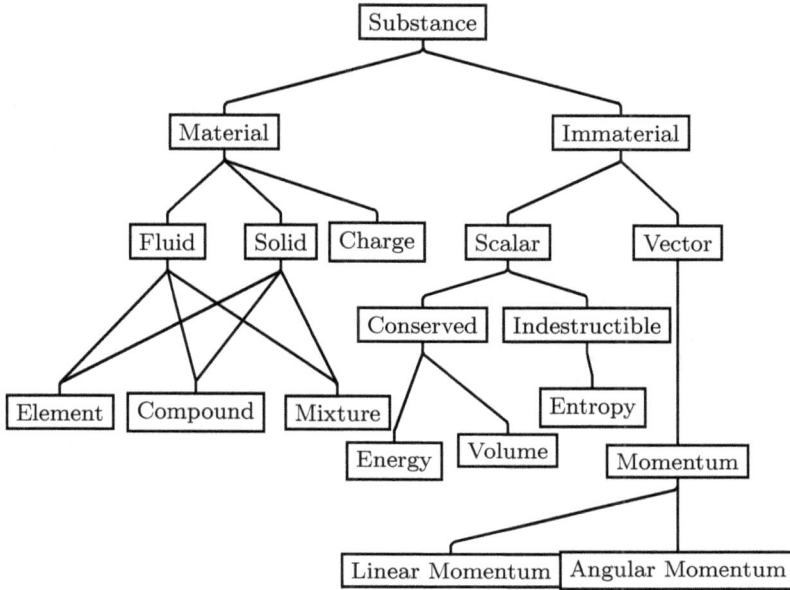

Fig. 8.1   Kinds of substances.

Physical properties of a body often but not always refer to the material substance in the body. For example, the volume $V_B(u)$ of the spatial region $B(u)$ is an immaterial extensive property of B.

Aside 8.3.1. The property of a body that distinguishes between a fluid that is either a gas or a liquid, and between a fluid or a solid, might be expressed in terms of the strength of cohesion between its "particles" – atoms or molecules. However, that approach is anathema to my determination that classical thermodynamics, and even chemical thermodynamics, be described without reference to particles (see Aside 1.10.1). Fortunately, continuum mechanics offers Cauchy's Fundamental Theorem of Elastic Bodies, see Appendix E. A modern development from this Theorem and technical definitions of the *stress tensor* in fluids and solids is available in [Romano *et al.* (2006)]. Modern intuitive discussion of stress and momentum currents in bodies is offered for example by [Herrmann and Schmid (Year not available)],[Herrmann and Job (2006)].

Definition 8.7. If $A, B \in \mathcal{X}$ and $u \in \Sigma$ then B **is enclosed by (or "contained in")** A **at** u if the spatial region of B is a subset of the spatial region of A in state u, that is, if $B(u) \subseteq A(u)$. B **is immersed in** A

at u if the boundary of B is a subset of the boundary of A, that is, if $\partial B(u) \subset \partial A(u)$.

If B is enclosed by A then all the substance $X := |B|$ is also substance of A, so $X_B \leq X_A$. If $V_B < V_A$ then X may or may not be uniformly distributed in A. Indeed, if X is not uniformly distributed in B then it is not uniformly distributed in A, but X could be uniformly distributed in B without being uniformly distributed in A.

Axiom 8.3.3. $\boxed{\text{Conductivity}}$

Let A, B, C be bodies, X any substance, and $u \in \Sigma$. Then

$$X_A^B(C)(u) = -X_B^A(C)(u) ,$$

and

$$X_B^A(C)(u) > 0 \tag{8.1}$$
$$\text{if and only if}$$
$$X_A(u) > 0 \text{ and } K_C^{(X)}(A, B)(u) > 0 \text{ and } \varphi_A^{(X)}(u) > \varphi_B^{(X)}(u) .$$

If X is a material substance then $X_A(u) \geq 0$, and

$$X_A(u) > 0 \text{ if and only if } \varphi_A^{(X)}(u) > 0 . \tag{8.2}$$

In words, the transport rate in the reverse direction is just the negative of the transport rate in the forward direction. Equation (8.1) declares there can be transport of a substance from one body to another if and only if there is substance to transport *and* there is a body with positive conductivity which can conduct the substance between the two bodies *and* the potential of the first body is strictly greater than the potential of the second body. Equation (8.2) asserts that a body may contain *material substance* if and only if it has positive substance potential.

Note that there is no presumption that the conductivity $K_C^{(X)}(B, A)(u)$ of X from A to B through C equals the (reverse) conductivity from B to A. For example, a rectifier or diode in an electrical circuit allows electricity to flow in only one direction.

Remark 8.8. In Theory of Substances the Conductivity Axiom is a discrete version of several hallowed "constitutive laws of physics":

$$J_x = -D\frac{\partial c}{\partial x} \qquad \text{Fick's First Law of Diffusion}$$

$$\frac{dQ}{dt} = kA\frac{\partial T}{\partial x} \qquad \text{Fourier Heat Conduction Law}$$

$$I = \frac{1}{R}V \qquad \text{Ohm's Law of Electricity}$$

$$V = \frac{(p_1 - p_2)\pi r^4}{8\eta L} \qquad \text{Poiseuille's Law of Liquid Flow .}$$

Definition 8.9. The notation $A \xrightarrow[X]{C} B$ signifies that there is a conduit C between stores A and B such that in any state the X-conductivity of C is positive. Formally, $K_C^{(X)}(A, B)(u) > 0$ for any state u.

Therefore, if $A \xrightarrow[X]{C} B$ then for any state u, $X_B^A(C)(u) > 0$ if and only if $\varphi_A^{(X)}(u) > \varphi_B^{(X)}(u) > 0$. If $A \xrightarrow[X]{C} B$ then the conductivity of C could vary from state to state even though it is always positive. If, however, every positive conductivity is independent of the state then it is as though there are fixed walls between pairs of storage bodies, but walls "permeable" to different substances to possibly different degrees. For example, a partition between two bodies may conduct heat very well and electricity somewhat, and if it is freely movable – say it is a frictionless piston sliding in a cylinder with different gases on its two sides – then it also conducts volume very well.

Definition 8.10. For any substance $X \in \mathcal{X}$ the storage bodies and their conduit arrows $A \xrightarrow[X]{C} B$ form a directed graph for which the notation shall be $\mathbb{K}^{(X)}$.

Since any directed graph is the disjoint union of its connected components, each storage body belongs to a set of storage bodies between any two of which there exists a chain of X conduits in one direction or the other. And between bodies of two different connected components there exist no conduits of X whatsoever – connected components are isolated with regard to the flow of X.

As seen in Fig. 8.2 the positive conductivities of a system form a multilevel graph. Any body may conduct more than one substance, and any body may be a conduit between more than one pair of other bodies.

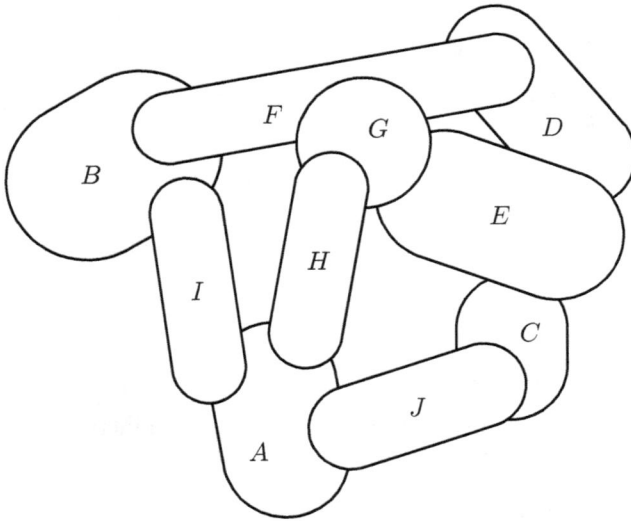

Fig. 8.2 A system $\mathcal{X} = [A\,B\,C\,D\,E\,F\,H\,I\,J]$ involves various levels of conductivities. For example, G is a conductor between E and F, and at a higher level H is a conductor between G and A.

Example 8.11. For preparing tea, say, an electric heating element in a cup of water is a conduit for electricity between a higher potential and a lower potential (which may alternate). The surface of the heating element is a conduit for entropy (the "heat substance") between the interior of the heating element and the water, which is a body partially enclosed by the cup, another body. In this example the heating element, water, and cup are three-dimensional bodies, but the surface interface between the heating element and the water is a two-dimensional body. Entropy is generated by the flow of electricity through the heating element, whose resistance is the reciprocal of its conductivity. There is also a surface interface between the water and the atmosphere, another body, and entropy may flow through it from the water to the atmosphere.

Definition 8.12. A **constraint** on a universe is a map

$$\mathcal{S} \times \mathcal{X} \times \mathcal{X} \times \mathcal{X} \xrightarrow{k} \mathbb{R}$$

such that $k_C^{(X)}(A, B) \geq 0$ for all states u. A path $[0, r] \xrightarrow{\alpha} \Sigma$ is **constrained by** k if $K_C^{(X)}(A, B)(\alpha(t)) = k_C^{(X)}(A, B)$ for $0 \leq t \leq r$.

It is conventional to refer to a wall that is impermeable to the flow of heat as *adiabatic*; whereas a wall that permits the flow of heat is termed *diathermal*. If a wall allows the flux of neither work nor heat, it is *restrictive with respect to energy*. A system enclosed by a wall that is restrictive with respect to energy, volume, and all the mole numbers is said to be *closed* [Callen (1985)].

Generally speaking, there are very many possibilities for constraints, starting with just two storage bodies and the conduits between them. If there are no conduits between them at all, so that there can be no flow of any substance between them, then they are **completely isolated** bodies. Or, there may be some conduits between two bodies for some substances, in which case they are **partially isolated**. The same terminology applies to two subsystems of the universe wherein if any two bodies one from each are completely isolated, then the subsystems are completely isolated. Or, some bodies of one subsystem and some of the other may be partially isolated, and so on.

A universe with bodies $\mathcal{X}$ and a distinguished system $\mathcal{B} \hookrightarrow \mathcal{X}$ determines the complement $\mathcal{E} := \mathcal{X} \backslash \mathcal{B}$. In this case $\mathcal{E}$ may be called the **environment** or the **surroundings** of $\mathcal{B}$, and this system may be completely or partially isolated from its environment.

Definition 8.13. Let u be a state. A path $[0, r] \xrightarrow{\alpha} \Sigma$ **goes through** u_* **at** T_* if $0 < T_* < r$ and $\alpha(T_*) = u_*$. If α is also constrained by k then it is called a **constrained virtual displacement** (with respect to k) at u_*.

Definition 8.14. Let u_* be a state, k a constraint and Y a quantity. Say Y is **independently variable** if there exists a constrained virtual displacement α_Y through u_* at T_* such that

$$\partial_0 X \alpha_Y (T_*) = 0 \text{ for all quantities } X \text{ distinct from } Y \qquad (8.3)$$
$$\partial_0 Y \alpha_Y (T_*) = 1 . \qquad (8.4)$$

Otherwise put, a quantity is independently variable with respect to a constraint if some constrained path can definitely vary the quantity without varying any other quantities at all.

Theorem 8.15. *(Independent Variability) Let X and $Y_1, \ldots, Y_n$ be quantities such that $X = X(Y_1, \ldots, Y_n)$ and $Y_1, \ldots, Y_n$ are independently variable.*

If

$$P_i := \frac{\partial X}{\partial Y_i} \ for \ i = 1, \ldots, n \tag{8.5}$$

then

$$\overset{\bullet}{X} = \sum_{i=1}^{n} P_i \cdot \overset{\bullet}{Y_i} \ . \tag{8.6}$$

Conversely, if there are quantities $P_1, \ldots, P_n$ such that Eq. (8.6) holds, then Eq. (8.5) is true.

Proof.   Equation (8.5) implies Eq. (8.6) by the Chain Rule. Conversely, let $\alpha_1, \ldots, \alpha_n$ be k-constrained virtual displacements through u_* at T_* for $Y_1, \ldots, Y_n$, respectively.

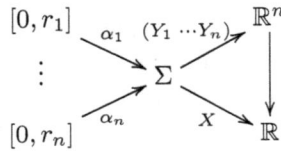

By the Chain Rule,

$$\overset{\bullet}{X} = \sum_{i=1}^{n} \frac{\partial X}{\partial Y_i} \cdot \overset{\bullet}{Y_i} \tag{8.7}$$

so subtraction from Eq. (8.6) yields

$$0 = \sum_{i=1}^{n} \left(P_i - \frac{\partial X}{\partial Y_i} \right) \cdot \overset{\bullet}{Y_i} \tag{8.8}$$

Therefore, at T_* and for $j = 1, \ldots, n$,

$$0 = \left(\sum_{i=1}^{n} \left(P_i \alpha_j - \frac{\partial X}{\partial Y_i} \alpha_j \right) \overset{\bullet}{Y_i} \alpha_j \right)(T_*)$$

$$= \sum_{i=1}^{n} \left(P_i \alpha_j(T_*) - \frac{\partial X}{\partial Y_i} \alpha_j(T_*) \right) \overset{\bullet}{Y_i} \alpha_j(T_*)$$

$$= \sum_{i=1}^{n} \left(P_i(u_*) - \frac{\partial X}{\partial Y_i}(Y(u_*)) \right) \overset{\bullet}{Y_i} \alpha_j(T_*)$$

$$= \sum_{i=1}^{n} \left(P_i(u_*) - \frac{\partial X}{\partial Y_i}(Y(u_*)) \right) \partial_0(Y_i \alpha_j)(T_*)$$

$$= P_j(u_*) - \frac{\partial X}{\partial Y_j}(Y(u_*)) \ . \qquad \square$$

Axiom 8.3.4. Additivity

For any bodies A, B, C, D and substance X,

$$X_{A+B} = X_A + X_B \tag{8.9}$$

$$\dot{X}_{A+B} = \dot{X}_A + \dot{X}_B \tag{8.10}$$

$$X^A_{B+C}(D) = X^A_B(D) + X^A_C(D) \tag{8.11}$$

$$X^{B+C}_A(D) = X^B_A(D) + X^C_A(D) \tag{8.12}$$

As defined above, for any body A and substance X the net transport rate X^A_A of X from A to itself is called the rate of appearance or disappearance of X in A according as $X^A_A > 0$ or $X^A_A < 0$.

Definition 8.16. For a body B say substance X is

uncreatable in B if $X^B_B \leq 0$

conserved in B if $X^B_B = 0$.

indestructible in B if $X^B_B \geq 0$

For any subsystem $[\,A\,B\,]$ and for any substance X by Additivity

$$
\begin{aligned}
X^{A+B}_{A+B} &= X^{A+B}_A + X^{A+B}_B \\
&= X^A_A + X^B_A + X^A_B + X^B_B \\
&= X^A_A + X^B_B \, .
\end{aligned}
$$

Thus for any subsystem the rate at which substance is created in the subsystem is the sum of the rates at which it is created in each of its constituent bodies, and likewise for destruction of substance.

Axiom 8.3.5. Circuitry

An important distinction exists between field theory and system theory. Field theory is based on infinitely small domains with distributed elements and calculates, using

mostly partial differential equations, the distribution or field of interesting variables. They can be the fluid flow around the hull of a ship or the electric field around a high voltage insulator. Field theory is indispensable for such problems. System theory on the other hand uses discrete elements of finite size. They are built up by interconnecting such elements, a proceeding usually called reticulation of a system [Thoma (1976)].

For each substance X the bodies of the universe are divided into two disjoint sets,

$$\mathcal{X} = \mathcal{C}^{(X)} \cup \mathcal{B}^{(X)} .$$

A body C in $\mathcal{C}^{(X)}$ is called an X-**conduit** and a body B in $\mathcal{B}^{(X)}$ is called an X-**store**. An X-store B satisfies $X_B \geq 0$ and $K_B^{(X)}(A, D) = \infty$ (where ∞ is an infinite hyperreal). An X-conduit C satisfies $X_C = 0$ and $K_C^{(X)}(A, B) > 0$.

Aside 8.3.2. This axiom definitely aligns the Theory of Substances in this book with such systems theories as the "network thermodynamics" of [Oster *et al.* (1973)], the tradition of "bond graphs" as in [Thoma (1976)] and [Cellier (1992)], and the "dynamical systems" approach of [Haddad *et al.* (2005)]. Much of my intuition about substances derives from familiarity with the electricity substance in electrical circuits among which the simplest are the networks of resistors and capacitors and batteries.

Axiom 8.3.6. $\boxed{\text{Process}}$

For any process $u \xrightarrow{\alpha} v \in \mathbb{U}$ there exists an equation

$$\dot{X}_A \, \alpha = \partial_0(X_A \alpha) .$$

Remark 8.17. This "adjointness" equation is what really explains the meaning of $\dot{X}_A$. The analogy is in particle mechanics, where $\dot{q}$ – a variable interpreted as "generalized velocity" – is a derivative with respect to time only in terms of an equation $\dot{q} = \dot{q}(t)$.

The composition of a path α and a quantity X yields a real-valued function of a real variable, as in

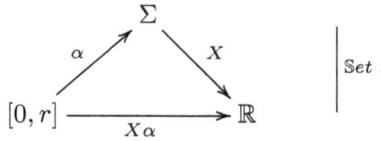

$$
\begin{array}{ccc}
 & \Sigma & \\
{}^{\alpha}\nearrow & & \searrow{}^{X} \\
[0,r] \xrightarrow[X\alpha]{} & & \mathbb{R}
\end{array}
\qquad \Big| \; Set \; .
$$

Consequently all the tools of the infinitesimal calculus are applicable to quantities depending on paths or processes. In particular, the derivative of $X\alpha$ with respect to time $t \in [0,r]$ is defined if $X\alpha$ is differentiable. The most common notation for that derivative is the quotient of infinitesimals,

$$
\frac{d(X\alpha)}{dt} \; ,
$$

but in Theory of Substances the preferred notation is $\partial_0 X\alpha$, so

$$
[0,r] \xrightarrow{\;\;\partial_0 X\alpha\;\;} \mathbb{R} \tag{8.13}
$$

$$
t \qquad\qquad \partial_0 X\alpha(t) := \frac{d(X\alpha)}{dt}(t) \; .
$$

Definition 8.18. Let α be a path and X a quantity. The **change in the quantity X corresponding to the path α** is defined by

$$
\Delta_\alpha X := X\alpha(r) - X\alpha(0) = X(y) - X(x) \; .
$$

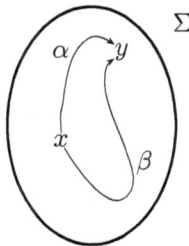

The change in X corresponding to α is independent of the path α except for its initial and final states. Formally, for any paths $x \xrightarrow{\alpha} y$ and $x \xrightarrow{\beta} y$ then $\Delta_\alpha X = \Delta_\beta X$ for any quantity X. Also, even if $\Delta_\alpha X = 0$, the value $X\alpha(t)$ might vary during the process.

Definition 8.19. Let X be a substance conducted from storage body A to storage body B via conduit C. Formally, $A \xrightarrow[B]{C} X$. Hence the transport rate $X_B^A(C)$ may be non-zero along a path $[0,r] \xrightarrow{\alpha} \Sigma$. The **total transport of X from A to B through C along** α is defined by

$$X_\alpha(A \xrightarrow{C} B) := \int_\alpha X_B^A(C) = \int_0^r X_B^A(C)(t)dt \,.$$

When C is understood from the context the notation is $X_\alpha(A \to B)$.

Theorem 8.20. *For any path* $x \xrightarrow{\alpha} y$ *and quantity* X *the following equation holds:*

$$\int_\alpha \frac{\partial_0 X}{X} = \ln \frac{X(y)}{X(x)} \,. \tag{8.14}$$

Proof. Integrating by the Change-of-Variable Rule,

$$\int_\alpha \frac{\partial_0 X}{X} = \int_0^r \frac{\partial_0 X \alpha(t)}{X \alpha(t)} dt = \int_0^r \frac{1}{X\alpha} \frac{dX\alpha}{dt} dt = \int_{(X\alpha)(0)}^{(X\alpha)(r)} \frac{du}{u} = \ln \frac{X(y)}{X(x)} \,.$$
$\square$

Axiom 8.3.7. $\boxed{\textbf{Balance Equation}}$

For every substance X and X-storage body B,

$$\overset{\bullet}{X}_B = \sum_{A \xrightarrow[X]{C} B} X_B^A(C) + X_B^B \quad (A, C \text{ are bound}) \,.$$

This Axiom is the Theory of Substances version of the "equations of continuity" in physics. It equates two quantities that depend on the state of the system. In words it says that the rate of change of substance in a body that can store the substance equals the sum of the rates at which the substance is transported via conduits to or from the body, plus (or minus) the rate at which the substance is created (or destroyed) in the body. The Balance Equation is analogous to Kirchoff's Current Law for the electrical substance, but is extended to all substances, and allows additionally for the creation or destruction of substance.

Example 8.21. Imagine a delimited region of the country with incoming and outgoing roads for automobiles, an automobile manufacturing plant, and an automobile junkyard with a giant crusher. At any given time there is a certain rate at which cars are arriving to the region and a certain rate at which cars are leaving from the region. At some initial time there is some definite number of cars in the region. At any subsequent time the total number of cars in the region depends on the rates of arrival and departure, and on the rates of manufacture and destruction of cars over the intervening interval of time. This analogy guides all thoughts about the "car substance" in the region, which is like a body in a system of such regions. "Equilibrium" in the system with respect to cars would be the situation in which no cars are moving on the connecting roads; "steady state" would correspond to the dynamic circumstance in which cars are moving, possibly being created, and possibly being destroyed, but at such rate that the total number changing is zero – equivalently, the total count of cars is constant.

For any isolated subsystem $[\,A\ B\ C\,]$ and for any substance X conducted by C from A to B, by Balance

$$\dot{X}_A = X_A^B(C) + X_A^A$$

$$\dot{X}_B = X_B^A(C) + X_B^B \qquad \text{hence}$$

$$\dot{X}_{A+B} = \dot{X}_A + \dot{X}_B = X_A^A + X_B^B = X_{A+B}^{A+B}\,.$$

The condition $\dot{X}_A + \dot{X}_B = 0$, that is, the sum of the rates of change of substance X in $[\,A\ B\,]$ is 0, may arise in different ways.

(1) $X_B^A(C) = X_A^A = X_B^B = 0$ There is no flow of X between A and B, and X is conserved in A and B.

(2) $X_B^A(C) = 0$ and $X_A^A = -X_B^B \neq 0$ There is no flow of X between A and B, but the rate of creation of X in A is exactly balanced by the rate of destruction of X in B.

(3) $X_B^A(C) > 0$ and $X_A^A = X_B^B = 0$ There is flow of X from A to B but X is conserved in both A and B.

(4) $X_B^A(C) > 0$, $X_A^A \neq 0$ and $X_B^B \neq 0$ There is some flow, some creation, and some destruction but these rates happen to balance out to 0.

This concludes the basic definition and axioms of a universe. However, further definitions and axioms will articulate intuitions about energy, entropy, and chemistry.

8.4 Reservoir & Capacity

For any body $B \in \mathcal{X}$ and immaterial substance (entropy, energy, volume, momentum) X the net exchange of X between B and other bodies A connected to B by X-conductors $A \overset{C}{\underset{X}{\to}} B$ depends on the conductivities $K_C^{(X)}(A, B)$. How the change of X_B is related to the change of potential $\varphi_B^{(X)}$ is by definition the "capacity" of B for X.

Definition 8.22. The **X-capacity of B in state u** is

$$C_B^{(X)}(u) := \frac{\dot{X}_B(u)}{\dot{\varphi}_B^{(X)}(u)} \quad [\mathbf{AMT}]^2[\mathbf{NRG}]^{-1} . \tag{8.15}$$

The **molar X-capacity of B in state u** is the "per-amount" capacity

$$c_B^{(X)} := \frac{C_B^{(X)}}{X_B} \quad [\mathbf{AMT}][\mathbf{NRG}]^{-1} .$$

In general there exists a diagram

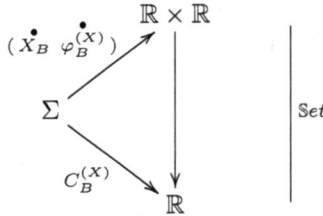

but if the capacity is independent of state except maybe for the magnitude of the potential then there exists a diagram

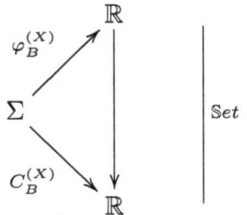

Definition 8.23. Let X be a substance. A body A is an **X-reservoir at potential** φ if for any X-conduit from A to some body B and non-equilibrium state $u \in \Sigma$ such that $\varphi_B^{(X)}(u) \neq \varphi_A^{(X)}(u) = \varphi$, the spontaneous process $u \overset{\alpha_u}{\longrightarrow} u_*$ satisfies $\varphi_B^{(X)}(u_*) = \varphi_A^{(X)}(u_*) = \varphi$.

Example 8.24. According to this definition, if there were an entropy reservoir then it would be an inexhaustible supply of entropy at a fixed temperature; in the thermodynamics literature such is called a "heat reservoir." Likewise, a volume reservoir would supply volume without losing pressure, and an electricity reservoir would deliver an unlimited amount of charge at a fixed voltage. Typically, an ocean is cited as being sufficiently large to serve for all intents and purposes as an entropy reservoir, and the atmosphere of the Earth as a volume reservoir. A very, very large electrochemical battery is close to being an electricity reservoir, and so on.

For a reservoir the ratio of the change in amount of substance to the corresponding change in potential is infinite. In that sense a reservoir has infinite capacity.

8.5 Equilibrium & Equipotentiality

> For an isolated system to be at stable equilibrium, the entropy must have a maximum value with respect to any allowed variations. Thus, to test whether or not a given isolated system is, in fact, at equilibrium and stable, we propose virtual displacement processes to evaluate certain variations.
>
> [Tester and Modell (2004)]

Definition 8.25. Let k be a constraint. A quantity $\Sigma \xrightarrow{Y} \mathbb{R}$ is at **equilibrium in a state** $u_* \in \Sigma$ **relative to** k if there exists a neighborhood of u_* of constrained states among which Y has an extreme value (maximum or minimum) at u_*.

A standard result of advanced calculus provides a *sufficient* condition for equilibrium:

Theorem 8.26. *[Simon and Blume (1994)] Let $F : U \to \mathbb{R}$ be a C^2 function whose domain is an open set U in $\mathbb{R}^n$. Suppose that x^* is a critical point of F in that it satisfies $DF(x^*) = 0$.*

(1) *If the Hessian $D^2F(x^*)$ is a negative definite matrix, then x^* is a strict local maximum of F;*

(2) *if the Hessian $D^2F(x^*)$ is a positive definite matrix, then x^* is a strict local minimum of F;*

(3) *if the Hessian $D^2F(x^*)$ is an indefinite matrix, then x^* is neither a local maximum nor a minimum of F.*

However, in thermodynamics *necessary* conditions for a *strict local maximum* are almost always paramount.

Theorem 8.27. *If quantity $\Sigma \xrightarrow{Y} \mathbb{R}$ is at a maximal equilibrium in a constrained state $u_* \in \Sigma$ relative to a constraint, then*

$$\partial_0 Y\alpha(T_*) = 0$$
$$\partial_0^2 Y\alpha(T_*) < 0$$

for any constrained virtual displacement $[\,0, r\,] \xrightarrow{\alpha} \Sigma$ through u_* at T_*.

Proof.   Any constrained virtual displacement must pass through a neighborhood of u_*. Hence, the quantity must achieve a strict local maximum at T_* as a real-valued function of a real variable. $\square$

Theorem 8.28. *Let $\mathcal{B} := [\,B_1 \cdots B_m\,]$ be a completely isolated system in which substance Y is conserved in every body, so that $Y_i^i := Y_{B_i}^{B_i} = 0$ for $i = 1, \ldots, m$. If $\overset{\bullet}{Y_i} := \overset{\bullet}{Y}_{B_i} \neq 0$ then there exists a conduit $B_i \xrightarrow[Y]{C} B_k$ for some C and k such that $1 \leq k \leq m$ and $k \neq i$.*

Proof.   By Y-Balance and Y-conservation

$$0 \neq \overset{\bullet}{Y_i} = \sum_{B_k \xrightarrow[Y]{C} B_i} Y_i^k(C) + Y_i^i$$

$$= \sum_{B_k \xrightarrow[Y]{C} B_i} Y_i^k(C) \quad (k, C \text{ are bound})$$

hence at least one term $Y_i^k(C) \neq 0$ for some k and C. $\square$

Put into words, this Theorem asserts that the only way the amount of substance in a body can be changing – in an isolated system in which the substance is conserved in all bodies – is to be connected by a conduit for that substance to some other body in the system.

Corollary 8.29. *If for each $i = 1, \ldots, m$ there exists a state u_i such that $\overset{\bullet}{Y}_i \neq 0$, then $\mathcal{B}$ is connected.*

Axiom 8.5.1. | **Virtual Displacement** |

Let $\mathcal{B} = [\, B_1 \cdots B_m \,]$ be a connected system of bodies in each of which X is conserved, and $u_* \in \Sigma$. For any conduit $B_i \overset{C}{\underset{X}{\to}} B_k$ with $i \neq k$ there exists a constrained virtual displacement $[\,0, r\,] \overset{\alpha}{\to} \Sigma$ through u_* at T_* such that $\overset{\bullet}{X}_i(u_*) \neq 0$ but $\overset{\bullet}{X}_j(u_*) = 0$ for $j = 1, \ldots, \widehat{i}, \ldots, \widehat{k}, \ldots, m$. Hence, $\overset{\bullet}{X}_i = -\overset{\bullet}{X}_k$ at u_*.

Put into words, this Axiom claims that in an X conserved, connected, isolated system, the amount of X in one body with a conduit to another body can vary up or down while being compensated by variation in the other body, without any changes throughout the rest of the system.

> [T]he system-wide uniformity of temperature, pressure and other intensive properties is obtained from the Gibbs criterion of equilibrium as a *deduction*, not an assumption.
>
> ---
>
> [Weinhold (2009)]

Lemma 8.30. *Let $\mathcal{B} = [\, B_1 \cdots B_m \,]$ be a connected system of bodies and $u_* \in \Sigma$. Let $Y_1, \ldots, Y_n$ be independently variable conserved substances, and suppose $X = X(Y_1, \ldots, Y_n)$, so that there exists a diagram*

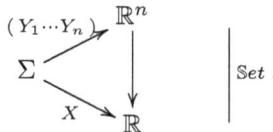

$$
\begin{array}{ccc}
& \overset{(Y_1 \cdots Y_n)}{\nearrow} \mathbb{R}^n & \\
\Sigma & \Big\downarrow & \text{Set} . \\
& \underset{X}{\searrow} \mathbb{R} &
\end{array}
$$

If X has an extremum at u_* and $B_i \overset{C}{\underset{Y_j}{\to}} B_k$ then

$$
\frac{\partial X_i}{\partial Y_{ji}}(Y(u_*)) = \frac{\partial X_k}{\partial Y_{jk}}(Y(u_*)) . \tag{8.16}
$$

Proof.   Let $X_i := X_{B_i}$ and $Y_{ji} := (Y_j)_{B_i}$ for $i = 1, \ldots, m$ and $j = 1, \ldots, n$. Then

$$X_\mathcal{B} = X_1(Y_{11}, \ldots, Y_{1m}) + \cdots + X_m(Y_{n1}, \ldots, Y_{nm})$$

since X is Additive. Hence by the Chain rule

$$
\begin{aligned}
\partial_0 X_\mathcal{B}\alpha(T_*) = \quad & \tfrac{\partial X_1}{\partial Y_{11}}\partial_0 Y_{11}\alpha(T_*) + \cdots + \tfrac{\partial X_1}{\partial Y_{n1}}\partial_0 Y_{n1}\alpha(T_*) \\
+ \qquad & \qquad\qquad\qquad\qquad + \\
\vdots \qquad & \qquad\qquad\qquad\qquad \vdots \\
+ \tfrac{\partial X_m}{\partial Y_{1m}}\partial_0 Y_{1m}\alpha(T_*) & + \cdots + \tfrac{\partial X_m}{\partial Y_{nm}}\partial_0 Y_{nm}\alpha(T_*)
\end{aligned}
\tag{8.17}
$$

for any path $[\,0, r\,] \xrightarrow{\alpha} \Sigma$. By the independent variability of $Y_1, \ldots, Y_n$ there exists for each $j = 1, \ldots, n$ a path such that all columns in Eq. (8.17) are zero except for column j, hence

$$\partial_0 X_\mathcal{B}\alpha(T_*) = \frac{\partial X_1}{\partial Y_{j1}}\partial_0 Y_{j1}\alpha(T_*) + \cdots + \frac{\partial X_m}{\partial Y_{jm}}\partial_0 Y_{jm}\alpha(T_*) \,. \tag{8.18}$$

By the Virtual Displacement Axiom, for any conduit $B_i \xrightarrow[Y_j]{C} B_k$ with $i \neq k$ there exists a constrained virtual displacement $[\,0, r\,] \xrightarrow{\alpha} \Sigma$ through u_* at T_* such that $\overset{\bullet}{Y}_{ji}\,\alpha(T_*) \neq 0$ but $\overset{\bullet}{Y}_{jl}\,\alpha(T_*) = 0$ for $l = 1, \ldots, \widehat{i}, \ldots, \widehat{k}, \ldots, m$. Hence, $\partial_0 Y_{ji}\alpha(T_*) = \overset{\bullet}{Y}_{ji}\,\alpha(T_*) = -\overset{\bullet}{Y}_{jk}\,\alpha(T_*) = -\partial_0 Y_{jk}\alpha(T_*)$ by the Process Axiom.

$$
\begin{aligned}
\partial_0 X_\mathcal{B}\alpha(T_*) &= \frac{\partial X_i}{\partial Y_{ji}}\partial_0 Y_{ji}\alpha(T_*) + \frac{\partial X_k}{\partial Y_{jk}}\partial_0 Y_{jk}\alpha(T_*) \tag{8.19}\\[6pt]
&= \frac{\partial X_i}{\partial Y_{ji}}\partial_0 Y_{ji}\alpha(T_*) - \frac{\partial X_k}{\partial Y_{jk}}\partial_0 Y_{ji}\alpha(T_*) \\[6pt]
&= \left(\frac{\partial X_i}{\partial Y_{ji}} - \frac{\partial X_m}{\partial Y_{jk}} \right) \partial_0 Y_{ji}\alpha(T_*)
\end{aligned}
$$

The left side of Eq. (8.19) equals 0 by the hypothesis that X has an extremum at u_*, and the right factor of the right side is not 0, hence the conclusion of the Theorem. $\qquad\square$

Aside 8.5.1. The Virtual Displacement Axiom with regard to one substance, and the concept of "independently variable" for a list of substances, to my mind capture and generalize "arbitrary and independent values" of energy and volume substances in the proof by ([Callen (1985)] pp. 49–51) of uniform temperature and uniform pressure resulting from "mechanical equilibrium."

Theorem 8.31. *(Equipotentiality) Let $\mathcal{X}$ be a completely isolated system and $Y_1, \ldots, Y_n$ conserved substances such that $Y_1, \ldots, Y_n$ are independently variable, $X = X(Y_1, \ldots, Y_n)$, and there are quantities $P_1, \ldots, P_n$ such that*

$$\overset{\bullet}{X} = \sum_{i=1}^{n} P_i \overset{\bullet}{Y}_i \ .$$

If X has an extremum at u_* then P_j is constant on each Y_j-connected component of $\mathbb{K}^{(Y_j)}$.

Proof. Within each Y_j-connected component Theorems 8.15 and 8.30 imply for each conduit $B_i \overset{C}{\underset{Y_j}{\to}} B_k$ that

$$P_{ji}(Y(u_*)) = \frac{\partial X_i}{Y_{ji}}(Y(u_*)) = \frac{\partial X_k}{\partial Y_{jk}}(Y(u_*)) = P_{jk}(Y(u_*)) \ .$$

$\square$

Remark 8.32. By the Conductivity Axiom there can be transport of a substance from one body to another if and only if there is a conduit for the substance between the two bodies, *and* the potential of the first body is strictly greater than the potential of the second body. In the case of an X extremum – for example an equilibrium with respect to a constraint – Theorem 8.31 signifies that there is no flow of any substance Y_j for $j = 1, \ldots, n$.

8.6   Entropy & Energy

> Entropy can be imagined as a grey paste which is indestructible, but can be generated, "out of nothing" by all kinds of frictions.
>
> ---
> [Thoma (1976)]

Axiom 8.6.1. $\boxed{\textbf{Entropy Axiom}}$

There exists a creatable but indestructible substance called **entropy** and denoted by S. For any storage body $B \in \mathcal{X}$ and any process $x \xrightarrow{\alpha} y \neq x$ with $x \neq y$ there is an increase of entropy in the body, $S_B(y) > S_B(x)$. Moreover, for any non-equilibrium state u there exists a unique process $u \xrightarrow{\alpha_u} u_*$ such that u_* is a maximal equilibrium state with respect to S. The duration of α_u is called the **relaxation time** of u ([Callen (1985)] p. 99), and α_u is called the **spontaneous process** of u.

Theorem 8.33. *A necessary condition for entropy S_B to be at equilibrium in a state $u_* \in \Sigma$ relative to constraints k is just that $\dot{S}_B(u_*) = 0$.*

Proof. Definition (8.25), Theorem 3.25 and the Entropy Axiom. $\qquad\square$

Remark 8.34. The Entropy Axiom is intimately associated with what is conventionally called "The Second Law of Thermodynamics." However, it is not so easy to pin down exactly one generally accepted statement of The Second Law. Indeed, there are at least twenty-one formulations of it [Cápek and Sheehan (2005)].

Theorem 8.35. *(Spontaneity Theorem) If in some state there exists a potential difference between two connected bodies then the spontaneous process ends in an equilibrium state in which the two potentials are equal.*

Proof. By the contrapositive of the Equipotentiality Theorem, the potential difference implies the system is out of equilibrium. By the Entropy Axiom the spontaneous process asserted exists. $\qquad\square$

Axiom 8.6.2. $\boxed{\textbf{Energy Axiom}}$

Aside 8.6.1. The word "energy" has been a source of confusion to me because it seems to come in so many different forms. I have encountered – in no particular order – free energy, ordered energy, chemical energy, thermal energy, kinetic energy, potential energy, stored energy, electrical energy, mechanical energy, latent energy, bound energy, released energy, psychic energy, biological energy, microscopic energy, multidirectional energy, external energy, internal energy, dark energy, nonthermal energy, expendable energy, First-Law energy, Second-Law energy, chemical potential energy, actual energy, and total energy. In Theory of Substances *there is only one concept of energy.*

There exists a substance called **energy** denoted by E that is conserved in every completely isolated system of bodies. For every body $B \in \mathcal{X}$ there exists a quantity

$$\Sigma \xrightarrow{E_A} \mathbb{R} \quad [\mathbf{NRG}] .$$

The rate of change

$$\Sigma \xrightarrow{\dot{E}_A} \mathbb{R} \quad [\mathbf{POW}]$$

of energy in a body is the sum of the rates at which energy is carried to or from A by the transport of substance to or from A during a process. The rate of change of energy with unit of measurement $[\mathbf{POW}] :=$ $[\mathbf{NRG}][\mathbf{TME}]^{-1}$ in a body is called the **power** exchanged with the body. By the Conductivity Axiom the transport of a substance depends on a potential difference across a conduit. Let the bodies $\mathcal{X}$ of the universe include two non-void non-overlapping subsystems $\mathcal{A}$ and $\mathcal{B}$ and think of $\mathcal{A}$ as the "system" and $\mathcal{B}$ as the "environment" of $\mathcal{A}$. All other bodies of the universe are considered to be conduits in circuits connecting bodies of $\mathcal{A} + \mathcal{B}$ to one another.

$$\dot{E}_A = \mathrm{Th}_A + \mathrm{Vo}_A + \mathrm{Ch}_A + \mathrm{Me}_A + \mathrm{El}_A \tag{8.20}$$

$$\mathrm{Th}_A = T_A S_A^A + \sum_{B \xrightarrow[s]{C} A} \left\{ T_A S_A^B(C) : B \in \mathcal{B} \right\} \tag{8.21}$$

$$\mathrm{Vo}_A = \sum_{B \xrightarrow[V]{C} A} \left\{ -P_A^{(|A|)} V_A^B(C) : B \in \mathcal{B} \right\} \tag{8.22}$$

$$\mathrm{Ch}_A = \sum_{X \in \mathcal{M}} \mu_A^{(X)} X_A^A + \sum_{X \in \mathcal{M}} \left\{ \mu_A^{(X)} X_A^B : B \in \mathcal{B} \right\} \tag{8.23}$$

$$\mathrm{Me}_A = \sum_{B \xrightarrow[p]{C} A} \left\{ \langle \vec{v}_A | \vec{p}_A^B(C) \rangle : B \in \mathcal{B} \right\} \tag{8.24}$$

$$\mathrm{El}_A = \sum_{B \xrightarrow[q]{C} A} \left\{ v_A q_A^B(C) : B \in \mathcal{B} \right\} \tag{8.25}$$

Equation (8.20) declares that the rate of change of energy in the system – that is, its power – is the sum of five terms. These terms change the power of A by five independent types of processes: thermal, shape, chemical, mechanical, and electrical processes. In greater detail,

Thermal The first term in Eq. (8.21) sums all power resulting from the creation of entropy in the bodies of $\mathcal{A}$, each of which is at a uniform temperature. For body $A \in \mathcal{A}$ at temperature T_A the power is its temperature multiplied by the rate $S_A^A \geq 0$ of entropy creation in A. Assuming this term is equal to 0 during a process formalizes absence of friction of all kinds. In that case the process is often called "reversible" and in some thermodynamic literature is represented by $\Delta S = 0$.

The second term in Eq. (8.21) sums the powers carried by entropy transport into or out of the system from the environment through thermal conductors – entropy conduits. Assuming this term is equal to 0 during a process formalizes thermal isolation of the system from the environment. In that case the process is often called "adiabatic."

Shape Changing the shape of a body does not necessarily change its volume, but changing its volume definitely implies change in shape, and requires work. By definition a rigid body does not change shape, hence does not change volume. Otherwise, a flexible body such as a liquid or gas fluid body enclosed in a container body may change volume by changes in shape of the container. The standard conduit of volume is a cylinder – a pipe – with a frictionless piston sliding within it. In a circuit the ends of a pipe C connect to bodies A and B of liquid or gas fluid enclosed in rigid container bodies. Movement of the piston from A towards B – expansion of A – adds to the volume of A what is lost in volume of B – total volume substance is conserved. The power delivered to A by the motion of the piston from B towards A – compression of A – is the product of the pressure $P_A^{(|A|)}$ of the substance $|A|$ in A and the rate $V_B^A = -V_A^B$ at which volume is leaving A. Equation (8.22) sums all the power exchanged between the system and its environment by volume conduits. Assuming this term is equal to 0 during a process formalizes the absence of volume changes in the system, in which case the process is sometimes called "isochoric" and $dV = 0$.

Chemical The first term in Eq. (8.23) represents power produced or consumed by creation or destruction of chemical substances in the bodies A of the system. Specifically, the chemical potential $\mu_A^{(X)}$ of substance X in A is multiplied by the creation rate X_A^A (which is negative for destruction). In other words, chemical reactions in A release or bind energy at a certain rate, which is the power of the reactions. The second term formalizes the chemical power exchanged between the system and its environment due to the diffusion or flow of chemical substances through membranes or pipes.

Mechanical Equation (8.24) represents power exchanged between the system and its environment due to frictional contact that results in transport of momentum between bodies at different velocities. This term includes both translational and rotational contributions to the power of the system. Since force is identified with momentum current, the unit of measurement of the inner product $\langle \overrightarrow{v}_A | \overrightarrow{p}_A^B(C) \rangle$ is

$$[\mathbf{DST}][\mathbf{TME}]^{-1} \cdot [\mathbf{FRC}] = [\mathbf{FRC}][\mathbf{DST}] \cdot [\mathbf{TME}]^{-1}$$
$$= [\mathbf{NRG}][\mathbf{TME}]^{-1}$$

that is to say, power.

Electrical Finally, Eq. (8.25) calculates the electrical power exchanged between the system and its environment via electrical conduits, say, wires. The power delivered to A is its electric potential v_A multiplied by the rate of transport q_A^B of electricity, that is, the electrical current. Along with Ohm's Law $I = E/R$ one learns $P = EI = E^2/R$.

Notation for basic substances quantities associated with these processes are given in Table 8.1. The importance of analogy in the creation of physics and in mathematical science education cannot be over-emphasized [Muldoon (2006)].

There are other energy-carrying substances – such as magnetism – and associated basic quantities ([Alberty (2001)] Table I), but in this book only those listed here may enter further discussion.

The Energy Carrier Axiom has many special cases.

Theorem 8.36. *(One Substance)([Alberty (2001)] Eq. (1.2-1)) If a body A is composed of exactly one substance X, is isolated except for one other body B, and undergoes a process that transports energy only by thermal (through an entropy conductor C) and compression transport (through a piston D) sub-processes and no electrical, hydraulic, rotational, or translational components, then*

$$\dot{E}_A = T_A S_A^B(C) - P_A^{(X)} V_A^B(D) + \mu_A^{(X)}(X_A^A + X_A^B(C)) . \qquad (8.26)$$

Table 8.1   The Grand Analogy.

	Substance	Current	Potential
	X_A	X_B^A	$\varphi_A^{(X)}$
PROCESS	[AMT]	[AMT][TME]$^{-1}$	[NRG][AMT]$^{-1}$
Thermal	S_A	S_B^A	T_A
Compression	G_A	G_B^A	$P_A^{(G)}$
Chemical	X_A	X_B^A	$\mu_A^{(X)}$
Translational	$\overrightarrow{p}_A$	$\overrightarrow{p}_B^A$	$\overrightarrow{v}_A$
Rotational	H_A	H_B^A	ω_A
Electrical	q_A	q_B^A	v_A

Substances, currents, and potentials corresponding to basic physical processes in this Theory of Substances.

Remark 8.37. The assertion that energy is conserved – that $\overset{\bullet}{E}_A = 0$ in any completely isolated system – is conventionally called "The First Law of Thermodynamics."

Remark 8.38. The idea that "energy is a substance" is not without its detractors. For example,

> **This hypothesis that energy is a fluid substance is in conflict with its inherent nature, but more significantly it does not contribute to the scientific use of the concept. Thus it is an unnecessary additional hypothesis, the introduction of which violates Ockham's principle [Warren (1983)].**

Naturally in this book such an allegation is without substance.

8.7   Fundamental Equation

The fundamental equation of thermodynamics for the internal energy U may include terms for various types of work and involves only differentials of extensive variables. The fundamental equation for U yields intensive variables as partial derivatives of the internal energy with respect to other extensive properties. In addition to the

terms from the combined first and second laws for a system involving PV work, the fundamental equation for the internal energy may involve terms for chemical work, gravitational work, work of electric transport, elongation work, surface work, work of electric and magnetic polarization, and other kinds of work. Fundamental equations for other thermodynamic potentials can be obtained by use of Legendre transforms that define these other thermodynamic potentials in terms of U minus conjugate pairs of intensive and extensive variables involved in one or more work terms. The independent variables represented by differentials in a fundamental equation are referred to as natural variables. The natural variables of a thermodynamic potential are important because if a thermodynamic potential can be determined as a function of its natural variables, all of the thermodynamic properties of the system can be obtained by taking partial derivatives of the thermodynamic potential with respect to the natural variables. The natural variables are also important because they are held constant in the criterion for spontaneous change and equilibrium based on that thermodynamic potential. By use of Legendre transforms any desired set of natural variables can be obtained. The enthalpy H, Helmholtz energy A, and Gibbs energy G are defined by Legendre transforms that introduce P, T, and P and T together as natural variables, respectively. Further Legendre transforms can be used to introduce the chemical potential of any species, the gravitational potential, the electric potentials of phases, surface tension, force of elongation, electric field strength, magnetic field strength, and other intensive variables as natural variables [Alberty (2001)].

Without numerical coordinates for states it is not possible to apply infinitesimal calculus to problems in mathematical science. Crucial are equations that relate theoretical and especially empirically measurable quantities $\Sigma \to \mathbb{R}$. In particular relations between fundamental quantities such as energy, entropy, temperature, volume, pressure, and so on, are assumed or derived.

Definition 8.39. Let $\mathcal{U}$ be a universe with states Σ and let $\Sigma \xrightarrow{E} \mathbb{R}$, $\Sigma \xrightarrow{S} \mathbb{R}$ be the energy and entropy substances. Let $\Sigma \xrightarrow{Y} \mathbb{R}^n$ be a list of quantities.

A **Fundamental Equation for energy** is a diagram

$$(8.27)$$

which in the thermodynamic literature would be represented simply by the equation $E = E(S, Y)$.

A **Fundamental Equation for entropy** is, likewise, represented by the equation $S = S(E, Y)$ corresponding to the diagram

$$(8.28)$$

Translation between the energy and entropy Fundamental Equations is straightforward if they are equivalent in the sense that the equation $E = E(S, Y)$ is solvable for S in terms of E and the equation $S = S(E, Y)$ is solvable for E in terms of S, both for specified Y.

Axiom 8.7.1. | **Fundamental Equivalence** |

The equation $E = E(S, Y)$ is solvable for S in terms of E and the equation $S = S(E, Y)$ is solvable for E in terms of S, both for specified Y.

Remark 8.40. By the Solvability Theorem 3.50 a sufficient condition for Fundamental Equivalence is that the partial derivative of, say, S with respect to E is positive,

$$\left(\frac{\partial S}{\partial E}\right)_Y > 0 \,, \qquad (8.29)$$

which is part of **Postulate III** in ([Callen (1985)] pp. 28–9).

The notation in Eq. (8.29) is very common in thermodynamics literature and bears explanation: it is equivalent to the assertion that $S = S(E, Y)$ and the definition that

$$\left(\frac{\partial S}{\partial E}\right)_Y := \frac{\partial S}{\partial E}(E, Y) .$$

The reason for this notation is that functional relationships for the same dependent variable in terms of alternative (lists of) independent variables are not usually given names. Thus, it might happen that $A = A(B, C)$ and later $A = A(D, C)$, meaning that there are understood but un-named functions f, g such that $A = f(B, C)$ and $A = g(D, C)$. In infinitesimal calculus there is no problem distinguishing $\frac{\partial f}{\partial C}$ from $\frac{\partial g}{\partial C}$. But in thermodynamics the notation $\frac{\partial A}{\partial C}$ is ambiguous, so a subscript is added to signal which dependency is intended, for example as in

$$\left(\frac{\partial A}{\partial C}\right)_B := \frac{\partial f}{\partial C} .$$

It is a standard result of the infinitesimal calculus that a necessary and sufficient condition for a dependent variable $S = S(E, Y)$ to have a maximum value for given E – but varying Y – is that the equations

$$\left(\frac{\partial S}{\partial Y}\right)_E = 0 \tag{8.30}$$

$$\left(\frac{\partial^2 S}{\partial Y^2}\right)_E < 0 \tag{8.31}$$

are true.

Dually, the same assertion is true with "maximum" replaced by "minimum" and "<" replaced by ">".

Theorem 8.41. *([Callen (1985)] pp. 134–5) If $S = S(E, Y)$ and $E = E(S, Y)$ are equivalent then Eqs. (8.30)–(8.31) are equivalent to the equations*

$$\left(\frac{\partial E}{\partial Y}\right)_S = 0 \tag{8.32}$$

$$\left(\frac{\partial^2 E}{\partial Y^2}\right)_S < 0 . \tag{8.33}$$

In other words, if the energy and entropy Fundamental Equations are equivalent then entropy maximization for a given energy is equivalent to energy minimization for a given entropy.

Proof. The Cyclic Rule and assumption Eq. (8.30) yield

$$\left(\frac{\partial E}{\partial Y}\right)_S = -\frac{\left(\frac{\partial S}{\partial Y}\right)_E}{\left(\frac{\partial S}{\partial E}\right)_Y} = 0 \, .$$

Appealing to the Cyclic Rule again, calculate the second derivative:

$$\left(\frac{\partial^2 E}{\partial Y^2}\right)_S = \frac{\partial}{\partial Y}\left(\frac{\partial E}{\partial Y}\right)_S$$

$$= \frac{\partial}{\partial Y}\left(-\frac{\left(\frac{\partial S}{\partial Y}\right)_E}{\left(\frac{\partial S}{\partial E}\right)_Y}\right)$$

$$= -\left(\frac{\left(\frac{\partial S}{\partial E}\right)_Y\left(\frac{\partial^2 S}{\partial Y^2}\right)_E - \left(\frac{\partial S}{\partial Y}\right)_E\frac{\partial^2 S}{\partial Y \partial E}}{\left(\frac{\partial S}{\partial E}\right)_Y^2}\right)$$

$$= -\frac{\left(\frac{\partial^2 S}{\partial Y^2}\right)_E}{\left(\frac{\partial S}{\partial E}\right)_Y} + \left(\frac{\partial S}{\partial Y}\right)_E\frac{\frac{\partial^2 S}{\partial Y \partial E}}{\left(\frac{\partial S}{\partial E}\right)_Y^2}$$

$$= -\left(\frac{\partial E}{\partial S}\right)_Y\left(\frac{\partial^2 S}{\partial Y^2}\right)_E + \left(\frac{\partial S}{\partial Y}\right)_E\frac{\frac{\partial^2 S}{\partial Y \partial E}}{\left(\frac{\partial S}{\partial E}\right)_Y^2}$$

which must be positive since $\left(\frac{\partial E}{\partial S}\right)_Y > 0$, $\left(\frac{\partial^2 S}{\partial Y^2}\right)_E < 0$ and $\left(\frac{\partial S}{\partial Y}\right)_E = 0$
by hypothesis. $\square$

8.8   Conduction & Resistance

Theorem 8.42. *Assume* $\mathcal{X} = [\, A\, B\, C\,]$ *is a system such that*

$$0 = \overset{\bullet}{S}_B = S_B^A + S_B^B + S_B^C \quad \textit{steady-state entropy, entropy balance} \quad (8.34)$$

$$0 = \overset{\bullet}{U}_B = Q_B^A + Q_B^C \qquad\qquad \text{steady-state energy, energy conserved}$$
(8.35)

$$Q_C^B = T_C \cdot S_C^B \text{ and } Q_B^A = T_B \cdot S_B^A \qquad\qquad \text{entropy carries energy}$$
(8.36)

Then

$$S_B^B = -\frac{1}{T_B} S_C^B (T_C - T_B) . \qquad (8.37)$$

Proof.

$$S_B^B = -S_B^A - S_B^C$$

$$= -\frac{Q_B^A}{T_B} + \frac{Q_C^B}{T_C}$$

$$= Q_C^B \left(\frac{1}{T_C} - \frac{1}{T_B} \right)$$

$$= \frac{Q_C^B}{T_C} \left(\frac{T_B - T_C}{T_B} \right) .$$

$\square$

Remark 8.43. This result is a discrete version of "the generation of entropy in conduction" equation

$$\pi_s = -\frac{1}{T} j_s \frac{dT}{dx}$$

at ([Fuchs (1996)] p. 362, Eq. (106)).

Definition 8.44. For a quantity X with potential $\varphi^{(X)}$, and two bodies A, C, define the **potential difference** by

$$_A X_C := \varphi_A^{(X)} - \varphi_C^{(X)} .$$

For a process α define

$$R := \frac{_A X_C \alpha}{X_C^A \alpha} \qquad\qquad \text{resistance} \qquad (8.38)$$

$$K := \frac{\Delta_\alpha X_A}{\Delta_\alpha \varphi_A^{(X)}} \qquad\qquad \text{capacitance} \qquad (8.39)$$

Remark 8.45. Note that resistance is defined at each time during the process, and capacitance depends only on the initial and final states of the process.

Remark 8.46. Capacitance is conventionally defined in electrical engineering by $dq = K\,dv$ or $i := \dfrac{dq}{dt} = K\dfrac{dv}{dt}$. Similarly, In thermodynamics *entropic capacitance* would be defined by $K(T) = \dfrac{dS}{dT}$ assuming entropy $S = S(T)$ ([Fuchs (1996)] p. 157).

Theorem 8.47. *If $\mathcal{X} = [\,A\ B\ C\,]$ is a system and X is a conserved substance flowing from A to C through $B = \left(X_C^A \right)$ which has resistance R, then $\overset{\bullet}{X}_C = -\,\overset{\bullet}{X}_A$ and*

$$\overset{\bullet}{X}_A = \frac{1}{R}\left(\frac{X_A}{K_A^X} - \frac{X_C}{K_C^X} \right). \tag{8.40}$$

Proof. The first assertion follows from $\overset{\bullet}{X}_A = X_A^C + X_A^A = X_A^C$ by the X-Balance Equation and the conservation of X. The second equation follows from $R X_C^A = \varphi_A^{(X)} - \varphi_C^{(X)} = \left(\dfrac{X_A}{K_A^X} - \dfrac{X_C}{K_C^X} \right)$ by the definitions of resistance and capacitance. $\qquad\square$

The theorem provides a system of coupled ordinary differential equations for conserved substance flow. If initial conditions and parameters are given then these equations are easy to simulate. Based on The Grand Analogy such a simulation applies to various physical situations such as coupled fluid tanks, sliding friction, electrical capacitors, and thermal conduction [Fuchs (1996)].

Chapter 9

Clausius, Gibbs & Duhem

9.1   Clausius Inequality

[W]hen [the system] is brought from one state to the other ... the difference of entropy is the limit of all possible values of the integral $\int \frac{dq}{t}$ denoting the element of the heat received from external sources, and t the temperature of the part of the system receiving it ([Gibbs (1957)] Volume I, p. 55).

For our purposes we may regard the content of the ordinary second law of thermodynamics as given by the expression

$$\Delta S \geq \int \frac{\delta Q}{T} \, ,$$

which states that the increase in the entropy of a system, when it changes from one condition to another, cannot be less than the integral of the heat absorbed divided for each increment of heat by the temperature of a heat reservoir appropriate for supplying the increment in question. The equality sign in this expression is to be taken as applying to the limiting case of reversible changes ([Tolman (1979)] pp. 558–9).

Let us consider a system S that undergoes a cyclic transformation. We suppose that during the cycle the system receives heat from or surrenders heat to a set of sources having the temperatures $T_1, T_2, \ldots, T_n$. Let the amounts of heat exchanged between the system and these sources be $Q_1, Q_2, \ldots, Q_n$, respectively; we take the $Q's$ positive if they represent heat received by the system and negative in the other case. We shall now prove that:

$$\sum_{i=1}^{n} \frac{Q_i}{T_i} \leq 0,$$

and that the equality sign holds ... if the cycle is reversible ([Fermi (1956)] p. 46).

In conduction, the current of energy entering a system at temperature T is given by the product of the current of entropy entering the system and the temperature of the system: $I_{E,th} = TI_s$ ([Fuchs (1996)] p. 88).

It seems that the view of Fermi is aligned with that of Tolman. Nevertheless, in Theory of Substances the view of Gibbs is adopted. Therefore,

Definition 9.1. The letter Q is conventionally associated with energy carried by an entropy current, so the power conveyed by "the flow of heat from B to A" is by definition

$$Q_A^B := T_A S_A^B \qquad [\textbf{NRG}][\textbf{TME}]^{-1}. \tag{9.1}$$

Theorem 9.2. *For any system* $\mathcal{X} = [A\,B]$ *and process* α,

$$\int_\alpha \frac{Q_A^B}{T_A} \leq \Delta_\alpha S_A, \tag{9.2}$$

and equality hold if and only if $S_A^A \alpha = 0$, *that is, if and only if no entropy is generated in* A *by the process.*

Proof. By the Entropy Balance Equation

$$\partial_0 S_A \alpha = S_A^B \alpha + S_A^A \alpha, \tag{9.3}$$

where $S_A^A \alpha \geq 0$ by the Indestructibility of Entropy. Integrating this equation, applying the Fundamental Rule, and by the definition of Q,

$$\Delta_\alpha S_A = \int_\alpha \partial_0 S_A \alpha = \int_\alpha S_A^B \alpha + \int_\alpha S_A^A \alpha \geq \int_\alpha \frac{Q_A^B}{T_A} \qquad (9.4)$$

and equality holds exactly when $S_A^A \alpha(t) = 0$ for $0 \leq t \leq r$. □

Corollary 9.3. ([Fermi (1956)] Eq. (61), Eq. (109))

(1) $\int_\alpha \dfrac{Q_A^B \alpha}{T_A \alpha} \leq 0$ if α is cyclic;

(2) $\underset{\substack{\alpha : x \to y \\ \partial_0 T_A \alpha = 0}}{\sup} \ Q_\alpha(B \to A) \leq T_A(x)(S_A(y) - S_A(x))$.

Proof. (1) If α is cyclic then $\Delta_\alpha S_A = 0$, so this result follows immediately from the Theorem. (2) $\partial_0 T_A \alpha = 0$ means temperature is constant throughout the process α, hence the constant value $T_A \alpha(0) > 0$ may be factored out of the denominator in Eq. (9.2). Consequently for any process α,

$$Q_\alpha(A \to B) = \int_\alpha Q_A^B \leq T_A(x)(S_A(y) - S_A(x)) , \qquad (9.5)$$

The result follows by definition of *supremum*, since the quantity $T_A(x)(S_A(y) - S_A(x))$ is independent of the choice of α. □

Remark 9.4. The right-hand side of inequality Eq. (9.5) depends only on the initial and final states of the processes entering into the formation of the *supremum*. Thus, $T_A(x))(S_A(y) - S_A(x))$ is an upper bound for all isothermal processes between those states. There is the question whether this is the *least* upper bound. In other words, might there be processes approaching arbitrarily close to the maximum possible amount of energy transportable by thermal means between A and the bodies of B? An answer to this question involves a discussion of "infinitely slow" processes and shall be deferred for the time being.

Aside 9.1.1. Actually, the concept "infinitely slow" process is surrounded by confusion due to association with related concepts such as "nearly continuous equilibrium," "maximal heating," "second-order infinitesimal," "dissipation-free," "isentropic," "equilibrium approximation," "reversible," "quasi-static," "invertible," and "relaxation allowing" processes, in no particular order. The best technical discussion I have found is Chapter 5, "Reversibility" in [de Heer (1986)].

Remark 9.5. The quantities $Q_\alpha(A \to B)$ and $T_A(x))(S_A(y) - S_A(x))$ are conventionally denoted by ΔQ and $T\Delta S$, so the corollary avers that

$$\Delta Q \leq T\Delta S .$$

9.2 Gibbs-Duhem Equation

Theorem 9.6. *(TPμ Theorem)([Alberty (2001)] Eqs. (1.1-1) to (1.1-5))*
For any body A if the entropy S_A, volume V_A, and amount X_A of a chemical substance are independently variable, and $E_A = E_A(S_A, V_A, X_A)$, then in an isolated system and for any process without dissipation

$$\frac{\partial E_A}{\partial S_A} = T_A$$

$$\frac{\partial E_A}{\partial V_A} = -P_A^{(X)} \tag{9.6}$$

$$\frac{\partial E_A}{\partial X_A} = \mu_A^{(X)} .$$

Proof. By the special case One Substance Theorem Eq. (8.26) of the Energy Axiom for a body A

$$\overset{\bullet}{E_A} = T_A \overset{\bullet}{S_A} - P_A^{(X)} \overset{\bullet}{V_A} + \mu_A^{(X)} \overset{\bullet}{X_A} ,$$

hence Eq. (9.6) follows from the Independent Variability Theorem 8.15.
$\square$

By definition any substance X is additive in the sense that for disjoint bodies A and B the equation $X_{A+B} = X_A + X_B$ is true. In particular, entropy S, volume V, and chemicals X are substances – and so is energy E. If it so happens that $E = E(S, V, X)$ and A, B are distinct bodies but identical in every way, then

$$\begin{aligned} 2E_A = E_{A+B} &= E(S_{A+B}, V_{A+B}, X_{A+B}) \\ &= E(S_A + S_B, V_A + V_B, X_A + X_B) \\ &= E(2S_A, 2S_A, 2S_A) . \end{aligned} \tag{9.7}$$

The observation Eq. (9.7) motivates

Axiom 9.2.1. | **Energy Homogeneity** |

Energy is a homogeneous function of order 1 depending on the entropy, volume, and chemical substances in a body, that is to say,

$$E(\lambda S, \lambda V, \lambda X) = \lambda E(S, V, X) . \tag{9.8}$$

It follows immediately by the Homogeneity Rule Theorem 3.54 that

$$E_A = \frac{\partial E_A}{\partial S_A} S_A - \frac{\partial E_A}{\partial V_A} V_A + \frac{\partial E_A}{\partial X_A} X_A . \tag{9.9}$$

Theorem 9.7. *(Euler Homogeneity Theorem) For an isolated body A if the entropy S_A, volume V_A, and amount X_A of a chemical substance are independently variable, and $E_A = E_A(S_A, V_A, X_A)$, then*

$$E_A = T_A S_A - P_A^{(X)} V_A + \mu_A^{(X)} X_A . \tag{9.10}$$

Proof. Since $\overset{\bullet}{S}_A = S_A^A$ and $\overset{\bullet}{X}_A = X_A^A$ for an isolated body, by the Energy Axiom

$$\overset{\bullet}{E}_A = T_A \overset{\bullet}{S}_A - P_A^{(X)} \overset{\bullet}{V}_A + \mu_A^{(X)} \overset{\bullet}{X}_A \tag{9.11}$$

or, in differential form,

$$dE_A = T_A dS_A - P_A^{(X)} dV_A + \mu_A^{(X)} dX_A . \tag{9.12}$$

(This is the "fundamental equation" for a "one-phase system with one species" as in ([Alberty (2001)] p. 1355).)But the differential of E_A is also given by the Chain Rule,

$$dE_A = \frac{\partial E_A}{\partial S_A} dS_A + \frac{\partial E_A}{\partial V_A} dV_A + \frac{\partial E_A}{\partial X_A} dX_A . \tag{9.13}$$

Therefore after subtraction of Eq. (9.13) from Eq. (9.12),

$$0 = \left(T_A - \frac{\partial E_A}{\partial S_A}\right) dS_A + \left(P_A^{(X)} - \frac{\partial E_A}{\partial V_A}\right) dV_A + \left(\mu(X)_A - \frac{\partial E_A}{\partial X_A}\right) dX_A \,.$$

The conclusion follows from Independent Variability Theorem 8.15 and Eq. (9.9). □

Corollary 9.8. *(Gibbs-Duhem Equation)*

$$X_A d\mu_A^{(X)} = -S_A dT_A + V_A dP_A^{(X)} \,.$$

Proof. After suppressing A and (X) for consistency with the thermodynamic literature the differential form of Eq. (9.10) is

$$\begin{aligned} dE &= T dS + S dT - P dV + V dP + X d\mu + \mu dX \\ &= T dS - P dV + \mu dX + S dT - V dP + X d\mu \\ &= dE + S dT - V dP + X d\mu \end{aligned}$$

by Eq. (9.13). Subtracting dE from both sides yields the assertion of the theorem. □

Remark 9.9. Although the extensive variables S, V, X are independently variable, the Corollary declares that – ultimately because of the Energy Homogeneity Axiom – the intensive variables T, P, μ are not.

Chapter 10

Experiments & Measurements

Repeatable experiments can lead to new concepts, measurements, and instruments. For example, experiments with rods upon fulcrums and objects upon the ends of the rods can lead to the concepts of balance and "relative weight" and then to a choice of a standard object. Then the standard is not forgotten, but is no longer mentioned, and the result is the concept of "weight" and the instrument for measuring it – by metonymy called a "balance."

Mathematics students are often encouraged to "read the masters," however technically archaic such readings may seem. In so doing students might gain intuitions beyond the rigor embodied in abstract, terse definitions, and see *why* things are so defined. The same advice might apply even more to physics and chemistry students (also see Section 1.5.1).

This Chapter introduces a handful of physics experiments – some recounted by Masters – and then recasts them in terms of the Theory of Substances.

10.1 Experiments

10.1.1 *Boyle, Charles & Gay-Lussac Experiment*

If a gas is expanded or contracted while the temperature is held constant, it is found experimentally that the pressure varies inversely as the volume. in algebraic form we say that $pV = C$, where C is a constant for a given temperature and a fixed quantity of gas. This empirical relation is known as Boyle's Law. ... There remains the question of the actual functional relationship between the

pV **product and the temperature. ... the** pV **product is found to be a *linear* function of the Celsius temperature** τ**; this linear variation is referred to as the "law of Charles and Gay-Lussac** [Arons (1965)]**.**

Definition 10.1. For any substance X and body B the **concentration of** X **in** B is denoted by $[X]_B$ and defined by

$$[X]_B := \frac{X_B}{V_B} \quad [\mathbf{AMT}][\mathbf{VLM}]^{-1} \,.$$

Definition 10.2. A substance X is a **gas** if there exists

$$c^{(X)} \in \mathbb{R} \quad [\mathbf{NRG}][\mathbf{AMT}]^{-1} \tag{10.1}$$

and there exists for each body B a quantity

$$\Sigma \xrightarrow{P_B^{(X)}} \mathbb{R} \quad [\mathbf{FRC}][\mathbf{ARA}]^{-1}$$

such that

$$P_B^{(X)} = c^{(X)} \cdot [X]_B \tag{10.2}$$

The quantity $P_B^{(X)}$ is called the **pressure** of gas X in B.

Theorem 10.3. *If X is a gas in a body B and there exists a process $x \xrightarrow{\alpha} y$ such that $\partial_0 T_B \alpha = 0$ – that is to say, if α is an isothermal process – then*

$$\mu_B^{(X)}(y) = \mu_B^{(X)}(x) + c^{(X)} \ln \frac{[X]_B(y)}{[X]_B(x)} \,. \tag{10.3}$$

Proof. In an isothermal process $dT_A = 0$ hence by the Gibbs-Duhem Equation

$$X_A d\mu_A^{(X)} = V_A dP_A^{(X)}$$

so

$$d\mu_A^{(X)} = \frac{V_A dP_A^{(X)}}{X_A} = \frac{c^{(X)} d[X]_A}{[X]_A} = c^{(X)} \frac{d[X]_A}{[X]_A} \,. \tag{10.4}$$

Integrating both sides of Eq. (10.4) along α eliminates α from the story, and the conclusion follows by appeal to the Process Axiom, the Fundamental Rule, and the Change-of-Variables Rule. $\qquad \square$

Remark 10.4. It follows immediately from Theorem 9.7 that the chemical potential of the substance in a body satisfies the equation

$$\mu_A^{(X)} = \frac{E_A}{X_A} + P_A^{(X)}\frac{V_A}{X_A} - T_A\frac{S_A}{X_A} . \tag{10.5}$$

If a "standard state" $o \in \Sigma$ is chosen, then a gas X has a standard potential $\mu_B^{(X)}(o)$ and a standard concentration $[X]_B(o)$ so that in any other state x the theorem provides a value of the potential

$$\mu_B^{(X)}(x) = \mu_B^{(X)}(o) + c^{(X)}\ln\frac{[X]_B(x)}{[X]_B(o)}$$

relative to the standard state. By the proof of Theorem 10.3 this value does not depend on the choice of process α leading from the standard state to x.

Definition 10.5. A gas X is an **ideal gas** if there exists $R^{(X)} \in \mathbb{R}$ such that

$$c^{(X)} = c^{(X)}(T_B) = R^{(X)}T_B$$

for any body B.

Axiom 10.1.1. $\boxed{\text{Universal Gas Constant}}$

There exists $R \in \mathbb{R}$ such that

$$R^{(X)} = R \quad [\mathbf{NRG}][\mathbf{AMT}]^{-1}[\mathbf{TMP}]^{-1}$$

for any ideal gas X.

Definition 10.6. The equation

$$P_A^{(X)}V_A = X_A R T_A$$

satisfied by any ideal gas $X := |A|$ is called the **Ideal Gas Law**.

Theorem 10.7. *(Chemical Potential Theorem) (1) For any ideal gas X the potential of a state $x \in \Sigma$ relative to standard state $o \in \Sigma$ is*

$$\mu_B^{(X)}(x) = \mu_B^{(X)}(o) + RT_B(X)\ln\frac{[X]_B(x)}{[X]_B(o)} . \tag{10.6}$$

(2) There exists an equation

$$P_B^{(X)}V_B = X_B R T_B . \tag{10.7}$$

Proof. Theorem 10.3 implies (1). (2) follows from the definition Eq. (10.2) of concentration, the definition of ideal gas, and the Universal Gas Constant Axiom. □

Remark 10.8. In the thermodynamics literature Eq. (10.7) is simply $PV = nRT$ and is called the **Ideal Gas Law**, where n stands for the number of moles of gas and R is in units of Joules per mole per degree Kelvin. Pressure and temperature are intensive variables conjugate to extensive variables volume and entropy, respectively. The product nR **[NRG][TMP]**$^{-1}$ has the same unit of measurement as entropy. The Ideal Gas Law equates a "mechanical energy" PV on the left to a "thermodynamic energy" nRT on the right. More precisely, an ideal gas relates energy transported by a mechanical process (change in pressure or volume) to energy transported by a chemical or thermal process (change in amount of substance or temperature). Such facts partially account for the fundamental role of the Ideal Gas Law – involving just four variables and a constant – in the history of thermodynamics.

One important property of the chemical potential has not yet been mentioned. This quantity displays a universal behavior when the molar density decreases. Indeed, when n/V is sufficiently small, the chemical potential as a function of the amount of substance is

$$\mu(n) - \mu(n_0) = RT \ln \left(\frac{n}{n_0} \right) \quad \text{for } V, T = \text{constant} . \tag{10.8}$$

This relation holds for whatever the substance, be it a gas (considered a solute in a vacuum), a solute in a liquid or a solute in a solid. It also holds for electromagnetic radiation (photons) and sound (phonons). Equation (10.8) is usually derived from the [Ideal Gas Law]. This seems reasonable since it is easy to verify the [Ideal Gas Law] experimentally. However, we believe that it is more appropriate from a conceptual point of view to conceive Eq. (10.8) as a basic law. Numerous other laws can be derived from Eq. (10.8). Examples are the law of mass action, the [Ideal Gas Law], Raoult's law, Henry's law, Nernst's distribution law, the vapour pressure equation, van t'Hoff's law or Boltzmann's distribution law [Job and Herrmann (2006)].

10.1.2 *Rutherford-Joule Friction Experiment*

Let us take, for example, a system composed of a quantity of water. We consider two states A and B of this system at atmospheric pressure; let the temperatures of the system in those two states be T_A and T_B, respectively, with $T_A < T_B$. We can take our system from A to B in two different ways. *First way*: We heat the water by placing it over a flame and raise its temperature from the initial value T_A to the final value T_B. The external work performed by the system during this transformation is practically zero. It would be exactly zero if the change in temperature were not accompanied by a change in volume of the water. Actually, however, the volume of the water changes slightly during the transformation, so that a small amount of work is performed ... We shall neglect this small amount of work in our considerations. *Second way*: We raise the temperature of the water from T_A to T_B by heating it by means of friction. To this end, we immerse a small set of paddles attached to a central axis in the water, and churn the water by rotating the paddles. We observe that the temperature of the water increases continuously as long as the paddles continue to rotate. Since the water offers resistance to the motion of the paddles, however, we must perform mechanical work in order to keep the paddles moving until the final temperature T_B is reached. Corresponding to this considerable amount of positive work performed by the paddles on the water, there is an equal amount of negative work performed by the water in resisting the motion of the paddles [Fermi (1956)].

The two ways Enrico Fermi describes for heating a body of water B are presented as Theory of Substances diagrams in Fig. 10.1. The "First Way" uses an entropy reservoir H – a heater – at some fixed temperature, with entropy conductor C from H to B. At equilibrium – by the Equipotentiality Theorem and the Entropy Axiom –B will have the same temperature as H.

The "Second Way" uses angular momentum current – torque – provided by a horse to rotate paddles D in contact via entropy conductor C – the common surface of the paddles and the water – with the water B. The

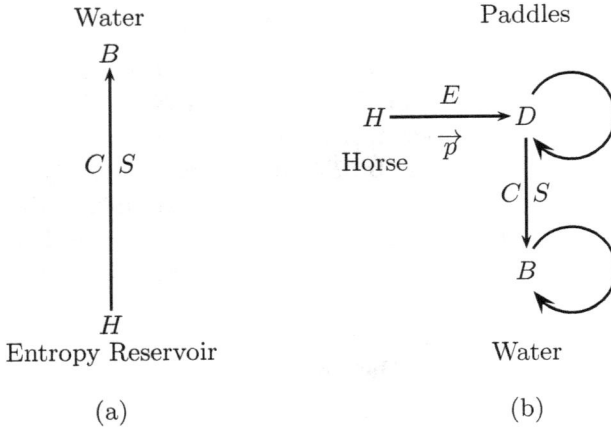

Fig. 10.1 (a) "First Way" to heat water, using a source of heat; (b) "Second Way" to heat water, using mechanical friction.

mechanical friction generates entropy in both D and B, hence the temperatures rise, by the Energy Axiom and the assumption that the system $[B\,C\,D\,E\,H]$ is isolated. If the horse is instructed to stop walking around the huge vat of water so that the paddles stop moving just when the temperature of the water reaches the same temperature as the entropy reservoir in the "First Way," the conclusion is that the total energy released mechanically by the horse to rotate the paddles in the "Second Way" is *equivalent* to the total energy absorbed from the heater in the "First Way."

10.1.3 *Joule-Thomson Free Expansion of an Ideal Gas*

Into a calorimeter Joule placed a container having two chambers, A and B, connected by a tube Fig. 10.2. He filled the chamber A with a gas and evacuated B, the two chambers having first been shut off from each other by a stopcock in the connecting tube. After thermal equilibrium had set in, as indicated by a thermometer placed within the calorimeter, Joule opened the stopcock, thus permitting the gas to flow from A into B until the pressure everywhere in the container was the same. He then observed that there was only a very slight change in the reading of the thermometer. This meant that there had been practically no transfer of heat from the calorimeter to the chamber or vice versa. It is assumed that if this experiment could be performed with an ideal gas, there would be no temperature change at all ([Fermi (1956)], p. 22).

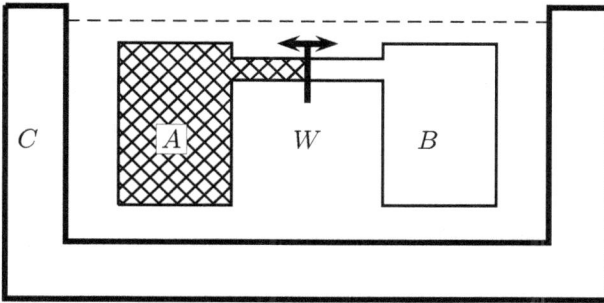

Fig. 10.2 *C* is Joule's calorimeter, a solid thermally insulated container. *W* is the body of water in which are immersed the two bodies. Initially *A* contains a quantity of gas substance, and initially *B* is void.

Experience 10.1.1. Given an isolated system $\mathcal{X} = [\,A\ B\ D\ E\ K\ W\,]$ as in Fig. 10.3, let *A* be a spatial region containing a gas, *B* a spatial region devoid of substance, *W* a body of, say, "water," let *K* be a volume conduit from *A* to *B* and suppose *D* and *E* are thermal conduits from *W* to *A* and

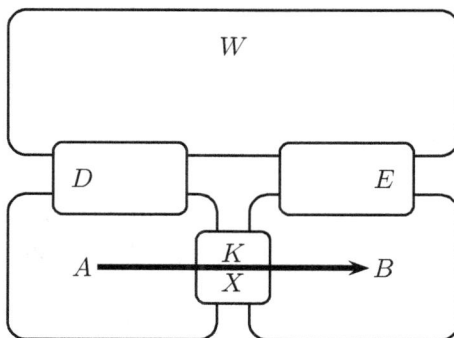

Fig. 10.3 Theory of Substances model of the Joule-Thomson gas expansion process. *W* is a body of fluid, say "water" whose temperature T_W is closely monitored during the process. *D* and *E* are thermal conduits. *K* is a conduit for gas substance *X*.

B, respectively. Formally,

$$A \overset{K}{\underset{V}{\to}} B$$

$$W \overset{D}{\underset{S}{\to}} A$$

$$W \overset{E}{\underset{S}{\to}} B$$

Assume state $u \in \Sigma$ satisfies $T_A(u) = T_W(u)$, that is, W and A are in thermal equilibrium. If A is a gas body with $X_A(u) > 0$ then $P_A^{(X)}(u) > 0$, and since B is void, $P_B^{(X)}(u) = 0$. Therefore u is not an equilibrium state by the contrapositive of the Equipotentiality Theorem. Hence by the Entropy Axiom there exists a (unique) process $u \overset{\alpha}{\to} u_*$ such that u_* is an equilibrium state. By the Equipotentiality Theorem, $\mu_B^{(X)}(u_*) = \mu_A^{(X)}(u_*)$. The empirical observation is that throughout the process $T_W(\alpha(t)) = T_W(u) = T_A(u)$.

Let $E = E_A, T = T_A, V = V_A$. If there exists a fundamental equation $E = E(T, V)$ the observation implies that $\left(\dfrac{\partial T}{\partial V}\right)_E = 0$, that is to say, since during the process the temperatures of W and A are always equal, there is no exchange of entropy between W and A, so the energy of A is invariant along the process. Consequently by the Cyclic Rule, $\left(\dfrac{\partial E}{\partial V}\right)_T = 0$, and so

$$dE = \left(\frac{\partial E}{\partial T}\right)_V dT + \left(\frac{\partial E}{\partial V}\right)_T dV = \left(\frac{\partial E}{\partial T}\right)_V dT \,.$$

Conventionally $C_V := \left(\dfrac{\partial E}{\partial T}\right)_V$ is called the **heat capacity at constant volume** of the gas X, and $dE = C_V dT$.

10.1.4 *Iron-Lead Experiment*

If we plunge a piece of iron and a piece of lead, both of equal weight and at the same temperature (100°C), into two precisely similar vessels containing equal quantities of water at 0°C we find that, after thermal equilibrium has been established in each case, the vessel containing the iron has increased in temperature much more than that containing the lead. Conversely, a quantity of water at 100° is cooled to a much lower temperature by a piece of iron at $0°$, than by an equal weight of lead at the same temperature. This phenomenon leads to a distinction between *temperature* and *quantity of heat* [Planck (1926)].

Experience 10.1.2. The result of the spontaneous process $u \xrightarrow{\alpha_u} u_*$ given the initial conditions

a piece of iron	$\lvert A \rvert = Fe$
a piece of lead	$\lvert B \rvert = Pb$
of equal weight	$M_A = M_B$
at same temperature	$T_A(u) = T_B(u) = 100°$
precisely similar vessels	$\lvert C \rvert = \lvert D \rvert = H_2O$
equal quantities of water	$(H_2O)_C = (H_2O)_D$
at same temperature	$T_C(u) = T_D(u) = 0°$
A immersed in B	$\partial A \subset \partial C$
B immersed in D	$\partial B \subset \partial D$

are

by the Spontaneity Theorem	$T_A(u_*) = T_C(u_*)$
and likewise	$T_B(u_*) = T_D(u_*)$
with the *observation* that	$T_C(u_*) > T_D(u_*)$.

On the face of it, the observed difference

$$T_A(u_*) - T_B(u_*) = T_C(u_*) - T_D(u_*)$$

depends on many choices, namely

$$A, B, C, D, Fe, Pb, M_A, M_B, T_A(u), H_2O, (H_2O)_C \text{ and } T_C(u) .$$

However, repetition of the experience leads to the conclusion that the result does not depend on the spatial regions $A(u), B(u), C(u), D(u)$ but only on the material substances pervading those regions, so the list of independent choices reduces to

$$Fe, Pb, M_A, M_B, T_A(u), H_2O, (H_2O)_C \text{ and } T_C(u) .$$

Furthermore, the difference is determined by repetition of the same conditions except for the choice of iron, Fe, versus lead, Pb. Hence, the basic experimental choices are

$$Fe, M_A, T_A(u), H_2O, (H_2O)_C \text{ and } T_C(u) .$$

These six choices are reduced further by standardization. First, instead of *mass* M_A switch to a standard *amount* of material substance, $X_A = 1$, where X is a variable representing the test substance, $X = Fe$ above. Second, use a standard substance, say W, in C, with a standard amount $W_C = 1$. $W = H_2O$ above. Third, use standard starting temperatures $T_A(u) = T_1$ and $T_C(u) = T_0$.

a test body	$\|A\| = X$
of standard amount	$X_A = 1$
standard quantity of standard material substance	$W_C = 1$
standard initial temperature	$T_A(u) = T_1$
standard initial temperature	$T_C(u) = T_0$
A immersed in B	$\partial A \subset \partial C$

The choices are reduced to X, T_1 and T_0 and

experiment	$A \xrightarrow[S]{\partial A} C$
spontaneous process	$u \xrightarrow{\alpha} u_*$

yields

$$\Delta_\alpha T_A = \Delta_\alpha T_A(X, T_1, T_0) \ .$$

Recall that the entropy capacity of A in state u is by definition

$$C_A^{(S)}(u) = \frac{\dot{S}_A(u)}{\dot{T}_A(u)} \quad [\mathbf{AMT}]^2[\mathbf{NRG}]^{-1} \ .$$

Hence, along the spontaneous process α,

$$C_A^{(S)}\alpha = \frac{\dot{S}_A(u)}{\dot{T}_A(u)}$$

$$= \frac{\partial_0 S_A \alpha}{\partial_0 T_A \alpha}$$

$$= \frac{1}{T_A \alpha} \frac{\partial_0 E_A \alpha}{\partial_0 T_A \alpha}$$

$$\approx \frac{1}{T_A \alpha} \frac{\Delta_\alpha E_A}{\Delta_\alpha T_A}$$

by the Process Axiom and the Energy Axiom (since the system is isolated). As recounted in ([Fuchs (1996)] pp. 160–161) the quantity

$$\frac{\Delta_\alpha E_A}{\Delta_\alpha T_A}$$

is conventionally called the **heat capacity** of X at $T_A(u)$, and he emphasizes that "it definitely cannot be thought of as a capacity in the ordinary sense of the word, since *heat* in this context cannot be thought of as residing in bodies." In [Bent (1965)] the definition of heat capacity is "the energy *absorbed by A from* its thermal surroundings C divided by A's change in temperature:

$$C_{Bent} = -\frac{dE_C}{dT_A}$$

where $dT_A = T_1 - T_0$." He goes on to point out that – as in the scenario depicted so far – if the volume V_A is constant during α then $dE_C = -dE_A$ (by isolation and the Energy Axiom). Hence, according to Bent, the **heat capacity at constant volume** is

$$C_V := \left(\frac{dE_A}{dT_A}\right)_{V_A} .$$

On the other hand, if the pressure P_A is constant during α, then $dE_C = -(dE_A + P_A dV_A) = -dH_A$, the enthalpy increment of A rather than the energy increment as in the constant volume scenario. Therefore, Bent goes on, the **heat capacity at constant pressure** is

$$C_P := \left(\frac{dE_C}{dT_A}\right)_{P_A} = -\left(\frac{dH_A}{dT_A}\right)_{P_A}$$

$$= -\left(\frac{dE_A}{dT_A}\right)_{P_A} - P_A \left(\frac{dV_A}{dT_A}\right)_{P_A} \tag{10.9}$$

But, by the Joule-Thomson experience *for an ideal gas* the energy of A depends only on its temperature, not on pressure or volume. Hence, the first term in Eq. (10.9) is C_V. As for the second term, since the pressure is constant by the Ideal Gas Law the increment of volume is

$$dV_A = \frac{X_A R}{P_A} dT_A ,$$

hence the second term in Eq. (10.9) equals $X_A R$. This proves that for an ideal gas the heat capacities at constant volume and constant pressure are related by

$$C_P - C_V = X_A R .$$

10.1.5 *Isothermal Expansion of an Ideal Gas*

A standard thermodynamic experiment is an isolated system

$$\mathcal{X} = [\,A\ B\ C\ D\ E\ H\,]$$

in which A is an ideal gas body in a rigid enclosure except for a volume conduit – a piston C – connecting to another gas body B. Expansion of A exerts force on the piston, also known as a momentum current through conduit E from A to C. E is just the two-dimensional interface between the gas and the piston, namely the intersection of the boundary of A with the boundary of C.

The only entropy conduit D connects body H to A. That is to say, energy may be carried by an entropy current from H to A in a thermal process. The Theory of Substances model of system $\mathcal{X}$ is depicted in Fig. 10.4.

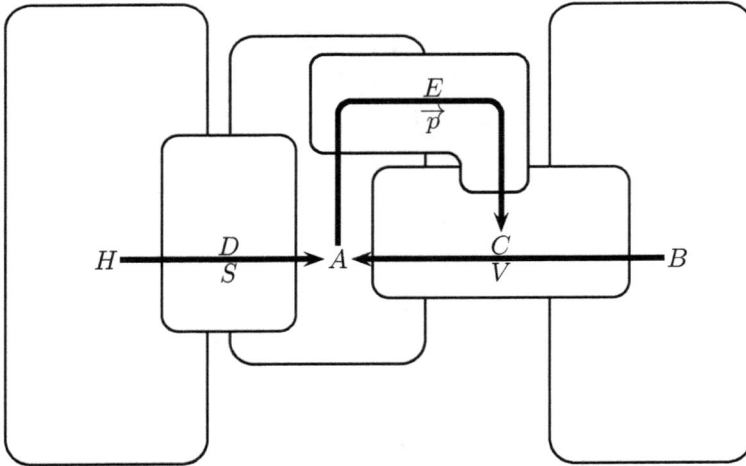

Fig. 10.4 Theory of Substances model of isothermal gas expansion. H is a "thermal reservoir" at temperature T_H with thermal conduit D to gas body A. C is a piston – a volume conduit – between A and B. The pressure in A induces momentum current – force – through momentum conduit E to piston C.

Theorem 10.9. *(Ideal Gas Isothermal Expansion [Bent (1965)]) Let* $x \xrightarrow{\alpha} y$ *be a process in the isolated system* $\mathcal{X}$ *such that no entropy is generated in* $\mathcal{X}$*, so that* $\dot{S}_{\mathcal{X}} = 0$ *along* α*, and that the temperature of* A *is constant along* α*:* $\partial_0 T_H = 0$ *and* $T_A = T_H$*. Assume that there exists a Fundamental*

Equation $E_A = E_A(T_A)$. Then

$$\Delta_\alpha \frac{S_A}{X_A} = R \ln \frac{V_A(y)}{V_A(x)} .$$ (10.10)

Proof. Since entropy is conserved in this isolated system and the only entropy conduit is between H and A, it follows that

$$\dot{S}_A = - \dot{S}_H .$$

Disregarding energy change in B, energy is conserved in the sub-system $[A \, C \, H]$, so by Additivity of Energy

$$0 = \dot{E}_{A+C+H} = \dot{E}_A + \dot{E}_C + \dot{E}_H$$

hence $\dot{E}_H = - \dot{E}_A - \dot{E}_C$. By the Joule-Thomson Experiment and the Fundamental Equation $E_A = E_A(T_A)$, the Chain Rule implies

$$\dot{E}_A = \frac{dE_A}{dT_A} \dot{T}_A = 0 .$$

Therefore, $\dot{E}_H = - \dot{E}_C$. Put into words, energy carried by entropy current from the heat reservoir H to the gas A is converted into movement of the piston C. Dividing by $T_A = T_H$ yields

$$-\frac{\dot{E}_C}{T_A} = \frac{\dot{E}_H}{T_H} = \dot{S}_H = - \dot{S}_A .$$

Therefore, $\dot{S}_A = \dfrac{\dot{E}_C}{T_A}$. Since the rate of energy transport from A to C is given by $\dot{E}_C = P^{(X)} V_A^B$ where $X = |A|$ is an ideal gas,

$$\dot{E}_C = X_A R T_A \frac{\dot{V_A^B}}{V_A} .$$

Since volume is conserved, $\dot{V}_A = \dot{V}_A^B$, hence

$$\dot{E}_C = P_A^{(X)} \dot{V}_A^B = X_A R T_A \frac{\dot{V}_A}{V_A} .$$

Dividing through by $X_A T_A$ yields

$$\frac{\dot{S}_A}{X_A} = R \frac{\dot{V}_A}{V_A} .$$

The result follows by integrating along α and appealing to the Fundamental Rule. $\square$

Corollary 10.10.

$$S_A(y) = S_A(x) - \ln \frac{P_A(y)}{P_A(x)} \, .$$

Proof. By the Ideal Gas Law and the assumption that temperature is constant in A,

$$\frac{P_A(x)}{P_A(y)} = \frac{V_A(y)}{V_A(x)}$$

from which the conclusion is immediate by appeal to the Theorem. $\square$

Putting this conclusion into words, isothermal expansion of an ideal gas body A within a fixed spatial region $A + B$ reduces the pressure and raises the volume and entropy of A.

10.1.6 *Reaction at Constant Temperature & Volume*

$\overset{\bullet}{E}_B$ – we could equally well speak about $\overset{\bullet}{S}_B$ – repre- sents the difference between the energies of the products of a reaction and the corresponding reactants. The energy of the reactants (the same remarks apply to the energy of the products) is the sum of the changes in the energy of B that occur as the reactants are added to B without any net change in the temperature of B or the volume of B. It is the last condition that causes difficulty. As each individual component is added to B at constant volume, the pressure in B will generally increase. This will gener- ally affect the molar energies of all the substances in B. How much these energies are affected will depend upon how much the pressure increases; this in turn, will depend upon the compressibilities of the components of B. This is an awkward situation. Effective energies and entropies for general use ... cannot easily be tabulated. A more useful equation would be one that permitted direct use to be made of the ordinary molar energies and entropies of chemical substances [Bent (1965)] (with slight changes in nota- tion for consistency with Theory of Substances).

Theorem 10.11. *If $\mathcal{X} = [\, B \ C \ H \,]$ is an isolated system at equilibrium such that C is an entropy conductor from entropy reservoir H at (fixed)*

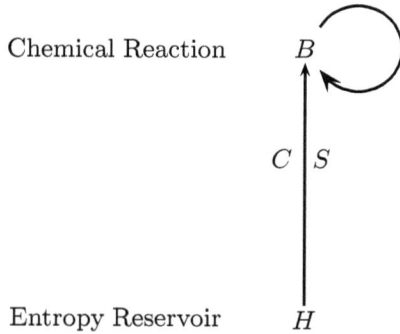

Fig. 10.5   Reaction at constant temperature and constant volume.

temperature T_H to reaction chamber B with fixed volume, then

$$\dot{S}_B - \frac{\dot{E}_B}{T_H} = 0 \ . \tag{10.11}$$

Proof.   By Entropy Additivity $0 = \dot{S}_{\mathcal{X}} = \dot{S}_H + \dot{S}_B$ since the system is at equilibrium (and conductor C is assumed not to generate entropy), hence $\dot{S}_B = -\dot{S}_H$. Also, calculate

$$\dot{S}_H = \frac{\dot{E}_H}{T_H} \qquad \text{Energy Axiom for Entropy Reservoir}$$

$$= -\frac{\dot{E}_B}{T_H} \qquad \text{isolation, Energy Additivity Axiom}$$

$$= -\frac{\dot{E}_B}{T_B} \qquad \text{equilibrium}$$

which yields the result.   □

10.1.7   *Reaction at Constant Pressure & Temperature*

A prototype chemistry experiment as in Fig. 10.6 is a reactive mixture of gases in a cylindrical chamber closed off by a massive piston free to move against the force of gravity. Outside the piston, atmospheric pressure is assumed negligible in comparison to the constant pressure exerted by the piston upon the reactive mixture.

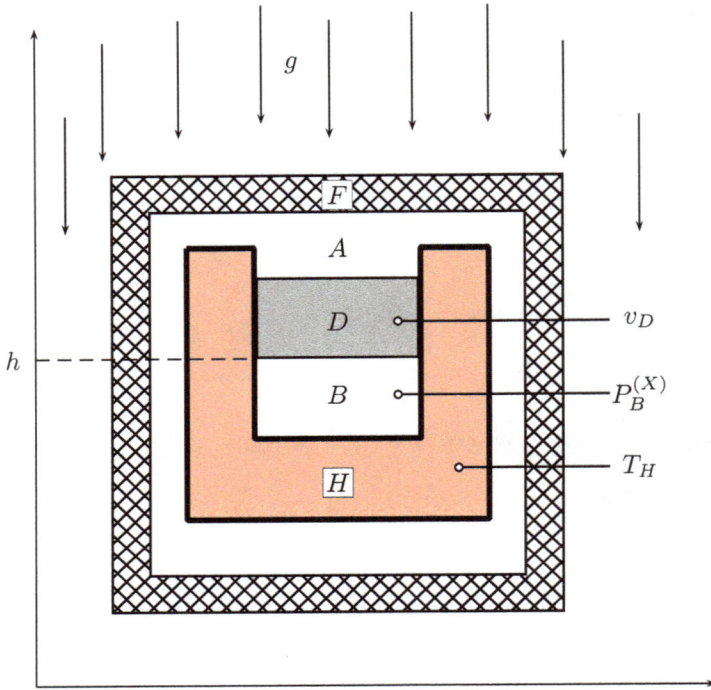

Fig. 10.6   Mental model of a typical chemical thermodynamics experiment: a gas re-action at constant pressure and temperature. The universe $\mathcal{U}$ includes the system $[\,A\ B\ D\ F\ H\,]$ in a uniform gravitational field (specified by gravitational acceleration constant g) in which the freely sliding piston body D with mass m_D at height h has velocity v_D. The body B with volume V_B beneath the piston is a mixture X_B of gases at pressure $P_B^{(X)}$. The body A with volume V_A is also a mixture Y_A of gases at pressure $P_A^{(Y)}$, so that the piston is a conduit for flow of conserved volume substance between A and B. Body H is a rigid entropy reservoir at temperature T_H. Body F is an isolating container of the entire apparatus.

Formally, the universe $\mathcal{U}$ consists of some large gravitational body in the vicinity of a system $\mathcal{X} = [\,A\ B\ C\ D\ E\ H\,]$ whose bodies are shown in Fig. 10.7. The piston D is a conduit for the transport of volume between the external gas body A and the mixture $X := |B|$. The equation

$$\overset{\bullet}{V}_{A+B} = 0 \tag{10.12}$$

expresses the Conservation of Volume. Conservation and Additivity of

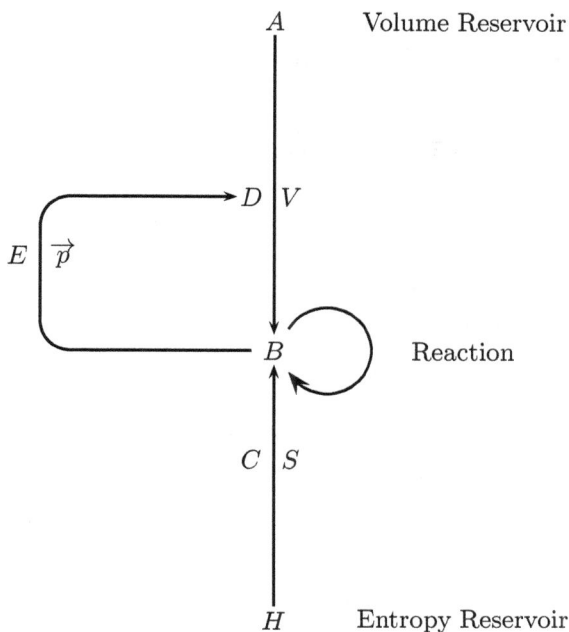

Fig. 10.7 Reaction at constant temperature and constant pressure.

Energy in $\mathcal{X}$ is expressed by the equation

$$\overset{\bullet}{E}_{\mathcal{X}} = \overset{\bullet}{E}_{A+B+C+D+E+H} = \overset{\bullet}{E}_A + \overset{\bullet}{E}_B + \overset{\bullet}{E}_C + \overset{\bullet}{E}_D + \overset{\bullet}{E}_E + \overset{\bullet}{E}_H = 0 \quad (10.13)$$

and assuming that atmosphere A and conductors C, E do not change in energy,

$$\overset{\bullet}{E}_{\mathcal{X}} = \overset{\bullet}{E}_B + \overset{\bullet}{E}_H + \overset{\bullet}{E}_D = 0 \, . \quad (10.14)$$

Remark 10.12. This equation corresponds exactly to the "First Law of Thermodynamics"

$$\Delta U_{\text{total}} = \Delta U_\sigma + \Delta U_\theta + \Delta U_{\text{wt}} = 0$$

in [Bent (1965)][Craig (1992)].

Thus, with focus on the rate of change of energy in the reactive mixture,

$$\overset{\bullet}{E}_B = -\overset{\bullet}{E}_H - \overset{\bullet}{E}_D \, . \quad (10.15)$$

The rate of change of energy $\overset{\bullet}{E}_D$ of the piston is equal to the rate of change of piston height multiplied by the force upon it due to (gases produced by) the reaction. That is, for

$$\overset{\bullet}{E}_D = v_D \cdot \overrightarrow{p}^B_D = \frac{dh}{dt} \cdot \left(a \cdot P^{(X)}_B \right) \tag{10.16}$$

where $v_D = \dfrac{dh}{dt}$, a is the area of the piston, and the force on it due to the reaction is the momentum current $\overrightarrow{p}^B_D$. Since $\overset{\bullet}{V}_B = \dfrac{dh}{dt} \cdot a$, by the Associative Law Eq. (10.15) is equivalent to

$$\overset{\bullet}{E}_B + P^{(X)}_B \overset{\bullet}{V}_B = - \overset{\bullet}{E}_H \ . \tag{10.17}$$

where now the focus is on the rate of change of energy of the reaction in B on the left versus the rate of change of energy of the heater H on the right.

Definition 10.13. The **enthalpy** of the reactive mixture in B is the quantity

$$H_B := E_B + P^{(X)}_B V_B \ ,$$

so that there exists a diagram

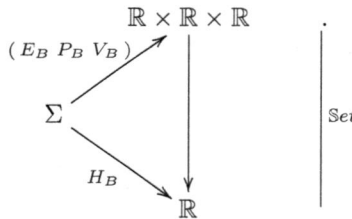

Theorem 10.14. *([Craig (1992)] p. 15) If $\partial_0 P^{(X)}_B \alpha = 0$ along a chemical process $x \overset{\alpha}{\to} y$ of the experiment in Fig. 10.7, then*

$$dH_B = dE_B + P^{(X)}_B dV_B \tag{10.18}$$

$$H_B(y) - H_B(x) = - (E_H(y) - E_H(x)) \ . \tag{10.19}$$

Proof. Equation (10.18) derives from the Product Rule and the constant pressure hypothesis, so $\partial_0 H_B \alpha = \partial_0 E_B \alpha + P_B \alpha \cdot \partial_0 V^{(X)}_B \alpha$. By Eq. (10.17) and the Process Axiom, $\partial_0 E_B \alpha + P_B \alpha \cdot \partial_0 V^{(X)}_B \alpha = -\partial_0 E_H \alpha$, so $\partial_0 H_B \alpha = -\partial_0 E_H \alpha$ from which Eq. (10.19) follows by the Fundamental Rule. $\square$

Theorem 10.15. *([Callen (1985)] p. 147, [Tester and Modell (2004)] p. 149) If a Fundamental Equation $E = E(S,V)$ exists for a body B of fluid then $P(S,V) = -\dfrac{\partial E}{\partial V}(S,V)$ by the Theorem 9.6. If this equation is solvable for V in terms of P so $V = V(S,P)$, then the Legendre Transform of E with respect to V is the enthalpy $H = E + PV$.*

Proof. In general, there exists a "y-intercept" variable $b = b(S,V)$ such that $E(S,V) = \dfrac{\partial E}{\partial V}(S,V) \cdot V + b(S,V)$, so

$$b(S,V) = E(S,V) - \frac{\partial E}{\partial V}(S,V) \cdot V = E(S,V) + P(S,V) \cdot V \ .$$

Allowing the substitutions $E(S,P) = E(S,V(S,P))$ and $b(S,P) = b(S,V(S,P))$ leads to

$$b(S,P) = E(S,P) + P \cdot V(S,P) = H(S,P) \ . \qquad \square$$

10.1.8 *Théophile de Donder & Chemical Affinity*

In Fig. 10.7 momentum current conductor E permits the piston to continuously equilibrate the pressure of B with the pressure of volume reservoir A (or, the weight of the piston), so that $\partial_0 V_A \alpha = 0$ even though volume may be conducted through D. Assume process α includes the chemical reaction

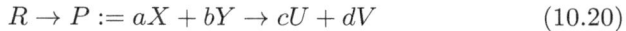

$$R \to P := aX + bY \to cU + dV \tag{10.20}$$

in B, and that no material substances are conducted from the environment $\mathcal{E}$ to B. Hence, by the Balance Axiom and the Definite Proportions Axiom there exists an advancement of reaction $\xi = \xi(\alpha(t))$ such that

$$X_B^B = \overset{\bullet}{X}_B = \nu_X \overset{\bullet}{\xi} \tag{10.21}$$

$$Y_B^B = \overset{\bullet}{Y}_B = \nu_Y \overset{\bullet}{\xi}$$

$$U_B^B = \overset{\bullet}{U}_B = \nu_U \overset{\bullet}{\xi}$$

$$V_B^B = \overset{\bullet}{V}_B = \nu_V \overset{\bullet}{\xi} \ .$$

Definition 10.16. The **affinity** of the reaction $R \to P$ in B is

$$\mathcal{A}_B := \mathcal{A}_B^{R \to P} := - \left(\mu_B^{(X)} \nu_X + \mu_B^{(Y)} \nu_Y + \mu_B^{(U)} \nu_U + \mu_B^{(V)} \nu_V \right) \ .$$

Remark 10.17. At http://en.wikipedia.org/wiki/Chemical_thermo dynamics it is observed that the minus sign in the definition of affinity "comes from the fact the affinity was defined to represent the rule that spontaneous changes will ensue only when the change in the Gibbs free energy of the process is negative, meaning that the chemical species have a positive affinity for each other."

Theorem 10.18. *If A is a volume reservoir and $P_B^{(X)} = P_A$ along α, then*

$$T_B S_B^H = \dot{H}_B \tag{10.22}$$

$$T_B S_B^B = \mathcal{A}_B \dot{\xi} \geq 0 \,. \tag{10.23}$$

Proof.   Let $X := |B|$ be the chemical composition of B. By the Energy Axiom and Entropy Balance in B,

$$
\begin{aligned}
\dot{E}_B &= \mathrm{Th_B} + \mathrm{Vo_B} + \mathrm{Ch_B} \\
&= T_B\,\dot{S}_B - P_B^{(X)} V_B^A(D) + \mu_B^{(X)}\,(\dot{X})_B + \mu_B^{(Y)}\,\dot{Y}_B + \mu_B^{(U)}\,\dot{U}_B + \mu_B^{(V)}\,\dot{V}_B \\
&= T_B S_B^H + T_B S_B^B - P_B^{(X)} V_B^A(D) \\
&\quad + \left(\mu_B^{(X)} \nu_X + \mu_B^{(Y)} \nu_Y + \mu_B^{(U)} \nu_U + \mu_B^{(V)} \nu_V \right) \dot{\xi} \\
&= T_B S_B^H + T_B S_B^B - P_B^{(X)} V_B^A(D) - \mathcal{A}_B\,\dot{\xi}
\end{aligned}
$$

hence

$$T_B S_B^H + T_B S_B^B = \left(\dot{E}_B + P_B^{(X)} V_B^A(D) \right) + \mathcal{A}_B\,\dot{\xi} \,. \tag{10.24}$$

The quantity in parentheses on the right-hand side of Eq. (10.24) is the rate of change of enthalpy H_B in B, which is the rate $T_B S_B^H$ of energy transport from heat source H. Thus, Eq. (10.22) holds. Therefore, Eq. (10.23) results after subtracting of Eq. (10.22) from Eq. (10.24) and recalling from the Entropy Axiom that entropy is indestructible.   □

This theorem has a significant background, beginning with the definition of "chemical affinity" by Théophile de Donder, whose early work on irreversible thermodynamics was developed further by Ilya Prigogine [Prigogine *et al.* (1948)][Oster *et al.* (1973)][De Groot and Mazur (1984)][Lengyel (1989)].

Let us limit ourselves to uniform systems (without diffusion) in mechanical and thermal equilibrium. The only irreversible phenomenon which we shall then have to consider is the chemical reaction. It is proved that the production of entropy per unit time, due to a chemical reaction, is

$$\frac{Av}{T} > 0 \tag{10.25}$$

where v is the rate of reaction, T the absolute temperature, and A the chemical affinity. Th. De Donder has shown that this afffinity can be easily calculated; for example, from the chemical potentials μ, we have

$$A = -\sum_{\gamma} \nu_{\gamma} \mu_{\gamma} \tag{10.26}$$

where ν_{γ} is the stoichiometric coefficient of the constituent γ in the reaction. Formula 10.25 gives us directly the fundamental inequality of De Donder:

$$Av > 0 \tag{10.27}$$

Affinity and reaction rate have therefore the same sign. At thermodynamic equilibrium we have simultaneously

$$A = 0, \quad v = 0 \tag{10.28}$$

Let us note that equations 10.27 and 10.26 are quite independent of the particular conditions in which the chemical reaction takes place (e.g., V and T constant or P and T constant) [Prigogine *et al.* (1948)].

The concept of chemical potential also has deep roots, especially in the work of Josiah Willard Gibbs.

> If to any homogeneous mass we suppose an infinitesimal quantity of any substance to be added, the mass remaining homogeneous and its entropy and volume remaining unchanged, the increase of the energy of the mass divided by the quantity of substance added is the *potential* for that substance in the mass considered [Gibbs (1957)].

Speaking to a conference of British chemists in 1876, [James Clerk] Maxwell distinguished between what we would today call "extensive" and "intensive" thermodynamic properties. The former scale with the size of the system. The latter, in Maxwell's words, "denote the intensity of certain physical properties of the substance." Then Maxwell went on, explaining that "the pressure is the intensity of the tendency of the body to expand, the temperature is the intensity of its tendency to part with heat; and the [chemical] potential of any component is the intensity with which it tends to expel that substance from its mass." The idea that the chemical potential measures the tendency of particles to diffuse is indeed an old one [Baierlein (2001)].

10.1.9 *Gibbs Free Energy*

Equation (10.22)

$$T_B S_B^H = \dot{H}_B$$

is equivalent to

$$\dot{H}_B - T_B S_B^H = 0$$

and since $\dot{S}_B = S_B^B + S_B^H = S_B^H$ at entropic equilibriumby Theorem 8.33, it is equivalent to

$$\dot{H}_B - T_B \dot{S}_B = 0 . \tag{10.29}$$

Definition 10.19. The **Gibbs free energy** of the reactive mixture $X := |B|$ is the quantity

$$G_B := H_B - T_B S_B = E_B + P_B^{(X)} V_B - T_B S_B .$$

Theorem 10.20. *If temperature and pressure are held constant then*

$$\dot{G}_B = -\mathcal{A}_B \dot{\xi}_R .$$

Proof. By definitions of Gibbs free energy and enthalpy and eliding subscripts for swiftness, $G = H - TS = E + PV - TS$ and so $\dot{G} = \dot{H} - \dot{T}S = \dot{H} - T\dot{S} - S\dot{T} = \dot{E} + P\dot{V} + V\dot{P} - T\dot{S} - S\dot{T}$. If temperature and pressure are held constant then $\overset{\bullet}{T} = 0$ and $\overset{\bullet}{P} = 0$, so $\overset{\bullet}{G} = \overset{\bullet}{E} + P\overset{\bullet}{V} - T\overset{\bullet}{S}$. By the Energy Axiom $\overset{\bullet}{E} = T\overset{\bullet}{S} - P\overset{\bullet}{V} + \langle \overrightarrow{\mu} | \overrightarrow{X} \rangle = T\overset{\bullet}{S} - P\overset{\bullet}{V} + \langle \overrightarrow{\mu} | \overrightarrow{\nu} \overset{\bullet}{\xi}_R \rangle = T\overset{\bullet}{S} - P\overset{\bullet}{V} + \langle \overrightarrow{\mu} | \overrightarrow{\nu} \rangle \overset{\bullet}{\xi}_R$. Adding equations yields the conclusion of the theorem by definition of chemical affinity. $\qquad\square$

Definition 10.21. The negative affinity $-\mathcal{A}_B$ reflects the change in Gibbs free energy per unit advance of the reaction ([McQuarrie and Simon (1997)] p. 965), and in the literature is denoted by

$$\Delta_R G_B := \frac{\overset{\bullet}{G}_B}{\overset{\bullet}{\xi}_R} = -\mathcal{A}_B [\mathbf{NRG}][\mathbf{AMT}]^{-1} .$$

Theorem 10.22. *For any state $u \in \Sigma(B)$,*

$$\Delta_R G_B(u) = \Delta_R G_B(o) - RT_B(u_*) Q_B^{(R)}(u) . \tag{10.30}$$

Proof. [Spencer (1974)]. $\qquad\square$

Theorem 10.23. *For the chemical reaction in reactor B the following are necessary conditions for the reaction network $\{R, -R\}$ to achieve chemical equilibrium at $u_* \in \Sigma(B)$:*

$$\frac{k_R}{k_{(-R)}} = -\frac{\prod\limits_{Y \in \underline{R}} Y_B^{-\nu_R^Y}}{\prod\limits_{Y \in \underline{-R}} Y_B^{-\nu_R^Y}} ; \tag{10.31}$$

$$\overset{\bullet}{\xi}_R = 0 ; \tag{10.32}$$

$$\overset{\bullet}{G}_B = -\mathcal{A}_B \overset{\bullet}{\xi}_R = 0 ; \tag{10.33}$$

$$\Delta_R G_B(o) = -RT_B(u_*) Q_B^{(R)}(u) . \tag{10.34}$$

Proof. Equation (11.20) is Theorem 11.4; Eq. (11.21) follows from the definition of stoichiometric coefficient in the Reaction Kinetics Axiom and the hypothesis that $\overset{\bullet}{X}_R = 0$. Equation (11.22) is immediate from the comment about entropic equilibrium at Eq. (10.29).

Aside 10.1.1. Equation (11.23) is bottom dollar for Biological Energetics ([Voet and Voet (1979)] p. 37). A major impetus for The Theory of Substances has been to master this equation after a sustained relentless search

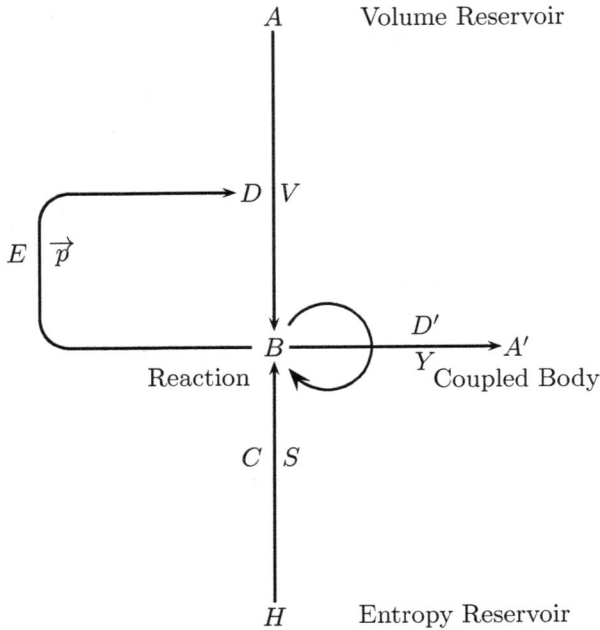

Fig. 10.8   Reaction at constant temperature and constant pressure coupled to another body by some energy-carrying substance.

for understanding. So it is ironic that although it follows immediately from Theorem 11.13, I refer the reader for that proof to the excellent literature. □
C'est la vie.

Theorem 10.24. *(Free Energy Theorem)([Craig (1992)] pp. 103–104) Let $\partial_0 P_B^{(X)} \alpha = 0$ along a chemical process $x \xrightarrow{\alpha} y$ of the experiment which adds to Fig. 10.7 coupling of energy via transport of substance Y through conduit D' to body A' as in Fig. 10.8. If no entropy is generated in A' then (for spontaneous process at constant temperature and pressure)*

$$\Delta_\alpha E_{A'} \leq -\Delta_\alpha G_B .\tag{10.35}$$

Proof.   By isolation of the entire system, Additivity of Energy, and the Energy Axiom,

$$0 = \dot{E}_{\mathcal{X}} = \dot{E}_B + \dot{E}_H + \dot{E}_D + \dot{E}_{A'}$$

hence

$$\Delta_\alpha E_H = -\Delta_\alpha H_B - \Delta E_{A'}$$

so

$$\Delta_\alpha S_H = \frac{\Delta_\alpha E_H}{T} = -\frac{\Delta_\alpha H_B}{T} - \frac{\Delta_\alpha E_{A'}}{T} \ .$$

Therefore, from $\Delta S_{A'} = 0$ and the Entropy Axiom

$$0 \le \Delta_\alpha S_\mathcal{X} = \Delta_\alpha S_B - \frac{\Delta_\alpha H_B}{T} - \frac{\Delta \alpha E_{A'}}{T}$$

hence

$$0 \le T \cdot \Delta_\alpha = -(\Delta_\alpha H_B - T \cdot \Delta_\alpha S_B) - \Delta_\alpha E_{A'}$$
$$= -\Delta_\alpha G_B - \Delta_\alpha E_{A'}$$

so the conclusion follows. $\qquad\square$

Thus, $-\Delta_\alpha G_B$ is the maximum amount of energy made available by a chemical reaction at constant temperature and constant pressure that can be stored in a useful form in an electrical system or its equivalent. For irreversible, spontaneous processes, $\Delta_\alpha S_\mathcal{X} > 0$ and $\Delta_\alpha E_{A'} < -\Delta_\alpha G_B$. ... A portion of the available, useful energy, $-\Delta_\alpha G_B$, enters the thermal reservoir through frictional, dissipative processes. The remainder is stored as $\Delta_\alpha E_{A'}$. In the worst case, as in Fig. 10.7, when no additional potential energy reservoir is coupled, all of $-\Delta_\alpha G_B$ is dissipated as thermal energy [Craig (1992)] (with slight changes in notation for consistency with Theory of Substances).

The transported coupling substance may be entropy, or electricity, or linear momentum, or – as in the case of muscle contraction – angular momentum, namely, rotation of the myosin head producing torque then mechanically translated into linear momentum current, also known as muscular force.

10.2 Measurements

The subsystem B of the system $\mathcal{X} := [\,A\ B\ C\ D\ E\ F\ H\,]$ with environment $\mathcal{E} := [\,A\ C\ D\ E\ F\ H\,]$ is the focus of interest in the typical chemical thermodynamic experiment in Fig. 10.6. Other important chemical thermodynamics experiments are performed with special cases of the generic system.

Aside 10.2.1. In the thermodynamics literature there is no more seemingly basic idea of "work" than lifting a weight. I have to say "seemingly" because hidden in this modest activity is a presumption that force is required to lift the weight to a certain height not only because imparting motion to a mass requires force, but also because there is a force resisting its upward motion. Of course, the word "upward" is implied by use of the word "lift," and the resisting force is that of "gravity." But gravity is a (scalar) potential field that influences the motion of bodies with mass because its negative gradient is a force field. All of this is implicit in "lifting a weight."

Definition 10.25. Let $R \hookrightarrow \mathbb{R}^3$ be a region of space. If for each point $p \in R$ there exists a diagram

$$\Sigma \xrightarrow{\phi_p} \mathbb{R} \qquad [\mathbf{NRG}][\mathbf{AMT}]^{-1}$$

then ϕ is called a **scalar potential field** in region R.

In words, there is a scalar potential field if merely the presence of some amount of a substance at a point in a region imbues it with energy. Of course, the magnitude of that energy depends on the state of the universe. An equivalent way to specify a scalar potential field – through adjointness – is by the existence of a diagram

$$R \xrightarrow{\phi_u} \mathbb{R} \qquad [\mathbf{NRG}][\mathbf{AMT}]^{-1} .$$

for each $u \in \Sigma$.

Definition 10.26. Let $R \hookrightarrow \mathbb{R}^3$ be a region of space. If for each point $p \in R$ there exists a diagram

$$\Sigma \xrightarrow{\phi_p} \mathbb{R}^3 \qquad [\mathbf{NRG}][\mathbf{AMT}]^{-1}[\mathbf{DST}]^{-1}$$

then ϕ is called a **vector force field** in region R. The force per unit amount of substance ϕ_p has three spatial components ϕ_{px}, ϕ_{py} and ϕ_{pz}.

An amount of a substance located in a region of space where a vector force field is defined experiences a force, since $[\mathbf{NRG}] = [\mathbf{FRC}][\mathbf{DST}]$. Any scalar potential field $\phi = \phi_u$ determines a vector force field by deriving its negative gradient $-\left(\dfrac{\partial \phi}{\partial x} \dfrac{\partial \phi}{\partial y} \dfrac{\partial \phi}{\partial z} \right)$. Thus, as the scalar potential increases in a certain direction, the force per unit amount of substance is in the opposite direction.

10.2.1 *Balance Measurements*

Example 10.27. Spring-loaded scales to weigh produce in a supermarket are based on balance between downward gravitational force and upward spring force. The resting position of a pointer along a line of equally spaced numbers points at a number, or more likely between two numbers. That number, or a visually interpolated in-between number, is taken to be the weight of the produce. The presumption is that the scale was calibrated by placing a standard weight on the scale and placing the number 1 at the corresponding pointer position. Assuming that the spring extends twice as far for twice as many standard weights, the positions of the remaining numbers are determined by duplicating the distance measured from the 0 point to the 1 point. How is it determined that two (standard) weights are equal, so that the proportionality of spring extension to weight can be confirmed? By deploying a gravitational balance with two pans connected by a rigid bar whose center point rests upon a fulcrum with little friction. If the two weights are placed in the two pans, and are at rest if the two pans are at the same height, then the two weights are considered to be equal.

Example 10.28. The simplest pressure measurement is analogous to a gravitational balance. Instead of two pans there are two chambers of gas, and instead of a rigid bar resting upon a fulcrum with little friction, there is a straight connecting tube between the chambers with a freely sliding piston. A pointer attached to the piston moves parallel to a scale of equally spaced numbers centered at 0. The instrument is calibrated by temporarily connecting the two chambers by a separate tube, and when the piston comes to rest, that is where the 0 is placed. After the temporary tube is disconnected, any change in the amount of gas in one chamber may result in movement of the piston. Thus, the pressure of the gas in that chamber is a number relative to the pressure of the gas in the other chamber. For example, the "other chamber" may be the Earth's atmosphere. In any case, pressure in a body of gas may be in balance with the pressure in some standard body.

Axiom 10.2.1. $\boxed{\text{Mass}}$

For each body B of the universe there exists a quantity

$$\Sigma \xrightarrow{\ m_B\ } \mathbb{R}$$

such that $m_B(x) > 0$ if there exists a chemical mixture X such that $X_B(x) > 0$. There exists a scalar potential field G defined throughout the universe

$$\Sigma \xrightarrow{\quad G_p \quad} \mathbb{R} \qquad [\textbf{NRG}][\textbf{AMT}]^{-1}$$

for $p \in \mathbb{R}^3$ and there exists a positive real number $g \in \mathbb{R}$ such that in a region of space where $G = g \cdot h$, the force on a body located at z coordinate $h > 0$ is $m_B \cdot g$.

Chapter 11

Chemical Reaction

The unit "mol rxn" stands for a single multiplier of
the stoichiometric coefficients of reactants *and* products
that tells how far a reaction has advanced in consumption
of reactants and in formation of products [Craig (1992)].

As conventionally conceived, a chemical reaction in a body – the *reactor* –
transforms a chemical mixture of compounds – the *reactants* – into another
mixture of compounds – the *products*. This means a given amount of one
mixture after a certain time is diminished while the amount of another
mixture is augmented. In any reaction the relative proportions of reactants
and products are characteristic of the reaction, independent of the reactant
and product amounts. The amount of transformation of reactant to product
– the *advancement* – depends on the state of the system, and so does the
rate of conversion. More precisely,

Axiom 11.0.2. Reaction Kinetics

There exists a non-void set $\mathcal{R}$ of **chemical reactions** and for each $R \in \mathcal{R}$
a diagram

$$\Sigma(B) \xrightarrow{\xi_R} \mathbb{F} \qquad\qquad \textbf{advancement of } R$$

$$\Sigma(B) \xrightarrow{k_R} \mathbb{F} \qquad\qquad \textbf{rate constant of } R$$

where B is the reactor. The unit of measurement for ξ_R is [**AMT**]. For
each compound $X \in \mathcal{C}$ the **stoichiometric coefficient** of X in R is by

definition a rational constant

$$\nu_R^X := \frac{\overset{\bullet}{X_B}}{\overset{\bullet}{\xi_R}} \in \mathbb{Q} \,. \tag{11.1}$$

Dimensionally a stoichiometric coefficient is

$$[\textbf{AMT}][\textbf{TME}]^{-1}([\textbf{AMT}][\textbf{TME}]^{-1})^{-1} = [\textbf{AMT}][\textbf{AMT}]^{-1} = [\ \ ] \,,$$

that is, dimensionless. If $\nu_R^X = 0$ then X is not transformed by R; if $\nu_R^X < 0$ then X is a **reactant** in R and is **consumed**; if $\nu_R^X > 0$ then X is a **product** that is **produced** by R. The sets of reactants $\underline{R}$ and products $\overline{R}$ are defined by

$$\underline{R} := \{\, X \in \mathcal{C} | \nu_R^X < 0 \,\}$$
$$\overline{R} := \{\, X \in \mathcal{C} | \nu_R^X > 0 \,\}$$

The conventional notation for a chemical reaction R exhibits the stoichiometric coefficients, as in

$$R : xX + yY \to uU + vV$$

where $-x, -y, u, v$ are the stoichiometric coefficients of X, Y, U, V in R, and most generally,

$$R : \sum_{X \in \underline{R}} \left(-\nu_R^X\right) X \to \sum_{X \in \overline{R}} \left(\nu_R^X\right) X \,.$$

An alternative representation of the same reaction is

$$R = \sum_{\nu_R^X \neq 0} \nu_R^X X \in \Sigma \mathbb{Q} \mathcal{C} \,.$$

According to the present axiom these are finite sums, that is to say, the sets of reactants and products of a reaction R are finite sets. Another convenient representation of the same reaction is

$$R : x_1 X_1 + \cdots + x_m X_m \to y_1 Y_1 + \cdots + y_n Y_n \tag{11.2}$$

where $x_1, \ldots, x_m, y_1, \ldots, y_n$ are the magnitudes of the reactant and product stoichiometric coefficients.[1] Furthermore, according to the present axiom,

---

[1] The x_i are negatives of the negative reactant stoichiometric coefficients.

(1) each reaction R is **balanced**, that is,

$$\sum_{X \in \underline{R}} \nu_R^X \Phi(X) + \sum_{X \in \overline{R}} \nu_R^X \Phi(X) = 0 \ ;$$

(2) if $R \in \mathcal{R}$ then the **reverse reaction** $-R := -\frac{1}{1}R \in \mathcal{R}$;
(3) if $X \in \mathcal{C}$ then the **decomposition reaction** $X \to \Phi(X) \in \mathcal{R}$.

It follows from (2) and (3) that for any compound there exists in $\mathcal{R}$ a **synthesis reaction** $\Phi(X) \to X$.

Definition 11.1. A **reaction network** is a finite set $N \subset \mathcal{R}$ of chemical reactions.

Axiom 11.0.3. $\boxed{\textbf{Mass Action}}$

For each body B, reaction network N, and species X,

$$\dot{X}_B = \sum_{\substack{R \in N \\ X \in \underline{R}}} k_R \nu_R^X \prod_{Y \in \underline{R}} Y_B^{-\nu_R^Y} + \sum_{\substack{R \in N \\ X \in \overline{R}}} k_R \nu_R^X \prod_{Y \in \underline{R}} Y_B^{-\nu_R^Y} \qquad (11.3)$$

Equation (11.3) has a long history and many variations, generically called something like **Mass Action Law** [Oster *et al.* (1973)][Érdi and Tóth (1989)][Aris (1989)][Gunawardena (2003)]. The simple version here considers the factor

$$\prod_{Y \in \underline{R}} Y_B^{-\nu_R^Y}$$

to model the likelihood of the reactants Y of a reaction R in network N to detract from – in the case of the first term – or contribute to – in the second term – the total amount of compound X in body B. The first term sums individual terms corresponding to each reaction in the network for which X is a reactant. Thus, the stoichiometric coefficient factors ν_R^X here are negative, and k_R is a proportionality constant called the **rate constant** of R. Dually, the second term sums individual terms corresponding to each reaction in the network for which X is a product. Thus, the stoichiometric coefficient factors there are positive.

Remark 11.2. The Mass Action Law for the set of the reactants and products in a reaction network is a system of non-linear ordinary differential equations. Thus, the right-hand sides of these equations define a vector field in the vector space $\mathbb{R}^C$ containing $\mathcal{M}$.

Since every reaction is reversible by the Reaction Axiom, for every reaction R there exists a reaction network $\{R, -R\}$. For this special reaction network the Mass Action Law is

$$\dot{X}_B = k_R \nu_R^X \prod_{Y \in \underline{R}} Y_B^{-\nu_R^Y} + k_{(-R)} \nu_R^X \prod_{Y \in \underline{-R}} Y_B^{-\nu_R^Y} . \qquad (11.4)$$

More concretely, for a reaction with two reactants and two products,

$$R : xX + yY \to uU + vV$$

the Mass Action Law for X comes down to

$$\dot{X}_B = -k_R x X^x Y^y + k_{(-R)} x U^u V^v .$$

The concept *chemical equilibrium* has multiple aspects, but for sure the first idea is that a reaction network might lead to a condition in which the amounts of compounds in the reactor do not change as time goes on. In the case of X this condition is formalized by

$$\dot{X}_B = 0 .$$

Definition 11.3. The network $\{R, -R\}$ is at **kinetic equilibrium** if $\dot{X}_B = 0$ for each species X in B.

Kinetic equilibrium is a dynamic condition: the point is that X may continue to be created by reaction R while being destroyed at the same rate by its reverse reaction. It all depends on amounts of species in B:

Theorem 11.4. *At kinetic equilibrium of the network* $\{R, -R\}$,

$$\frac{k_R}{k_{(-R)}} = -\frac{\prod\limits_{Y \in \underline{R}} Y_B^{-\nu_R^Y}}{\prod\limits_{Y \in \underline{-R}} Y_B^{-\nu_R^Y}} .$$

Proof. Taking for granted that the reverse reaction constant $k_{(-R)}$ is positive sanctions division by it, and the (magnitude of the) stoichiometric coefficient ν_R^X cancels itself. The negation sign arises from the negative stoichiometric coefficients of reactants, so the ratio $\dfrac{k_R}{k_{(-R)}}$ is positive. $\square$

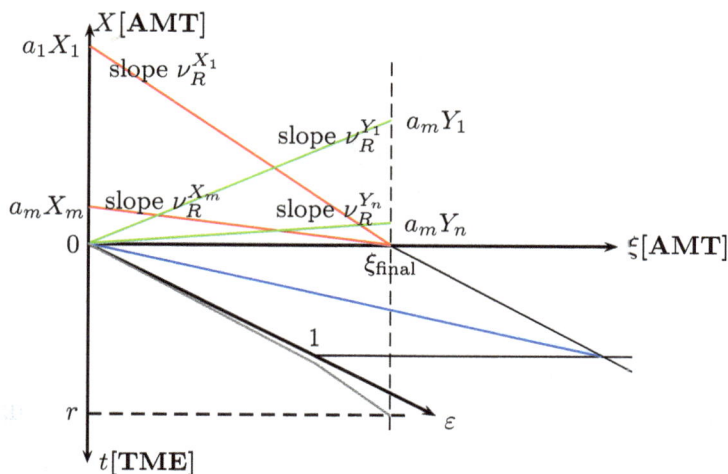

Fig. 11.1   There are four axes in this figure. Advancement of reaction R is measured along the $\xi[\textbf{AMT}]$ axis. Chemical process $[0, r] \xrightarrow{\alpha_R} \Sigma(B)$ – graphed in gray – duration is measured from 0 to r along the $t[\textbf{TME}]$ axis. The blue graph (coming out of the page along the ε axis) depicts completion ε from 0 to 1 as advancement progresses from 0 to ξ_{final}. Completion is forced by consumption of all of the reactants $X_1, \ldots, X_m$ with initial amount $a_1, \ldots, a_m$, graphed in red, resulting in products $Y_1, \ldots, Y_n$ in amounts $b_1, \ldots, b_n$ graphed in green.

11.1   Chemical Reaction Extent, Completion & Realization

If the initial amounts $a_1, \ldots, a_m$ of the reactants are a positive multiple $a_i = \lambda x_i$ of the magnitudes of the reactant stoichiometric coefficients, then total depletion of any one reactant implies that all reactants are depleted: let reaction Eq. (11.2) be realized by a chemical process $[0, r] \xrightarrow{\alpha_R} \Sigma(B)$.

Theorem 11.5. *If* $X_i \alpha_R(r) = 0$ *for some i then* $X_j \alpha_R(r) = 0$ *for all* $j = 1, \ldots, m$.

Proof.   The equation $\overset{\bullet}{X}_i = -x_i \overset{\bullet}{\xi}_R$ follows from Eq. (11.1). Hence by the Process Axiom, $\partial_0 X_i \alpha_R = -x_i \partial_0 \xi_R \alpha_R$. Integration yields

$$X_i \alpha_R(t) - a_i = X_i \alpha_R(r) - X_i \alpha_R(0)$$

$$= \int_0^t \partial_0 X_i \alpha_R dt$$

$$= -x_i \int_0^t \partial_0 \xi_R \alpha_R dt$$

$$= -x_i (\xi_R \alpha_R(t) - \xi_R \alpha_R(0))$$

$$= -x_i \xi_R \alpha_R(t) \ .$$

Thus

$$X_i \alpha_R(t) = a_i - x_i \xi_R \alpha_R(t) \ . \tag{11.5}$$

If $a_i = \lambda x_i$ then $0 = X_i \alpha_R(r) = \lambda x_i - x_i \xi_R \alpha_R(r)$ implies $\xi_{\text{final}} = \lambda$.
Therefore, $X_j \alpha_R(r) = a_j - x_j \xi_{\text{final}} = \lambda x_j - x_j \xi_{\text{final}} = \xi_{\text{final}} x_j - x_j \xi_{\text{final}} = \boxed{0}$

This theorem justifies the convergence of the red lines in Fig. 11.1 to 0 at
ξ_{final}. The blue curve is the graph of the **completion** $\varepsilon = \varepsilon(t)$ of R during
α_R, varying from 0 to 1 in direct proportion to the advancement ξ_R.

Remark 11.6. "Completion" is also known as "extent of reaction" but this
terminology shall be avoided in this book since it is sometimes confused with
"advancement" in the literature.

In general the amounts of reactants may not be the exact same multiple
of their stoichiometric coefficients. Thus one or more reactants may be
completely consumed before others. In any case, some positive amounts
of the requisite reactants provide input for the reaction to advance. If the
reaction advances then the amounts of reactants and products determine a
point in the vector space $\Sigma \mathbb{Q} \mathcal{C} \hookrightarrow \mathbb{R}^{\mathcal{C}}$ of formal first-order combinations of
species with rational coefficients, and the curve in this space is a solution
curve of the Initial Value Problem (IVP) in $\mathbb{R}^{\mathcal{C}}$ with initial conditions given
by those initial amounts, and vector field corresponding to the Mass Action
Law.

Definition 11.7. Let $\mathcal{M} \xrightarrow{\vec{R}} \mathbb{R}^{\mathcal{C}}$ denote the vector field corresponding to
reaction R determined by the Mass Action Law, Eq. (11.3). A **realization
of R at** $X = a_1 X_1 + \cdots + a_m X_m \in \mathcal{M}$ **of duration** r for some $r >$
0 of the solution curve of the (necessarily unique) solution curve of the
Mass Action Law differential equation Eq. (11.3) given the initial condition
$a_1 X_1 + \cdots + a_m X_m = \lambda x_1 X_1 + \cdots + \lambda x_m X_m$, where $\lambda \in \mathbb{F}$. In formal detail,
given $R \in \mathcal{R}$ and $r > 0$ the realization at X is the map $[0, r] \xrightarrow{\sigma(R,r)} \mathbb{R}^{\mathcal{C}}$

such that

$$\sigma(R,r)(0) = X \tag{11.6}$$

$$\partial_0 \sigma(R,r) = \vec{R} . \tag{11.7}$$

Axiom 11.1.1. $\boxed{\text{Reaction}}$

For each realization $\sigma(R,r)$ there exists a reactor B_R and a thermodynamic process $[0,r] \xrightarrow{\alpha} \Sigma(B_R)$ such that

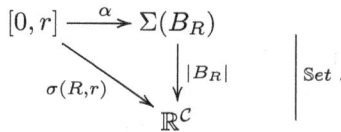

$$
\begin{array}{ccc}
[0,r] & \xrightarrow{\;\alpha\;} & \Sigma(B_R) \\
& & \downarrow |B_R| \\
\sigma(R,r) & \searrow & \\
& & \mathbb{R}^{\mathcal{C}}
\end{array}
\qquad Set \; .
$$

Since the initial reactant amounts are the same multiple λ of the stoichiometric coefficients x_i in the representation Eq. (11.2), it follows that in a realization the ratios of reactant amounts are invariant. That is, by Eq. (11.5)

$$\frac{X_i \alpha(t)}{X_j \alpha(t)} = \frac{a_i - x_i \xi_R \alpha(t)}{a_j - x_j \xi_R \alpha(t)} = \frac{\lambda x_i - x_i \xi_R \alpha(t)}{\lambda x_j - x_j \xi_R \alpha(t)} = \frac{x_i(\lambda - \xi_R \alpha(t))}{x_j(\lambda - \xi_R \alpha(t))} = \frac{x_i}{x_j} .$$

Also, the balance condition satisfied by R, re-expressed using Eq. (11.2), is

$$\sum_{i=1}^{m} x_i \Phi(X_i) = \sum_{j=1}^{n} y_j \Phi(Y_j) ,$$

which, upon multiplication by λ implies that the reactants and products are balanced all along a chemical process. In particular, if the yield of the reaction at the end of the realization is $Y := \sigma(R,r)(r)$ then $\Phi(X) = \Phi(Y)$.

11.2   Chemical Equilibrium

Let

$$R := aX + bY \to cU + dV \tag{11.8}$$

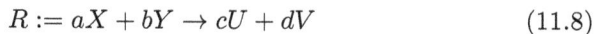

be a chemical reaction.

Definition 11.8. The **reaction quotient** of R is the quantity

$$\Sigma(B) \xrightarrow{Q_B^{(R)}} \mathbb{R}$$

defined by the equation

$$Q_B^{(R)}(u) = \frac{[U]_B^c [V]_B^d}{[X]_B^a [Y]_B^b} \, .$$

Definition 11.9. The reactor B may be held at a fixed temperature and pressure throughout the process. That is, a **standard temperature and pressure** (T_o, P_o) may be established by convention, and the environment of the system B may include machinery to maintain the standard temperature and pressure conditions ("**STP**"), so that $T_B\alpha(t) = T_o$ and $P_B\alpha(t) = P_o$ for $t \in [0, r]$. Thus, for example, pressure may be maintained at atmospheric pressure by connecting B to the atmosphere with a volume conductor – a barostat. Temperature may be maintained by cooling and heating apparatus in contact with B in a feedback loop including a temperature sensor inside B – a thermostat.

Axiom 11.2.1. | **Molar Entropy** |

For any temperature-pressure pair $a = (T, P)$ there exists a map

$$\mathcal{M} \xrightarrow{S_a} \mathbb{R} \qquad [\mathbf{NRG}][\mathbf{TMP}]^{-1}$$

such that

(1) $S_a(xX + yY) = xS_a(X) + yS_a(Y)$, and
(2) for any body B if $|B|(u) = X$ and $a = \left(T_B(u), P_B^{(X)}(u)\right)$ then

$$S_B(u) = S_a(X) \, .$$

This Axiom declares that the entropy of a body is determined by its temperature, pressure, and the chemical compounds of which it is composed. If $|B|(u) = X \in \mathcal{M}$ and $X_B(u) = 1[\mathbf{AMT}]$ then $S_B(u)$ is the entropy of 1 unit of X in state u. That number is denoted by $S_a^{(X)} := S_B(u)$ for $a = \left(T_B(u), P_B^{(X)}(u)\right)$ and is called the **molar entropy** of chemical mixture X. If $a = o$ is the standard state at **STP** then $S_o^{(X)}$ is the **standard molar entropy** of X.

Theorem 11.10. *(Entropy of Ideal Gas)([Gibbs (1957)] pp. 12–13) If $X \in \mathcal{C}$ is an ideal gas and B is a body such that $|B| = X$ then for states $x, y \in \Sigma(B)$ and a process $x \xrightarrow{\alpha} y$,*

$$[S_B]_x^y := \Delta_\alpha S_B = C_V \ln \frac{T_B(y)}{T_B(x)} + R \ln \frac{V_B(y)}{V_B(x)} . \tag{11.9}$$

Proof. Taking into account the hypotheses, recall the following equations:

$$dE_B = T_B dS_B - P_B^{(X)} dV_B \qquad\qquad \text{Energy Axiom}$$
$$\tag{11.10}$$

$$E_B = C_V T_B \qquad\qquad \text{heat capacity at constant volume}$$
$$\tag{11.11}$$

$$P_B^{(X)} V_B = R T_B \qquad\qquad \text{Ideal Gas Law}$$
$$\tag{11.12}$$

Calculate

$$T_B = \frac{E_B}{C_V} \qquad\qquad \text{by (11.11)}$$
$$\tag{11.13}$$

$$P_B^{(X)} = \frac{R E_B}{C_V V_B} \qquad\qquad \text{by (11.12,11.13)}$$
$$\tag{11.14}$$

$$dE_B = \frac{E_B}{C_V} dS_B - \frac{R E_B}{C_V V_B} dV_B \qquad\qquad \text{by (11.10,11.13,11.14)}$$
$$\tag{11.15}$$

$$\frac{dE_B}{E_B} = \frac{1}{C_V} dS_B - \frac{R}{C_V} \frac{dV_B}{V_B} \qquad\qquad \text{divide by } E_B$$
$$\tag{11.16}$$

$$[\ln C_V T_B]_x^y = \frac{[S_B]_x^y}{C_V} - \frac{R}{C_V} [\ln V_B]_x^y \qquad\qquad \text{by (11.11); integrate along } \alpha$$
$$\tag{11.17}$$

$$[S_B]_x^y = C_V \ln \frac{T_B(y)}{T_B(x)} + R \ln \frac{V_B(y)}{V_B(x)} \qquad\qquad \text{re-arrangement.}$$
$$\tag{11.18}$$

$$\square$$

Definition 11.11. A pair (M, N) of mixtures is **balanced** if $\Phi(M) = \Phi(N)$ and is **realizable by reaction** R if there exists a realization $\sigma(R, r)$

such that $\sigma(R,r)(0) = M$ and $\sigma(R,r)(r) = N$. The notation for such a scenario is $M \xrightarrow{R} N$, which is called the **transformation** of M to N by R (in process $x \xrightarrow{\alpha} y$). In case $M = \Phi(N)$ is the decomposition of N, $\Phi(N) \xrightarrow{R} N$ is called the **formation** of N.

Definition 11.12. The negative affinity $-\mathcal{A}_B$ reflects the change in Gibbs free energy per unit advance of the reaction ([McQuarrie and Simon (1997)] p. 965), and in the literature is denoted by

$$\Delta_R G_B := \frac{\overset{\bullet}{G}_B}{\overset{\bullet}{\xi}_R} = -\mathcal{A}_B [\mathbf{NRG}][\mathbf{AMT}]^{-1} \, .$$

Theorem 11.13. *For any state* $u \in \Sigma(B)$,

$$\Delta_R G_B(u) = \Delta_R G_B(o) - RT_B(u_*)Q_B^{(R)}(u) \, . \tag{11.19}$$

Proof. [Spencer (1974)]. □

Theorem 11.14. *For the chemical reaction in reactor B the following are necessary conditions for the reaction network* $\{ R, -R \}$ *to achieve chemical equilibrium at* $u_* \in \Sigma(B)$:

$$\frac{k_R}{k_{(-R)}} = -\frac{\prod\limits_{Y \in \underline{R}} Y_B^{-\nu_R^Y}}{\prod\limits_{Y \in \underline{-R}} Y_B^{-\nu_R^Y}} \, ; \tag{11.20}$$

$$\overset{\bullet}{\xi}_R = 0 \, ; \tag{11.21}$$

$$\overset{\bullet}{G}_B = -\mathcal{A}_B \overset{\bullet}{\xi}_R = 0 \, ; \tag{11.22}$$

$$\Delta_R G_B(o) = -RT_B(u_*)Q_B^{(R)}(u) \, . \tag{11.23}$$

Proof.   Equation (11.20) is Theorem 11.4; Eq. (11.21) follows from the definition of stoichiometric coefficient in the Reaction Kinetics Axiom and the hypothesis that $\overset{\bullet}{X}_R = 0$. Equation (11.22) is immediate from the comment about entropic equilibrium at Eq. (10.29).

Aside 11.2.1. Equation (11.23) is bottom dollar for biological energetics ([Voet and Voet (1979)] p. 37) and some aspects of muscle contraction. A major impetus for The Theory of Substances has been to master this equation after a sustained relentless search for understanding. So it is ironic that although it follows immediately from Theorem 11.13, I refer the reader for that proof to the excellent literature. *C'est la vie.*   □

Equation (11.23) is very important. It permits calculating the equilibrium constant for a chemical reaction without making a direct study of the reaction. $\Delta_R G_B(o)$ comes from tabulated values of standard Gibbs energies of formation or from $\Delta_R H_B(o)$ and $\Delta_R S_B(o)$ values through the relationship $\Delta_R G_B(o) = \Delta_R H_B(o) - T_B \Delta_R S_B(o)$ [Craig (1992)].

11.3 Chemical Formations & Transformations

Primary *thermodynamic* variables of interest to chemists are the temperature and pressure of the reactor, hence the Reaction Axiom diagram extends to

$$[0,r] \xrightarrow{\alpha} \Sigma(B_R) \xrightarrow{(T_B, P_B)} \mathbb{R} \times \mathbb{R} \qquad (11.24)$$

with $\sigma(R,r)$ diagonal and $|B_R|$ mapping to $\mathbb{R}^C$, Set.

Equation (11.24) commences the conceptual blending of chemical kinetics on the left with thermodynamics on the right [Fauconnier and Turner (2002)].

Definition 11.15. The enthalpy change $H_R^o := [H_B]_x^y$ induced by a transformation $X \xrightarrow{R} Y$ at **STP** is called the **enthalpy of transformation at STP**. The enthalpy change induced by a formation $\Phi(Y) \xrightarrow{R} Y$ per amount of Y is called the **enthalpy of formation of Y at STP** and is defined by

$$H_Y^0 := \frac{H_R^0}{Y_B(y)} . \qquad (11.25)$$

Remark 11.16. In the literature H_R^0 is called "enthalpy of reaction" or "heat of reaction," and alternatively denoted by something like $\Delta H^0 R$. The ratio Eq. (11.25) is called the "molar enthalpy (or heat) of formation." If instead of amount in moles the denominator is the mass of Y produced in the formation reaction, then the quantity is called the "specific enthalpy (or heat) of formation."

11.4 Monoidal Category & Monoidal Functor

Formations and transformations of chemical mixtures by reactions accompanied by their enthalpy changes may be assembled into an interesting algebraic structure.

(1) By the Reaction Axiom, for any reaction $R \in \mathcal{R}$ there exists its reverse reaction $-R \in \mathcal{R}$. If the transformation $X \xrightarrow{R} Y$ at **STP** has a reverse transformation $Y \xrightarrow{-R} X$ at **STP** then the system $B_R + B_{(-R)}$ undergoes no change in amounts of material substance, just as though no chemical reaction at all has taken place. Therefore, $H^0_{(-R)} = -H^0_R$.

(2) If there is a transformation $X_1 \xrightarrow{R_1} Y_1$ at **STP** in B_1 and another transformation $X_1 \xrightarrow{R_2} Y_2$ at **STP** in B_2, and the system $B_1 + B_2$ is isolated, then by the Additivity Axiom for Energy and the Energy Axiom, $[H_{B_1+B_2}]^y_x = H^0_1 + H^0_2$.

(3) As a special case of (2), if there is a transformation $X \xrightarrow{R_1} Y$ at **STP** in B_1 and a transformation $Y \xrightarrow{R_2} Z$ at **STP** in B_2, the net result in the system $B_1 + B_2$ is a transformation of X to Z, and the net enthalpy change of the combined transformations is $[H_{B_1+B_2}]^y_x = H^0_1 + H^0_2$.

> **[C]hemical reactions are morphisms in a symmetric monoidal category where objects are collections of molecules** [Baez and Stay (2009)].

> **Hess's Law of Constant Heat Summation: The enthalpy change in a chemical reaction is the same whether it takes place in one or several stages. The enthalpy change in the overall chemical reaction is the algebraic sum of the enthalpy changes of the reaction steps** [Bothamley (1923)].

Definition 11.17. A category $\mathbb{C}$ is a **monoidal category** if the following diagrams exist in $\mathbb{C}at$:

$$\mathbb{C} \times \mathbb{C} \xrightarrow{\;\square\;} \mathbb{C} \xrightarrow{\;1\;} \mathbf{1} \tag{11.26}$$

$$
\begin{array}{ccc}
\mathbb{C} \times \mathbb{C} \times \mathbb{C} & \longrightarrow & \mathbb{C} \times \mathbb{C} \\
\downarrow & & \downarrow \\
\mathbb{C} \times \mathbb{C} & \longrightarrow & \mathbb{C}
\end{array}
\tag{11.27}
$$

$$\mathbb{C} \xrightarrow{(1_\mathbb{C}\ 1)} \mathbb{C} \times \mathbb{C} \xleftarrow{(1\ 1_\mathbb{C})} \mathbb{C} \qquad (11.28)$$

where Eq. (11.26) introduces a functor $\square$ that defines a binary operation on objects and morphisms of $\mathbb{C}$ and a distinguished object 1, such that by Eq. (11.27) the operation $\square$ satisfies the Associative Law, and by Eq. (11.28) it satisfies the Left and Right Identity Laws.

Aside 11.4.1. In the higher reaches of category theory I have been given to understand that the definition of monoidal category can be abbreviated to the declaration that a monoidal category is a monoid in the category of categories. Compare the diagrams above to the definition:

A **monoid** is a diagram $M := A \times A \xrightarrow{*} A \xleftarrow{e} 1$ such that $A \times A \xrightarrow{*} A$ is a semigroup, $1 \xrightarrow{e} A$ is a pointed set, and there exists a diagram

$$A \xrightarrow{(1_A\ e)} A \times A \xleftarrow{(e\ 1_A)} A \qquad \qquad (11.29)$$

$$\mathcal{Set}\ .$$

What distinguishes a monoidal category from a monoid is that there are more items in play. That is to say, not only can objects be squared together by $\square$, also morphisms can be squared together. Thus, since $\square$ is a functor, given objects a, b the squaring together of their identity morphisms satisfies the equation $1_a \square 1_b = 1_{a \square b}$. Importantly, there is also an Exchange Law: for morphisms $a \xrightarrow{f} b \xrightarrow{g} c$ and $a' \xrightarrow{f'} b' \xrightarrow{g'} c'$ the equation $gf \square g'f' = (g \square g')(f \square f')$ is also satisfied. An important resource situating monoidal categories in the context of theoretical physics is [Coecke and Éric Oliver Paquette (2009)]. It starts by discussing "cooking with vegetables."

Definition 11.18. A monoidal category $\mathbb{C}$ is a **symmetric monoidal category** if there exist for every two objects a, b and morphisms $a \xrightarrow{f}$

$c, b \xrightarrow{g} d$ of $\mathbb{C}$, these diagrams:

$$a\square b \xrightarrow{\sigma_{ab}} b\square a \qquad\qquad (11.30)$$

with $1_{a\square b}$ and σ_{ba} arrows to $a\square b$, labeled $\mathbb{C}$

$$a\square b \xrightarrow{\sigma_{ab}} b\square a \qquad\qquad (11.31)$$

with $f\square g$ down to $c\square d$, $g\square f$ down to $d\square c$, and $c\square d \xrightarrow{\sigma_{ba}} d\square c$, labeled $\mathbb{C}$

The "symmetry" of a monoidal category is analogous to commutativity of a monoid, except that isomorphism as in Eq. (11.30) replaces equality of $a\square b$ and $b\square a$, and of course there is the new equation relating the isomorphisms to the squaring of morphisms in Eq. (11.31).

Example 11.19. Let G be a commutative group with binary operation $+$ and identity 0. Let $\overrightarrow{G}$ be the directed graph with dots the items $a \in G$, and arrows the lists $[\,a\ c\ b\,]$ such that $c = b - a$. The **domain** of $[\,a\ c\ b\,]$ is a and its **codomain** is b. In particular, for each $a \in G$ there is an **identity arrow** $[\,a\ 0\ a\,]$ in $\overrightarrow{G}$. Define a **law of composition** for composable morphisms $[\,a_1\ c\ a_2\,]$ and $[\,a_2\ d\ a_3\,]$ by

$$[\,a_1\ c\ a_2\,] \overrightarrow{\circ} [\,a_2\ d\ a_3\,] := [\,a_1\ c+d\ a_3\,] .$$

Then $\overrightarrow{G}$ is a category. Moreover, $\overrightarrow{G}$ is a symmetric monoidal category with squaring defined on objects by $a\square b := a + b$ and on arbitrary morphisms by

$$[\,a_1\ c\ a_2\,]\square[\,a_3\ d\ a_4\,] := [\,a_1 + a_3\ c+d\ a_2 + a_4\,]$$

which makes sense because $c+d = (a_2-a_1)+(a_4-a_2) = (a_2+a_4)-(a_1+a_3)$. The monoidal identity is $[\,0\ 0\ 0\,]$. The monoidal symmetry Eq. (11.31) is also an immediate consequence of the commutativity of G since, e.g., $\sigma_{a\square b} := 1_{a\square b}$.

Therefore, the additive group of vectors of any vector space forms a symmetric monoidal category. In particular, the additive group of real numbers forms a symmetric monoidal category, and this is denoted by $\overrightarrow{\mathbb{R}}$.

Example 11.20. Let $\mathbb{R}$xn be the **directed graph** whose dots are mixtures R of chemical compounds and whose arrows are lists $[\,R\ R \to P\ P\,]$ where

$R \to P$ is a chemical reaction with reactants R and products P. The **domain** of $[\,R\ R \to P\ P\,]$ is R and its codomain is P. For each mixture R there is an **identity arrow** $[\,R\ R \to R\ R\,]$. Define a **law of composition** for composable morphisms $[\,R\ R \to P\ P\,]$ and $[\,P\ P \to Q\ Q\,]$ by

$$[\,R\ R \to P\ P\,] \overrightarrow{\circ} [\,P\ P \to Q\ Q\,] := [\,R\ R \to Q\ Q\,]. \qquad (11.32)$$

Then $\mathbb{R}xn$ is a category. Moreover, $\overrightarrow{\mathbb{R}xn}$ is a symmetric monoidal category with squaring defined on objects by $R \square R' := R + R'$ and on arbitrary reactions by

$$R \to P \square R' \to P' := R + R' \to P + P'$$

which is well-defined because $-(R + R') + (P + P')$ is still balanced. the monoidal identity is the empty reaction $\emptyset \to \emptyset$.

Definition 11.21. Let $\mathbb{C}$ and $\mathbb{D}$ be monoidal categories and $\mathbb{C} \xrightarrow{F} \mathbb{D}$ a functor. Then F is a **monoidal functor** if it preserves the squaring operation, that is to say, if $F(a \square_{\mathbb{C}} b) = F(a) \square_{\mathbb{D}} F(b)$.

In this book the whole reason to define the abstractions of (symmetric) monoidal categories and their functors is that there exists an important monoidal functor from $\mathbb{R}xn$ to $\overrightarrow{\mathbb{R}}$.

11.5   Hess' Monoidal Functor

A set of reactions are said to be "added" when one follows the other with at least one product in each prior reaction serving as a reactant in a later one. Chemical equations, on the other hand, are conventional symbolic representations of reactions. The addition of chemical equations is closely analogous to the addition of algebraic equations, with the arrow replacing the equal sign. Any set of balanced equations can be "added", even if none represents a reaction that anyone has ever observed [Diemente (1998)].

Theorem 11.22. *At standard temperature and pressure, there exists a monoidal functor*

$$\mathbb{R}xn \xrightarrow{\mathcal{H}^0} \overrightarrow{\mathbb{R}}$$

*which assigns enthalpy of formation to mixtures and enthalpy of transformation to reactions. This functor may be called **Hess' monoidal munctor**.*

Example 11.23. The Combustion of Magnesium

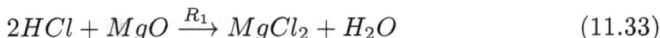

$$2HCl + MgO \xrightarrow{R_1} MgCl_2 + H_2O \tag{11.33}$$

$$2HCl + Mg \xrightarrow{R_2} MgCl_2 + H_2 \tag{11.34}$$

$$2H_2 + O_2 \xrightarrow{R_3} 2H_2O \tag{11.35}$$

$$2Mg + O_2 \xrightarrow{R} 2MgO \tag{11.36}$$

$$HCl \xrightarrow{1_{HCl}} HCl \tag{11.37}$$

$$O_2 \xrightarrow{1_{O_2}} O_2 \tag{11.38}$$

$$MgCl_2 \xrightarrow{1_{MgCl_2}} MgCl_2 \tag{11.39}$$

$$2R_1 = (2 \cdot 1_{MgCl_2} + R_3) \circ (2R_2 + 1_{O_2}) \circ (4 \cdot 1_{HCl} - R) \tag{11.40}$$

[W]e *are* adding equations but are *not* adding reactions. When a piece of burning magnesium is under consideration, the first three equations of this set cannot possibly represent reactions because hydrochloric acid, water, magnesium chloride, and hydrogen have absolutely nothing to do with the combustion of magnesium. Hess' Law is simply a convenience: if by some means we can determine $\Delta H_1, \Delta H_2$, and ΔH_3 and then invoke the law of conservation of energy, we can calculate ΔH_4. And it is a safe bet that this problem is solved in dozens of high-school chemistry labs across the country every school year. This is done by performing the reactions represented by the equations. The first two reactions of this set are easily run in familiar Styrofoam-cup calorimeters, making ΔH_1 and ΔH_2 experimentally accessible. The third reaction is of course an explosion. It may be performed as a demonstration, but ΔH_3 has to be looked up in a handbook or some other compilation. Then we use Hess' Law to calculate the heat of combustion of magnesium [Diemente (1998)].

Applying Hess' monoidal functor yields

$$2\mathcal{H}^0(R_1) = (2\mathcal{H}^0(1_{MgCl_2}) + \mathcal{H}^0(R_3)) \tag{11.41}$$

$$+ (2\mathcal{H}^0(R_2) + \mathcal{H}^0(1_{O_2})) \tag{11.42}$$

$$+ (4\mathcal{H}^0(1_{HCl}) - \mathcal{H}^0(R)) \tag{11.43}$$

which is easy to solve for $\mathcal{H}^0(R))$, as required.

PART 6

Muscle Contraction Research

This part of the book focuses on the muscle contraction research —
Chronology. It consists of excerpts – some lightly edited for consistency,
continuity and simplicity – selected from diverse publications. The choices
exhibited certainly do not form an encyclopedic survey of a deep, long and
complex research effort. Rather, they are a non-random sampling with a
bias towards revealing some twists and turns of theory as it collides with
experimental facts. There is also a strong bias towards relishing a "crisis"
with regard to the choice of theoretical interpretation of muscle contraction
data inexplicable without a cooperative molecular mechanism. A further
bias favors selections offering molecular-level computer simulations of muscle contraction.

Chapter 12

Muscle Contraction

12.1 Muscle Contraction: Chronology

12.1.1 *19th Century*

1864 Andrew G. Szent-Györgyi [Szent-Györgyi (2004)] A viscous protein was extracted from muscle with concentrated salt solution by Kühne, who called it **myosin** and considered it responsible for the **rigor** state of **muscle**.

12.1.2 *1930–1939*

1930 "Muralt and Edsall showed that the **myosin** in solution had a strong flow birefringence with indications that the particles were uniform in size and shape" [Szent-Györgyi (2004)].

1935 "Weber developed a new technique for the in vitro study of **contraction**. He squirted **myosin** dissolved in high salt into water where it formed threads that became strongly birefringent upon drying" [Szent-Györgyi (2004)].

1938 [In 1922 Archibald V. Hill received a half share of the Nobel Prize in Physiology or Medicine "for his discovery relating to the production of heat in muscle."]

> The isometric lever was adjusted by a separate rack and pinion, mounted on the same stand as the rest, and was disengaged when the main isotonic lever was in use. The chains and the isometric lever together permitted 0.93 mm. shortening when the muscle developed a force of 100 g.
>
> If a muscle is stimulated isometrically and then suddenly released under a small load, it shortens rapidly and during its

shortening the galvanometer gives a quick extra deflexion. This sudden extra deflexion of the record implies a sudden increment in the rate of heat production of the muscle. The increase of heat rate is proportional to the speed of shortening and stops when the shortening of the contractile component stops: the total extra heat is proportional to the total shortening.

We have seen that in shortening a distance x cm., extra heat ax g. cm. is set free. If P g. be the load lifted, the work done is Px g. cm. Thus the total energy, in excess of isometric, is $(P + a)x$ g. cm. The *rate* of extra energy liberation, therefore, is $(P + a)dx/dt$, or $(P + a)v$, if v be the velocity (cm./sec.) of shortening.

It is found experimentally that the rate of extra energy liberation, $(P + a)v$, is a rather exact linear function of the load P, increasing as P diminishes, being zero when $P = P_0$ in an isometric contraction and having its greatest value for zero load. We may write therefore

$$(P + a)v = b(P - P_0) \, .$$

[Hill (1938)]

Aside 12.1.1. This early work on production of heat by muscle contraction already connects animal motion to thermodynamics, and that is the primary motivation for my Theory of Substances in this book.

1939 "Engelhardt and Lyubimova reported in a careful study that **myosin** had **ATPase** activity" [Szent-Györgyi (2004)].

12.1.3　*1940–1949*

1941 "Engelhardt *et al.* also checked the effect of ATP on the **myosin** fibers of H.H. Weber and found that the fibers became more extensible. Engelhardt and Lyubimova's experiments represented the opening salvo in the revolution of muscle biochemistry."

"Albert Szent-Györgyi and colleagues then established that the **myosin** used by previous investigators consisted of two proteins. These were purified and shown to be necessary for the **contraction** elicited by **ATP**" [Szent-Györgyi (2004)].

1942 "**Actin** was discovered by Straub. Together with **myosin** and **ATP** it constitutes the **contractile system**. In the absence of salt, actin molecules are stable as monomers (G-actin); in the presence of salt, especially divalent cations, actin polymerizes. The high asymmetry of the polymerized actin (F-actin) is indicated by its high viscosity, thyxotropy and strong double refraction (Straub)."

"Albert Szent-Györgyi observed that exposure of ground **muscle** to high salt concentrations for 20 minutes extracted a protein of low viscosity (**myosin A**), whereas overnight exposure solubilized a protein with high viscosity (**myosin B**). The viscosity of **myosin B** was reduced by adding **ATP** while the viscosity of **myosin A** remained essentially unaffected."

"Needham *et al.* found that **ATP** reduced the viscosity and flow birefringence of **myosin**. These changes were reversed upon exhaustion of **ATP**. They proposed that **ATP** caused a reversible change in the asymmetry of the **myosin** molecule possibly due to the shortening of the molecule, or changes in the interaction between micellae formed by **myosin** molecules."

"Szent-Györgyi discovered that the threads prepared from **myosin B** using H. H. Weber's method shortened on addition of boiled muscle juice, but when fibers of **myosin A** were tested these remained unchanged. The shortening was apparently due to exclusion of water. The active material in the boiled extract was identified as **ATP**. In his autobiography, Szent-Györgyi (1963) describes that "to see them (the threads) contract for the first time, was perhaps the most thrilling moment of my life."

"Straub joined Szent-Györgyi about this time and it became clear that the difference between **myosin** B and A was due to the presence of another protein that they called "actin," which, when combined with **myosin**, was responsible for the high viscosity and for contractility" [Szent-Györgyi (2004)].

Aside 12.1.2. Unquestionably this observation by Albert Szent-Györgyi was an epochal event in the history of scientifically understanding muscle contraction.

1943 "**myosin** A was purified as paracrystals by Szent-Györgyi and retained the name **myosin**. In a very elegant series of experiments actin was purified by Straub. **myosin B** was renamed acto**myosin**."

"Straub showed that the newly discovered protein existed in two forms: globular actin (G-actin) that was stable in the absence of salt, and in

the presence of ions it polymerized to form fibrous actin (F-actin)."
"Szent-Györgyi demonstrated that ATP had a dual function that depended on ionic strength. At low ionic strength ATP induced contraction, at high ionic strength it dissociated actin from **myosin**. It was realized that the rigor state was due to the formation of acto**myosin** in the absence of ATP. In fact, rigor mortis was the result of the depletion of ATP" [Szent-Györgyi (2004)].

1946 "**Tropomyosin** was discovered and isolated by Bailey" [Szent-Györgyi (2004)].

1949 "The steady-state ATPase activity was increased during the contraction of **actomyosin** or of minced and washed muscle."
"The development and the behavior of the glycerol extracted psoas muscle preparation by Szent-Györgyi brought conclusive evidence that the interaction of ATP with **actomyosin** was the basic contractile event. The glycerol extracted psoas muscle preparation consists of a chemically skinned muscle fiber bundle that is permeable to ions. On addition of $Mg2^+$ ATP the preparation develops a tension that is comparable to the tension development of living muscle. Moreover, the preparation behaves somewhat like **actomyosin**" [Szent-Györgyi (2004)].

Aside 12.1.3. Again, Albert Szent-Györgyi adds monumentally to the experimental tools for studying muscle contraction.

12.1.4   *1950–1959*

1952 "Cross-striated muscle is organized in sarcomeres, repeating units 2–3 μm long. Hugh E. Huxley, in his Ph.D. thesis in 1952, observed that the basic meridional periodicities of muscle remain constant at various muscle lengths" [Szent-Györgyi (2004)].

1953 "Electron microscopy revealed the presence of two types of filaments: 1.6 μm long thick filaments located in the A-band and 1 μm long thin filaments stretching from the Z-band to the H-zone."
"Cross-striated muscle is organized in sarcomeres that extend from one Z-line to the next. The distance between Z-lines is 2–3 μm. The thin filaments contain actin and the thick filaments contain **myosin**. The thick filaments have bipolar symmetry with a central bare zone in which there are no cross-bridges. The actin fiber symmetry reverses in the Z-line. The area not penetrated by the thin filaments is variable, and

is known as the H-zone."

"Hasselbach also observed independently the removal of **myosin** from the A-band with pyrophosphate solution. **myosin** added to the ghost fibers bound to the thin filaments demonstrating that these contained actin. Light microscopic observations distinguished a zone of high refractive index, the A (anisotropic) band from the I (isotropic) band" [Szent-Györgyi (2004)].

1954 "The sliding filament theory[Huxley (1984)] was based on the observations of constancy of the length of the A-band and the shortening of the I band during a contraction. As pointed out by A. F. Huxley, this observation was made by applying interference microscopy to the most differentiated motile system available, namely intact frog muscle fibers [Huxley and Niedergerke (1954)]."

"A very similar observation was made on glycerol-extracted myofibrils using phase contract microscopy [Huxley and Hanson (1954)]. These authors were also able to associate actin with the thin filaments and **myosin** with the thick filaments. The sliding filament hypothesis was proposed to explain these observations" [Szent-Györgyi (2004)].

Isolated fibres from frog muscle, give satisfactory propagated twitches and tetani, in which activation is complete very early after the first stimulus, while tension develops more slowly even if the contraction is isometric, and changes of length, both during stimulation and in the resting muscle, can be controlled by holding the tendon ends.

An interference microscope in which the reference beam does not traverse the specimen would be expected to give a satisfactory 'optical section' of the fibre.

The contrast between the A-(higher refractive index) and I-bands could be controlled or reversed by changing the background path-difference between the two beams; the measured widths of the bands were independent of this adjustment. The fibre was photographed on moving film by a series of ten flashes from a discharge tube at intervals of about 20 msec., and could be stimulated by pulses of current synchronized with these flashes.

The similarity of the changes during passive shortening and during isotonic contraction, and the absence of change during isometric twitches, show that the changes in the ratio of widths of the A- and I-band depend simply on the length of the fibre, and are unaffected by the 'activation' or by tension development as such. The approximate constancy of A-band width under a wide range of conditions (including shortening within the physiological range) agree with the observations of H. E. Huxley and J. Hanson on separated myofibrils reported in the accompanying communication. The natural conclusion, that the material which gives the A-bands their high refractive index and also their birefringence is in the form of submicroscopic rods of definite length, was put forward by Krause, and receives strong support from the observations reported here. The identification of this material as myosin, and the existence of filaments (presumably actin) extending through the I-bands and into the adjacent A-bands, a shown in many electron microscope studies, makes very attractive the hypothesis that during contraction the actin filaments are drawn into the A-bands, between the rodlets of myosin. (This point of view was reached independently by ourselves and by H. E. Huxley and Jean Hanson in the summer of 1953) [Huxley and Niedergerke (1954)].

A possible driving force for contraction in this model might be the formation of actin-myosin linkages when adenosine triphosphate, having previously displaced actin from myosin, is enzymatically split by the myosin. In this way, the actin filaments might be drawn into the array of myosin filaments in order to presen to them as many active groups for actomyosin formation as possible; furthermore, if the structure of actin is such that a greater number of active groups could be opposed to those on the myosin by, for example, a coiling of the actin filaments, then even greater degrees of shortening could be produced by essentially the same mechanism [Huxley and Hanson (1954)].

"Previous theories took it for granted that contraction was the result of a length change in long polymer-like molecules. Until the epoch-making papers in 1954 the idea that movement might result from a process other than the shortening of molecular structures had just not been considered" [Szent-Györgyi (2004)].

1955 "Sedimentation data suggested that LMM and HMM were linearly attached in **myosin** (Lauffer and Szent-Györgyi) The division of roles between LMM and HMM then became clear. LMM is responsible for filament formation, whereas HMM contains the sites responsible for ATPase activity and also the sites for interacting with actin" [Szent-Györgyi (2004)].

1957 "Cross-bridges were clearly visualized by Huxley by electron microscopy of ultra-thin sections."

"A. F. Huxley investigated the idea that the entropy of cross-bridge attachment may be used to drive the cross-bridge cycle, which is still a relevant idea."

"A mechanistic relationship between possible cross-bridge movements and the mechanical properties of muscle first was proposed by Huxley. However, it turned out that an understanding of the structural changes that the **myosin** cross-bridge undergoes during a cycle necessitated the determination of the structures of actin and the **myosin** cross-bridge at atomic resolution. This took another 30 years" [Szent-Györgyi (2004)].

1958 "Shortly afterwards, Huxley proposed a mechanical cross-bridge cycle that is similar to present-day models"[Szent-Györgyi (2004)].

12.1.5 *1960–1969*

1962 "**Tropomyosin** can be removed from actin at low temperatures (Drabikowski and Gergely)."

"It also combines with **troponin**, the complex responsible for thin filament regulation by blocking the actin sites necessary for binding **myosin** in a calcium-dependent manner."

"The factor was later identified by Hasselbach and Makinose and by Ebashi and Lipmann as fragmented sarcoplasmic reticulum that acted as Ca^{++}-pump. The triggered release of sequestered Ca^{++} to 10 μm caused muscle to contract" [Szent-Györgyi (2004)].

Aside 12.1.4. More and more links between the mind and body were unveiled as the years went on.

1963 "2 μm long **myosin** containing thick filaments with cross-bridges and 1 μm long actin containing thin filaments are shown. As the **sarcomere** shortens, the **myosin** cross-bridges react with actin and propel the thin filaments toward the center of the sarcomere. Both filament types remain at constant lengths during contraction. The sliding of the filaments explains the constancy of the A-band and the changes of the I-band and the H-zone. During a contraction actin filaments move toward the center from both halves of the sarcomere. This necessitates a change in direction (orientation) of the actin filaments every half sarcomere. The directionality is built into the way actin and **myosin** assembled into filaments. Thick filaments are bipolar structures. Their assembly begins with the tail-to-tail association of the LMM fractions so that the heads come out pointing in opposite directions. Then filaments grow by addition of **myosin** molecules onto these bipolar nuclei. The overall result is a smooth central region 0.2 μm long that is free of **myosin** heads, while the molecules in the two halves of the filaments face in opposite direction (Huxley)."

"About 400 **myosin** molecules assemble to form a filament, which interacts with actin filaments containing about the same number of actin monomers (Hanson and Lowy; Huxley). Similar filaments form readily in vitro by self-assembly except that they display variable filament lengths (Huxley). **myosin** therefore has multiple functions: filament formation, ATPase activity, and reversible combination with actin. HMM is therefore a two-headed molecule connected to LMM via S2; LMM plus S2 forms the rod portion of the molecule, which for most of its length is a coiled-coil a-helix" [Szent-Györgyi (2004)].

Aside 12.1.5. Greater and greater detail on the actual shape of a specific molecule directly implicated in muscle contraction.

1964 "Ebashi and colleagues discovered that for the relaxing effect **tropomyosin** was required. However, only "native" **tropomyosin** was effective. This was due to an additional protein, named troponin (Ebashi; Ebashi and Ebashi)" [Szent-Györgyi (2004)].

1965 "The first direct evidence for a change in cross-bridge shape that might provide the basis for movement was obtained by Reedy *et al.*, who discovered that the angle of the **myosin** cross-bridge in insect muscles depended on the state of the muscle."

"The use of transient kinetics to explore the steps of the cross-bridge cycle was introduced by Tonomura and colleagues, who showed that there is an initial rapid liberation of phosphate by **myosin** (Kanazawa and Tonomura)" [Szent-Györgyi (2004)].

1966 "**Tropomyosin** is located in the thin filaments, lying on a flat surface formed by the two strands of actin. In studies using electron microscopy and small angle X-ray diffraction (Cohen and Longley) magnesium salts of **tropomyosin** form paracrystals that have a repeat period of 396 Å, indicating an elongated structure with an end-to-end overlap" [Szent-

Györgyi (2004)].

1967 "The important electron microscope studies of Slayter and Lowey established that a **myosin** molecule ended in two globules (heads). Troponin is arranged periodically, each **tropomyosin** binds one troponin molecule (Ohtsuki *et al.*)" [Szent-Györgyi (2004)].

1969 "Lowey *et al.* showed that, at low ionic strengths, papain or chymotrypsin splits **myosin** into S1 and rod. The S1 combined with actin and was a fully active ATPase."

"Electron microscopy combined with X-ray diffraction showed that at rest the cross-bridges extended at right angle from the thick filament (90°), whereas in rigor (no ATP present) the cross-bridges protruded at an acute angle (45°). Therefore, when Huxley put forward a swinging cross-bridge model, proposing that the **myosin** head attached to actin changes its angle during the contraction cycle, the idea was widely supported."

"Nevertheless, in point of fact it took many years to produce direct evidence in support of the swinging cross-bridge model" [Szent-Györgyi (2004)].

12.1.6 *1970–1979*

1971 "Lymn and Taylor provided evidence that hydrolysis of ATP occurs in the detached state when **myosin** is not bound to actin; they showed that the addition of ATP to **myosin** results in a burst of ATP hydrolysis that was nearly stoichiometric with the **myosin** heads. The burst occurred when the active site was unoccupied, so this finding indicated that the dissociation of ADP was the limiting reaction of the cycle."

"Greaser and Gergely showed that troponin consists of three different subunits. TroponinC (TnC) binds Ca^{++} and is related to calmodulin; troponinI (TnI) is an inhibitory subunit that binds to TnC and to actin and troponinT (TnT) in a Ca^{++}-dependent manner, and TnT binds to **tropomyosin**. It is thought that in the absence of Ca^{++} the affinity between TnC and TnI is strong so that **tropomyosin** is held over the **myosin**-binding site of actin. In the presence of Ca^{++} the binding between TnI and TnC weakens, **tropomyosin** is allowed to roll azimuthally around actin to open up the binding-site for **myosin**. Combination with S1 evidently leads to its further movement. This steric hindrance model of regulation was based on observations of the low angle X-ray diffraction patterns from muscle fibers. A detailed compendium of data on all aspects of muscle function available from antiquity until 1970 can be found in the monumental book by Needham (1971), which contains some 2,400 references" [Szent-Györgyi (2004)].

R. W. Lymn and E. W. Taylor 1971

The aim of enzyme kinetic studies is to provide a molecular mechanism and a quantitative model for the events which are presumed to take place in muscle contraction. A large body of evidence derived from electron microscopy and X-ray diffraction, has led to what may be called the sliding-filament-moving-bridge model (H. E. Huxley). The four steps in the cycle are: (1) the dissociation of the actin-bridge complex, (2) movement of the free **myosin** bridge, (3) recombination of the bridge with actin, and (4) the drive stroke.

The four main steps in the enzyme mechanism are: (1) the binding of ATP and very rapid dissociation of acto**myosin**, (2) the splitting of ATP on the free **myosin**, (3) recombination of actin with the **myosin**-products complex, and (4) the displacement of products.

The similarities of the two schemes are obvious. Steps 1 and 3 in both schemes involve the dissociation and recombination of actin and **myosin** and it appears quite reasonable to identify the corresponding steps. The chemical mechanism provides an explanation of how a cycle involving dissociation and recombination of **myosin** bridges could be coupled to ATP hydrolysis.

Our kinetic studies provide no evidence for a large change in the configuration of **myosin** associated with ATP binding or hydrolysis and there is no kinetic evidence of any kind for a configuration change that could be identified with a step in the contraction cycle. We consider that the type of model in which the configuration is modulated by binding and dissociation of substrate or products is too naive and apparently incapable of accounting for certain features of muscle.

The kinetic scheme could also be made to fit the general features of the model of A. V. Huxley in which the movement of free bridges is due to thermal energy rather than interaction with substrate [Lymn and Taylor (1971)].

1972 "In the thin filaments **tropomyosin** lies on the flat surface formed between the two strands of actin. **Tropomyosin**'s length somewhat exceeds the pitch of the long actin helix so that the **tropomyosin** molecules overlap when binding to actin. The presence of **tropomyosin** and the overlap between the **tropomyosin** molecules confers cooperativity to the regulatory system (Bremel and Weber)" [Szent-Györgyi (2004)].

1974 "The kinetic analysis with S1 indicated the existence of several ATP states and several ADP states (Bagshaw and Trentham). Kinetic analysis also demonstrated that the bound ATP was in equilibrium with the bound ADP and inorganic phosphate. An equilibrium constant of -7 indicated the reversibility between the states of the bound ATP and bound ADP and Pi. Therefore, hydrolysis of the ATP does not dissipate its energy while the nucleotide is bound" [Szent-Györgyi (2004)].

Terrell L. Hill 1974

The generally accepted view, at the present time, concerning the mechanism of contraction of vertebrate striated muscle includes the following assumptions: (a) the **myosin** cross-bridges act independently ; (b) force is generated by the cross-bridges, and a cross-bridge exerts a force only when it is attached to a site on an actin filament; and (c) each cross-bridge in the overlap zone makes use of a cycle (or cycles) of biochemical states, including attachment to actin and splitting of ATP. This is an oversimplified statement, but it suffices to indicate in a general way the class of models with which we shall be concerned in the present paper. Other types of mechanisms will not be considered at all, though some may be viable possibilities. Also, the activation of contraction will not be discussed.

Statistical mechanics provides an unambiguous and unique formal procedure that should be used to connect the biochemical (kinetic) assumptions of any particular model in the above class with the mechanical (and thermal) properties that are consequences of these assumptions. In a modest way, this is analogous to the well-known general applicability of the formalism of equilibrium statistical mechanics to an arbitrary molecular model (of a gas, solid, polymer solution, etc., etc.) in order to deduce the observable thermodynamic properties implicit in the model.

The basic principles to be developed here were outlined very briefly in earlier papers. This work amounted to a generalization, using statistical mechanics, of procedures first introduced by

A. F. Huxley (1957) in the treatment of a special case [Hill (1974)].

12.1.7 *1980–1989*

Jack A. Rall 1982

In 1923–1924 Wallace O. Fenn devised experiments to determine the relationship between energy (heat plus work) liberated during isotonic muscle contraction and work performance. Fenn studied afterloaded isotonic contractions. In this type of contraction initial or rest length is first fixed and then the muscle shortens against various afterloads (a load which the muscle does not support while at rest but is subjected to as it shortens during contraction). From these experiments Fenn concluded that (1) "whenever a muscle shortens upon stimulation and does work in lifting a weight, an extra amount of energy is mobilized which does not appear in an isometric contraction." (2) Further, "the excess energy due to shortening in contraction is very nearly equal to the work done...." These conclusions constitute what has become known as the Fenn effect, i.e., energy (E) liberated during a working contraction approximately equals isometric (I) energy liberation plus work (W) done or $E \cong I + W$.

To appreciate the significance of Fenn's results it is necessary to understand the prevailing view of muscle contraction in the 1920's. The viscoelastic (or new elastic body) theory of muscle contraction could be traced back to the 1840's. The view was held that, after a stimulus, muscle acted like a stretched spring released in a viscous medium. The stimulated muscle then would liberate, in an all-or-none fashion, an amount of energy that varied with initial length and which could appear as either heat or work. The amount of potential energy that could be converted into work depended on the skill of the experimenter in arranging levers, and thus work should bear no relation to total energy liberated. This theory predicts that the amount of energy liberated in an isotonic contraction would be independent of work or load and equivalent to energy liberated in an isometric contraction. Fenn's results clearly were inconsistent with this theory. A. V. Hill states that Fenn's conclusions "were obviously the death warrant of the visco-elastic theory" [Rall (1982)].

Aside 12.1.6. Evidence of one of the very first crises in the theory of muscle contraction.

1986 "Direct demonstration of the in vitro sliding of actin filaments over lawn of **myosin** molecules attached to a cover-slip (Kron and Spudich)" [Szent-Györgyi (2004)].
1988 "Measurements of the step size and tension induced by single **myosin** molecules acting on an actin filament attached to a very thin glass needle (Kishino and Yanagida)" [Szent-Györgyi (2004)].

12.1.8   *1990–1999*

Clarence E. Schutt and Uno Lindberg 1992

The central problem of muscle contraction is to account for the mechanism of force generation between the thick and thin filaments that constitute the sarcomere. In A. F. Huxley's analysis, the proportionality of isometric tension to the overlap of thick and thin filaments, as well as the other mechanical properties of muscle, are neatly explained with so-called independent force generators that are presumed to be distributed uniformly along the overlap zone. In most models, the individual force contributions of these generators are summed up by some relatively inextensible structural element to which each generator has a single point of attachment. In the case of the classical rotating crossbridge model of force generation, thin filaments themselves serve the role of the inextensible structural elements. In this note, we propose an alternative model in which repetitive length changes in segments of actin filaments, induced by **myosin** heads, generate forces that are summed and transmitted to the Z disc by **tropomyosin**.

To qualify as an acceptable independent generator model, a mechanism of skeletal muscle contraction must provide explanations for four experimental phenomena: (i) isometric tension is proportional to the degree of overlap between thick and thin filaments; (ii) stiffness is also proportional to filament overlap; (iii) recovery of tension following a quick release scales to overlap with a time course independent of overlap; (iv) speed of shortening is independent of overlap for an isotonic contraction. In the sarcomere, the regular array of **myosin** heads projecting from thick filaments uniformly subdivides actin into independent segments, each capable of developing a constant force as it contracts. If provision is not made for an inextensible element to sum the individual force contributions, the maximal force felt at the Z line would only be as great as that produced by any single segment of the thin filament, and tension would be independent of overlap. In skeletal muscle, adjoining **tropomyosin** molecules form a continuous rope-like structure through a head-to-tail association strengthened by troponin T. We propose that **tropomyosin** is the inextensible parallel component that sums the individual forces being developed over the length of an actin filament in the overlap zone and transmits them to the Z disc.

Tension is developed by a helicalizing segment of an actin filament having two points of attachment: to **tropomyosin** on one end and to a **myosin** head on the other. Thus, contracting actin segments pull on **tropomyosin** while being anchored via cross-bridges to thick filaments. A useful analogy is a tug-of-war in which each person (ribbon segment) pulls independently on the rope (**tropomyosin**) by digging their heels into the ground (**myosin**). The tension on the rope is the sum of the individual forces [Schutt and Lindberg (1992)].

Aside 12.1.7. I count this quotation as evidence of a major schism in the muscle contraction research community. On one side are those scientists who consider all the filamentary molecules involved in muscle contraction to remain fixed in length, versus those who do not.

Clarence E. Schutt and Uno Lindberg 1993

Any theory of muscle contraction must account for the striking fact that muscle fibers shortening against a load (and thereby performing work) are able to draw from biochemical sources significantly greater amounts of free energy than equivalent isometrically-contracting fibers. Originally discovered by W. O. Fenn, these observations imply that a muscle is not a spring that converts potential energy into mechanical work, but is rather a device in which mechanical events in the contracting fiber control the rates of the biochemical reactions that provide the energy used by the fiber to perform work.

There are, in principle, many types of cross-bridge mechanisms, but we will focus on the Huxley- Simmons model because it accords so well with the requirements of the independent force generator hypothesis. The basic feature of the model is that **myosin** heads are capable of attaching to actin filaments at a succession of sites of increasing binding energy. As the head moves through these binding states, perhaps by tilting, it transmits a force proportional to the gradient of the binding energy to an elastic element situated somewhere in the cross-bridge. The stretched elastic element can then pull on the attached actin filament as elastic energy is converted into the work of moving a load. The Huxley-Simmons proposal has been analyzed in great detail. It would appear from this analysis that the Fenn effect can be explained by this model as long as a **myosin** head remains attached while the hookean element discharges its stored energy.

The recent challenge to the cross-bridge theory comes from a number of experiments indicating that the distance through which an actin filament moves per ATP molecule hydrolyzed is at least 400 $\overset{\circ}{A}$ and may be over 1,000 $\overset{\circ}{A}$. Thus, the **myosin** power-stroke is longer than twice the physical length of the head so that, even lying on its back and transforming $180°$ to a position on its stomach, it would still fall far short of delivering its punch.

The field of muscle contraction is in a state of crisis. The prevailing paradigm, the rotating cross-bridge theory, which has served so well to guide the design of experiments, seems less credible than it did just a few years ago. This crisis comes at a time when X-ray crystallography is revealing images of the force-producing molecules at atomic resolution, and in vitro reconstitution systems and genetic engineering are providing the means to test the principal tenets of the theory. We believe that the source of the difficulty is that the mechanical role of **tropomyosin**, the third filament system comprising sarcomeres, has not been properly understood, nor has the actin ATPase been appreciated as a source of Gibbs free energy for muscle fibers performing work [Schutt and Lindberg (1993)].

Aside 12.1.8. There you have it, a declaration of "crisis." And as well, specific mention of Gibbs free energy in relation to muscle contraction, see Section 10.1.9.

Guangzhou Zou and George N. Phillips, Jr. 1994

[W]e simulate the behavior of individual molecules of muscle thin filaments during their regulation from the relaxed, or "off" state, to the contracting, or "on" state. A one-dimensional array of finite automata is defined based only on local interactions. The overall regulatory behavior of muscle thin filaments emerges from the collective behavior of these finite automata. The key assumption in this paper is that the state-transition rate constant of each constituent molecule of the thin filament is a function of the states of its neighboring molecules.

The transition rules for the constituent molecules of the muscle thin filament are constructed based on an understanding of the structure of the thin filament and experimental kinetic data on the local interactions between the molecules. Our goal is to build a computational machine that can mimic muscle filaments under a wide variety of experimental conditions rather than to propose a new data-fitting technique that can simply reproduce data curves. The transition rate constants in our model are fundamentally different from simple data-fitting parameters in the following ways: (1) the transition rate constants have clear physical meaning and correspond to well defined chemical processes; (2) the transition rate constants can be independently measured, in principle, without referring to any specific model; and (3) once the values of the rate constants are determined, either by direct experiment or by comparisons of the model with certain data sets, they are not allowed to have multiple values for comparison with different experimental data [Zou and George N. Phillips (1994)].

Aside 12.1.9. An early simulation effort adopts finite automata, which places this aspect of muscle contraction research squarely in the realm of stochastic timing machinery, see Aside 6.0.1.

M. P. Slawnych, C. Y. Seow, A. F. Huxley, and L. E. Ford 1994

It is generally believed that muscle contraction is generated by cyclical interactions between myosin cross-bridges and actin thin filaments, with ATP hydrolysis providing the energy. It is also well accepted that the cross-bridge cycle is composed of at least several physical and chemical steps [Lymn and Taylor (1971)]. Whereas many experiments are designed to study just one or two

of these transitions, the cyclical nature of the actomyosin interaction often precludes such experimental isolation. The exact
interpretation of the results therefore demands some prediction
of the effect of each intervention on the entire cycle. To make
these predictions we have developed a computer program that is
capable of determining the response to any length change of a
system that can have as many transitions as are known to occur
in muscle.

A distinction must be made between a program, which we
describe here, and a model. In the context of this paper, a model
is defined to be a specific cross-bridge scheme whose transitions
are all defined. In contrast, the program is a tool for deriving the
response of a particular model [Slawnych *et al.* (1994)].

Clarence E. Schutt and Uno Lindberg 1998

The allosteric transmission of conformational changes along
actin filaments, linked to ATP exchange and hydrolysis on both
actin and **myosin**, is the motivating idea behind viewing of muscle contraction as a Markov process. Although it is intrinsically
difficult to infer from the properties of the isolated parts of a
self-coordinating piece of machinery how it works, there does exist for the case of muscle several lines of evidence for the kind
of cooperativity required for such a model [Schutt and Lindberg
(1998)].

Thomas Duke 1999

I investigate how an ensemble of motor proteins generates sliding between a single pair of filaments, under conditions in which
an external force opposes the motion, and report a striking correspondence with a number of features that are familiar from
experiments on muscle. The Fenn effect and A. V. Hill's characteristic relation between force and velocity are reproduced, and
so are the deviations from Hill's law that have been detected in
single muscle fibers.

Although apparently a dynamical process, the relative motion of a pair of rigid filaments, generated by an ensemble of
$N \sim 100$ motor proteins, may be treated as a problem of mechanical equilibrium that is continually being adjusted as chemical reactions occur. This is because the viscous relaxation time

of the system is very rapid compared with the typical time between reaction events; whenever the chemical state of one head is altered, the filaments quickly shift position to maintain the equality of the sum of the forces acting in the ensemble of cross-bridges and the (constant) load. Simulation of this stochastic process is straightforward if a record is kept of the state and the strain of each **myosin** molecule. It involves the repetition of five steps: (i) Evaluate the mean time t until the next attachment/detachment event; (ii) determine the time step by drawing a random variable from an exponential distribution with mean t; (iii) choose, with probability proportional to the respective rates of the different events, which **myosin** molecule was involved and which transition occurred; (iv) change the state of the **myosin** head accordingly and readjust the mechanical equilibrium; (v) reequilibrate the power-stroke transition of all of the bound heads. This procedure permits the calculation of the sliding velocity as a function of the force opposing the motion.

Results are the time-averaged velocity of an indefinitely long thin filament being propelled by a thick filament comprising $N = 150$ **myosin** molecules (total number of chemical reaction events $= 10^8$).

A clear transition in behavior is apparent: at low load, the motion is smooth, but at high load the thin filament advances in discrete steps. Stepwise movement is due to the synchronization of the power strokes of the **myosin** molecules. At first sight, this is surprising, because the individual molecules are undergoing chemical reactions with stochastic kinetics. Coordination arises, despite this randomness, because when one head changes chemical state, the strain that is generated is communicated by means of the rigid thin filament to all of the other attached heads, thereby regulating their biochemistry [Duke (1999)].

Aside 12.1.10. Perhaps among the earliest published details on a muscle contraction simulation algorithm, this work is also notable for its stochastic framework, its direct mention of the classical work of Fenn and Hill, and its recognition of a feedback loop that yields cooperative behavior of molecules. To me what seems to be missing (see Aside 1.10.1) is thermodynamics.

12.1.9 *2000–2010*

Clarence E. Schutt and Uno Lindberg 2000

The paradigm for chemomechanical process in biology is the "sliding filament model of muscle contraction," in which cross-bridges projecting out of the **myosin** thick filaments bind to actin thin filaments and pull them towards the center of the sarcomeres, the basic units of contraction in muscle fibers. Actin is generally thought to be an inert rodlike element in this process. The **myosin** cross-bridges bind ATP as they detach from actin and hydrolyze it in the unattached state. Upon rebinding actin, the **myosin** head 'rotates' through several binding sites on actin of successively lower energy while stretching a molecular "spring" that then pulls on the actin filament. In this manner, converting bond energy into elastic energy, it is believed that the free energy of ATP hydrolysis is transduced into work. **myosin** is often called a "motor molecule," because the macroscopic forces generated by muscle fibers could be explained as the summed effect of hundreds of **myosin** heads independently pulling on each actin filament. That situation has changed very recently. New measurements on the extensibility of actin filaments, and reconsideration of the thermodynamics of muscle have cast doubts on the validity of the conventional cross-bridge theory of contraction. Attention is being increasingly focused on models that take into account the overall spatial and temporal organization in muscle lattices, and the possibility of cooperativity amongst the **myosin** motors [Schutt and Lindberg (2000)].

Josh E. Baker and David D. Thomas 2000

A. V. Hill treats muscle as a conventional chemical thermodynamic system that is externally held at a constant force. In such systems, forces equilibrate among the molecules that make up the system and chemical potentials are coupled to the ensemble force. In contrast, A. F. Huxley treats individual **myosin** cross-bridges as independent thermodynamic systems that are internally held at constant forces by a rigid muscle lattice. Huxley's description of muscle as a collection of mechanically isolated thermodynamic systems is often referred to as the independent force generator model.

The models of A. V. Hill and A. F. Huxley are two mutually exclusive and fundamentally different physical descriptions of mechanochemical coupling in muscle. Which model most accurately describes mechanochemical coupling in muscle depends on whether individual cross-bridges in muscle are mechanically coupled to each other through compliant muscle structures (A. V. Hill) or mechanically uncoupled from each other by rigid muscle structures (A. F. Huxley). While most molecular models of muscle are based on A. F. Huxley's independent force generator model, recent measurements of compliance in muscle filaments imply that **myosin** cross-bridges are mechanically coupled to each other through these filaments. Moreover, recent spectroscopic/mechanical studies provide biochemical support for and imply a molecular basis for A. V. Hill's thermodynamic model of muscle mechanochemistry.

While individual **myosin** heads generate force and motion in muscle, they do so as an ensemble of motors among which forces equilibrate, not as a collection of independent macroscopic machines that are mechanically isolated from each other. These observations are consistent with stochastic molecular approaches to modeling muscle, and they suggest a conventional biochemical thermodynamic approach (e.g. the Nernst equation) to modeling muscle in which chemical potentials are defined in terms of state variables of the chemical system, not state variables of the molecules in that system. In essence, these results suggest a chemical basis for A. V. Hill's muscle equations.

While the working stroke of an individual cross-bridge generates force and motility, it also performs internal work on other **myosin** heads, cooperatively biasing net work production by the reversible working strokes of these heads [Baker and Thomas (2000)].

Aside 12.1.11. I am particularly keen on this work which takes sides on controversy in muscle contraction research regarding the "independent force generator model" of Andrew F. Huxley. Emphasis on thermodynamics and Archibald V. Hill's model, with special regard to cooperativity in my mind is a complement – and compliment – to the work of Thomas Duke.

Joanna Moraczewska 2002

The presence of **tropomyosin** on the thin filament is both necessary and sufficient for cooperativity to occur.

Most probably the main function for **tropomyosin**'s ends is to confer high actin affinity. Since truncation lowered **tropomyosin** affinity without eliminating cooperativity with **myosin** $S1$, the source of the cooperativity must be a conformational change in actin that occurs upon $S1$ binding and is propagated to the neighboring actin molecules. This manifests itself by cooperative binding of **tropomyosin**. This idea is supported by the observation that binding of truncated **tropomyosin**s to actin fully decorated with **myosin** heads was non-cooperative. Once **myosin** changed actin structure, **tropomyosin** bound tightly and non-cooperatively [Moraczewska (2002)].

Josh E. Baker 2003

To the Editor: The data and analysis presented by Karatzaferi *et al.* support a new paradigm for muscle contraction, which if correct demands a fundamental reassessment of decades' worth of muscle mechanics studies. At issue is how mechanics and chemistry are coupled in muscle.

The conventional model of the past 40 years has mechanics and chemistry coupled within individual cross-bridges, with ATP, ADP, and Pi concentrations formally expressed as functions of a mechanical parameter (the molecular strain, x) of an isolated cross-bridge. In contrast, by expressing $[ADP]$ as a function of a mechanical parameter of the muscle system (the macroscopic muscle force, $PSL\text{-}ADP$), Eq. (1) in Karatzaferi *et al.* implicitly couples mechanics and chemistry at the level of an ensemble of cross-bridges. Equations of this form have been legitimized only within the context of a thermodynamic muscle model: a model originally developed to account for the first direct measurements of mechanochemical coupling in muscle. By demonstrating that Eq. (1) accurately describes their data, where conventional models fail, Karatzaferi *et al.* provide additional experimental support for a thermodynamic muscle model, not, as stated, "a molecular explanation for [it]". The conventional model of mechanochemical coupling uses rational mechanics to describe muscle force as a sum of well-defined **myosin** cross-bridge forces. In contrast, a thermodynamic model describes muscle force as an emergent property of a dynamic actin-**myosin** network, within which the force of a given cross-bridge stochastically fluctuates due to force-generating transitions of neighboring cross-bridges that are transmitted through compliant linkages. The "molecular explanation" for a thermodynamic muscle model is that, through these intermolecular interactions, the mechanics and chemistry of a given cross-bridge are mixed up with the mechanics and chemistry of its neighbors.

The above competing descriptions of muscle force (molecular reductionist vs. thermodynamic) are mutually exclusive. As Gibbs points out, "If we wish to find in rational mechanics an a priori foundation for the principles of thermodynamics, we must seek mechanical definitions of temperature and entropy". Thus the thermodynamic muscle model supported by Karatzaferi *et al.* represents a fundamental shift in our understanding of muscle mechanics. In essence, if this model is correct, then one must conclude that the successes of conventional muscle models are superficial, resulting from strain (x)-dependent rate constants that were artificially tuned to make individual **myosin** cross-bridges mimic the emergent properties (P) of dynamic actin-**myosin** networks in muscle. Although it remains to be determined which model most accurately describes muscle mechanics, growing support for a thermodynamic model of muscle from studies like that presented in Karatzaferi *et al.* suggests that this paradigm shift and its profound implications for our understanding of muscle contraction warrant careful consideration [Baker (2003)].

Aside 12.1.12. The emphasis on thermodynamics in the work of Josh E. Baker calls for a careful understanding of chemical thermodynamics and energy transduction. My Theory of Substances is supposed to be careful (see Aside 8.1.1).

Andrew G. Szent-Györgyi 2004

Since antiquity, motion has been looked upon as the index of life. The organ of motion is muscle. Our present understanding of the mechanism of contraction is based on three fundamental discoveries, all arising from studies on striated muscle. The modern era began with the demonstration that contraction is the result of the interaction of two proteins, actin and **myosin** with ATP, and that contraction can be reproduced in vitro with purified proteins. The second fundamental advance was the sliding filament theory, which established that shortening and power production are the result of interactions between actin and **myosin** filaments, each containing several hundreds of molecules and that this interaction proceeds by sliding without any change in filament lengths. Third, the atomic structures arising from the crystallization of actin and **myosin** now allow one to search for the changes in molecular structure that account for force production [Szent-Györgyi (2004)].

Gerald H. Pollack, Felix A. Blyakhman, Xiumei Liu, and Ekatarina Nagornyak 2004

Fifty years have passed since the monumental discovery that muscle contraction results from relative sliding between the thick filaments, consisting mainly of **myosin**, and the thin filaments, consisting mainly of actin [Huxley and Niedergerke (1954)][Huxley and Hanson (1954)]. Until the early 1970's, considerable progress have been achieved in the research field of muscle contraction. For example, A. F. Huxley and his coworkers put forward a contraction model, in which the myofilament sliding is caused by alternate formation and breaking of cross-links between the cross-bridges on the thick filament and the sites on the thin filament, while biochemical studies on acto**myosin** *ATPase* reactions indicated that, in solution, actin and **myosin** also repeat attachment-detachment cycles. Thus, when a Cold Spring Harbor Symposium on the Mechanism of Muscle Contraction was held in 1972, most participants felt that the molecular mechanism of muscle contraction would soon be clarified, at least in principle.

Contrary to the above "optimistic" expectation, however, we cannot yet give a clear answer to the question, "what makes the filaments slide?"

This paper has a dual goal. First it outlines the methods that have evolved to track the time course of sarcomere length with increasingly high precision. Serious attempts at this began roughly at the time the sliding filament theory was introduced in the mid-1950s, and have progressed to the point where resolution has reached the nanometer level. Second, and within the context of these developments, it considers one of the more controversial aspects of these developments: stepwise shortening.

True dynamic measurements based on optical diffraction were realized with the advent of linear photodiode arrays. Measurements using this approach were made in both cardiac and skeletal muscle.

The results of such experiments showed that shortening occurred in steps. That is, the shortening waveform was punctuated by a series of brief periods in which shortening was zero or almost zero, conferring a staircase-like pattern on the waveform.

In sum, stepwise shortening was observed in many laboratories using a variety of methods. Perhaps because of the unexpected level of synchrony implied by the results, skepticism was appreciable, and various papers along the way criticized certain aspects the earliest results – all of which inspired responses and additional controls. To the knowledge of these authors, the latter results have not been criticized; nor have the confirmations in other laboratories.

Maughan: "Is there enough flexibility in the thin (actin) filament to accommodate your model?"

Pollack: "The model does not rely on actin-filament flexibility. It is based on propagated local shortening along the actin filament. There is appreciable evidence for the actin filament shortening." [Pollack *et al.* (2005)]

Aside 12.1.13. This work comes down on the side of those researchers who hold that a major filament involved in muscle does indeed vary in length.

H. J. Woo and Christopher L. Moss 2005

The Brownian ratchet model, first described by Feynman as a demonstration of the inevitable reversibility of any macroscopic movements driven by equilibrium fluctuations, has provided conceptual guidelines of how free energy transductions could become possible far from equilibrium. A prototypical Brownian ratchet has external controls switching the potential of mean force that the Brownian particle feels along a reaction coordinate x (the

displacement of a motor head on the linear track) between a flat
profile and an asymmetric profile, resulting in the nonzero average
net flux of the particle.

A largely unresolved question, however, is how the key mechanisms of such theoretical models are implemented in reality within
their protein constituents. We argue that attentions to structural details should help resolve some of the issues that have
arisen within the studies of motor proteins. One of such issues is
whether the Brownian ratchet models should be regarded as representing an alternative mechanism superseding the traditional
power stroke description exemplified by the swinging lever arm
model of muscle contractions.

Within the unified point of view based on the general PMF
$G_i(x, y)$ and its one-dimensional projections, therefore, the distinction between the two seemingly different perspectives becomes a quantitative one. The difference in particular centers on
the extent to which the stochastic dynamics of the motor head
reaction coordinate, largely on $G_2(y)$, is affected by the thermal
diffusion or concerted relaxations toward the minimum in free energy. The question needs to be addressed based on considerations
of molecular structures, for example, by calculating the diffusion
coefficients of the motor head on the PMF, which can be obtained
using molecular simulation techniques. In this paper, we adopt
the version of description using the free energy landscape of the
conformational reaction coordinate y. The choice is appropriate
for nonprocessive motor protein systems such as **actomyosins**,
where a typical motor complex is expected to undergo more concerted movements on average in its conformational space than in
its positional displacement [Woo and Moss (2005)].

Ganhui Lan and Sean X. Sun 2005

We will show that there are important collective effects in
skeletal muscle dynamics. The geometrical organization of the
sarcomere and the kinematics of the constitutive parts play an
important role. We will provide an explanation for the observed
synchrony in muscle contraction and show how an increasing load
force leads to an increasing number of **myosins** working on actin.
We show that a force-dependent ADP release step can explain the
dynamics of skeletal muscle contraction.

At low load conditions, if there are many bound working heads, they must mechanically oppose each other. Thus, synchrony must exist among the motors and the number of actin-bound motors must change as a function of the external load. Electron micrographs of muscle under tension show increasing order in the cross-bridge arrangement as a function of load force. These measurements are consistent with the notion of synchrony among the motors. In our model, we show how synchrony is achieved during muscle contraction.

The work of Duke established the basic framework of understanding muscle contraction. Duke's model is also based on the swinging cross-bridge mechanism of Huxley and Simmons, which now is widely accepted as the basic explanation of the role of the **myosin** in muscle contraction. The model presented in this article builds upon Duke's and Huxley's earlier works. We show how the thin filament movement is connected with the conformational change in the **myosin** motors. Duke's work treated the chemical rate constants as fitting parameters. He contends that synchrony in muscle contraction is due to a slow phosphate (Pi) release step. Forces from other **myosin** motors can assist Pi release. Biochemical studies, however, suggest that Pi release is rapid. In our current work, experimentally measured chemical rate constants are used. Realistic geometrical arrangement of the mechanical elements in the sarcomere is included. Thus, the number of unknown parameters is limited to the mechanical constants of the **myosin** motor during its chemical cycle and the elastic modulus of the stalk protruding out of the thick filament.

There are ~ 150 **myosin** motors interacting with the hexagonal thin filaments. When an external load force is applied to the Z-disk, all six actin filaments are under the same amount of mechanical tension. If we simplify the problem and assume that the Z-disk can only move in the x direction, then it is equivalent to model 150 **myosin** motors interacting with a single actin filament. If the Z-disk is always held perpendicular to the x axis by other tissue, then the current assumption is a valid one.

The dynamics of molecular motors can be described by the coupled Langevin equations

$$\zeta \, \dot{\xi} = -\frac{\partial E(\zeta, \overrightarrow{s})}{\partial \xi} - F + f_B(t)$$

$$\frac{\partial \overrightarrow{s}}{\partial t} = \mathbf{K} \cdot s \, ,$$

where ξ is a dynamical observable of interest. The value ζ is the friction due to the surrounding medium. The value F is an external load force and $f_B(t)$ is the Brownian random force obeying the fluctuation dissipation theorem. The value $\overrightarrow{s}$ is the chemical state of the molecular motor and $E(\xi, \overrightarrow{s})$ is the elastic energy of the motor as a function of the dynamical observable and the chemical state. $\mathbf{K}$ is a matrix of kinetic transition rates describing the chemical reactions in the motor catalytic site. $\mathbf{K}$ is, in principle, a function of ξ also.

A single **myosin** motor binds and hydrolyzes ATP to generate force. The binding and hydrolysis is also coupled to **myosin**'s affinity for actin. Changes in the chemical state are coupled to conformational changes in the **myosin** motor domain.

Using 150 **myosin** motors, we have computed the force versus velocity curve for muscle contraction. The computational results are slightly different from the experimental data at large load forces. We argue that this is not surprising. In the experimental situation, the contraction is not only due to the **myosin** motors working along actin, but also due to the contraction of the passive force generator, titin. Thus a fraction of the applied force is balanced by titin, and the force along the actin filament is lower than the total applied force. Titin is also a nonlinear elastic object. At high load forces, the resorting force generated by titin can be quite substantial. Independent measurements of titin elasticity suggest that titin is responsible for $\sim 20\%$ of the contractile force. Thus, our computational results are consistent with experimental measurements.

We plot the average number of working heads as a function of F. We see that the number of working heads is very low when the load force is small. As the load force increases, the number of working heads increases gradually. When the load force is small, the rate-limiting step is actin binding. After a

myosin head is bound, it quickly releases Pi and makes a power stroke to reach the equilibrium conformation of the A.M.D state. At this equilibrium conformation, the ADP release rate is quick and the kinetic cycle proceeds without hindrance. If the load force is high, then the **myosin** head cannot complete its power stroke. The conformation is stuck in the ADP state before the equilibrium value. At this position, ADP release is slow and rate-limiting. Thus, the kinetic cycle is stopped until another **myosin** head binds to actin and makes a power stroke. If there are enough heads bound, the collective power stroke can overcome the load force and reach the equilibrium conformation. Thus, the con- formation-dependent ADP release step is the explanation for synchrony in muscle contraction [Lan and Sun (2005)].

Aside 12.1.14. In addition to the clear model of chemical to mechanical energy transduction in this muscle contraction simulation work, and the reference to the research of Thomas Duke , here is a direct reference to Langevin Equations (see Eq. (5.67) and Section B.2).

Leslie Chin, Pengtao Yue, James J. Feng, and Chun Y. Seow 2006

To further our understanding of muscle contraction at the molecular level, increasingly complex models are being used to explain the force-velocity relationship and how it can be changed under different conditions. In this study, we have developed a seven-state model to specifically address the question of how ATP and its metabolites alter the transitions within the cross-bridge cycle and hence modify the characteristics of the force-velocity relationship. In this study, we also investigated the biphasic behavior of velocity in muscles shortening at near-maximal load and explained the behavior in terms of velocity-dependence of transition rates. Another major component of this study is the development of a rapid computational method for obtaining exact solutions of simultaneous equations from a cyclic matrix of any size. This new tool is particularly suited for analyzing cyclic interactions or reactions, such as those found in the muscle cross-bridge cycle. One major advantage of this tool is that it allows investigators to formulate the cross-bridge cycle with virtually unlimited number of states and monitor the flux of cross-bridges in and out of each state in real-time, because of its high computational efficiency. Another advantage of this method compared to conventional numerical methods is that it does not have a nonconvergence problem where numerical iterations do not lead to a solution, and can handle stiff matrices (matrices with widely varying rate constants) that need to be used in the cross-bridge cycle simulation [Chin *et al.* (2006)].

Aside 12.1.15. This work introduces another muscle contraction simulation. It is not clear to me how chemical thermodynamics is reflected in the underlying theory.

Jeffrey W. Holmes 2006

A. V. Hill's paper "The heat of shortening and the dynamic constants of muscle" is a wonderful classic from a bygone era, 60 pages of detailed methods, experiments, and modeling representing years of work. In the first of three sections, Hill outlines the design and construction of his experimental system, with detailed circuit diagrams, the complete equations for a Wheatstone bridge amplifier, instructions on how to build a thermopile, and more. The second section presents the results of a series of experiments on mechanics and heat production in frog skeletal muscle. The final section presents the classic two-component Hill model with a contractile and an elastic element in series, develops the appropriate equations, and shows that this model explains many of the key experimental observations presented earlier in the paper.

MATLAB simulation is provided in the APPENDIX [Holmes (2006)].

Andrew D. McCulloch and Won C. Bae 2006

When a series of stimuli is given, isometric force rises to a plateau (unfused tetanus) which ripples at the stimulus frequency. As stimulus frequency is increased, the plateau rises and becomes a smooth fused tetanus.+

While Hill's equation was initially based on an incorrect thermodynamic derivation, it has been determined to be empirically accurate.

Fundamental Assumptions:

(1) Resting length-tension relation is governed by an elastic element in parallel with a contractile element. In other words,

active and passive tensions add. The **elastic element** is the passive properties.

(2) **Active contractile element** is determined by length-tension and velocity-tension relationships only.

(3) **Series elastic element** explains the difference between the twitch and tetanic properties.

Limitations of Hill Model Division of forces between parallel and series elements and division of displacements between contractile and elastic elements is arbitrary (i.e., division is not unique). Structural elements cannot be identified for each component.

Hill model is only valid for steady-shortening of tetanized muscle.

(1) For a twitch we must include the time-course of activation and hence define "active state";

(2) Transient responses observed not reproduced [McCulloch and Bae (2006)].

Scott L. Hooper, Kevin H. Hobbs, and Jeffrey B. Thuma 2008

All muscles contain thin filaments and thick filaments. Muscle thin filaments (diameter 6–10 nm) are a double helix of polymerized actin monomers, and have, with minor variation, a common structure across Animalia. The double helix repeats once every 28 monomers if the monomers from both strands are counted. Due to the helical nature of the filament, the molecule repeats every 14 monomers if the distinction between strands is ignored. Two important thin filament associated proteins in striated muscle are the globular protein troponin and the filamentous protein **tropomyosin**. Two troponin complexes (one for each helix) bind once every 14 monomers. **Tropomyosin** twists with the double helix and sterically blocks the **myosin** binding sites at rest but moves away from them in the presence of Ca^{++}.

Muscle thick filaments are composed of **myosin**. **myosin** is composed of three pairs of molecules, the heavy chain, the essential light chain, and the regulatory chain. The tails of the heavy chains form a coiled-coil tail and the other end of each heavy chain and one essential and one regulatory chain form one of the combined molecule's two globular heads (which engage the actin filament to produce force) The extended tails bind together to form the thick filaments.

In both types of filaments **myosin** heads possess an ATPase activity, and can bind to sites on the actin thin filaments. In their unbound state ADP-Pi is bound to the heads. The heads are in considerable disorder, but generally lie at obtuse angles relative to the **myosin** tail. Initial binding of the **myosin** head to the thin filament is weak with the head having a $\sim 45°$ angle relative to the thin filament long axis. As binding proceeds, the portion of the **myosin** heavy chain engaged with the actin retains its position and shape, but the region closer to the thick filament rotates toward the Z-line, which produces an M-line directed force on the thin filament. Force is thus not generated by rotation of the entire **myosin** head, but instead in a lever-like manner in which rotation of a more distant portion of the head uses the actin-binding portion to transfer force to the thin filament. At the end of the stroke the lever arm has a $\sim 135°$ angle relative to the thin filament long axis and points to the Z-line.

If this were the end of the process, the muscle would not contract further, and, furthermore, would become rigid, since the tight binding of the **myosin** head to the actin, and the inability of the **myosin** head to rotate back to its original angle, would lock the thin and thick filaments into a unyielding conformation (this is the basis of rigor mortis). However, **myosin** head rotation is accompanied by Pi unbinding and then ADP unbinding. ATP can then bind to the **myosin** head, which causes the head to detach from the thin filament. The ATP is then dephosphorylated, at which time it can again bind the thin filament. This cross-bridge cycling is the fundamental mechanism for generating force in all muscles.

The molecular basis of force generation (a lever arm magnifying relatively small changes at its base) is similar in invertebrates and vertebrates. Less well understood is how the cross-bridges function as a collective [Hooper *et al.* (2008)].

Aside 12.1.16. Of all the open questions about muscle contraction, that of "cooperativity" or "synchrony" or "collectivity" is most interesting to me because it is potentially a fundamental illustration of the phenomenon of *emergence* (see Aside 6.8.1).

Haruo Sugi, Hiroki Minoda, Yuhri Inayoshi, Fumiaki Yumoto, Takuya Miyakawa, Yumiko Miyauchi, Masaru Tanokura, Tsuyoshi Akimoto, Takakzu Kobayashi, Shigeru Chaen, and Seiryo Sugiura 2008

Although 50 years have passed since the monumental discovery that muscle contraction results from relative sliding between **myosin** and actin filaments produced by **myosin** cross-bridges, some significant questions remain to be answered concerning the coupling of cross-bridge movement to ATP hydrolysis. In the cross-bridge model of muscle contraction, globular **myosin** heads, i.e., the cross-bridges extending from the thick filament, first attach to actin in the thin filament, change their structure to produce relative myofilament sliding (cross-bridge power stroke), and then detach from actin.

Recent crystallographic, electron microscopic, and X-ray diffraction studies suggest that the distal part of the cross-bridge (M; called the catalytic domain because it contains a nucleotide binding site) is rigidly attached to the thin filament, while its proximal part acting as a lever (the lever arm region) is hinged to M; the lever arm movement around the hinge produces the power stroke. Based on biochemical studies of acto**myosin** $ATPase$, it is generally believed that ATP reacts rapidly with M to form a complex $M - ADP - Pi$, and this complex attaches to actin (A) to exert a power stroke, associated with release of Pi and ADP. After the end of the power stroke,M is detached from A upon binding of next ATP with M. After detachment from A, M performs a recovery stroke, associated with the formation of $M - ADP - Pi$. Can a change in lever arm orientation on the thick filaments take place in the absence of actin? Some of the evidence for the lever arm hypothesis is based on experiments with **myosin** fragments, not connected to the thick filament, and additional mechanisms may be involved in vivo.

The most direct way to study cross-bridge power and recovery strokes is to record the ATP-induced movement of individual cross-bridges in the thick filaments, using the hydration chamber (HC; or gas environmental chamber), with which biological macromolecules can be kept wet to retain their physiological function in an electron microscope with sufficiently high magnifications. With this method, Sugi *et al.* succeeded in recording ATP-

induced cross-bridge movement in the **myosin–paramyosin** core complex, although the results were preliminary and bore no direct relation to the cross-bridge movement in vertebrate skeletal muscle. In the present study, we attempted to measure the ATP induced cross-bridge movement in vertebrate thick filaments muscle with the HC and succeeded in recording images of the thick filaments, with gold position markers attached to the cross-bridges, before and after application of ATP. Here, we report that, in response to ATP, individual cross-bridges move for a distance (peak at 5–7.5 nm), and at both sides of the filament bare region, across which cross-bridges polarity is reversed, the cross-bridges are observed to move away from, but not toward, the bare region. After exhaustion of ATP, the cross-bridges returned toward their initial position, indicating reversibility of their ATP-induced movement. Because the present experiments were made in the absence of the thin filament, our work constitute a direct demonstration of the cross-bridge recovery stroke in vertebrate muscle thick filaments [Sugi *et al.* (2008)].

12.2 Conclusion

Muscle contraction is not completely understood but a very great deal has been mastered, even at the molecular level. Open questions about the relationship of entropy to motion, about the extensibility – or not – of basic building blocks of muscle, about the relationships between time, force, and velocity of muscle contraction, and how to simulate its salient features continue to be explored. A variety of pertinent mathematical and computational technologies surveyed in this book are intended to attract the interest of high school students and their teachers to the scientific study of muscle contraction.

PART 7
Appendices

Appendix A

Exponential & Logarithm Functions

Theorem A.1 ([R. L. Finney and Giordano (2003)]). *The formula*

$$\ln x = \int_1^x \frac{1}{t} dt$$

defines a function $\ln : (0, \infty) \to \mathbb{R}$.

Theorem A.2. *The following are true:*

(1)
$$\frac{d \ln x}{dx}(x) = \frac{1}{x};$$

(2) $\ln ax = \ln a + \ln x;$

(3) $\ln$ *is strictly monotonically increasing;*

(4) $\ln$ *has a strictly monotonically increasing inverse function.*

Definition A.3. Let $E : \mathbb{R} \to (0, \infty)$ denote the inverse function of $\ln$.

Theorem A.4. *The following are true:*

(1) $E(0) = 1$ *and*

$$\frac{d E(x)}{dx}(x) = E(x);$$

(2)

$$E(1) = \lim_{x \in \mathbb{R}, 0 \leftarrow x} (1 + x)^{1/x} = \lim_{n \in \mathbb{N}, n \to \infty} \left(1 + \frac{1}{n}\right)^n;$$

(3)

$$\lim_{x \in \mathbb{R}, x \to \infty} E(x) = \infty$$

329

and

$$\lim_{x \in \mathbb{R}, -\infty \leftarrow x} E(x) = 0.$$

Definition A.5. Let e:=E(1).

Theorem A.6 ([Herstein (1964)]). *e is transcendental.*

Theorem A.7 ([Ahlfors (1966)][Rudin (1966)]). *If* $F : \mathbb{R} \to (0, \infty)$ *and both* $F(0) = 1$ *and* $\dfrac{d\,F(x)}{dx}(x) = F(x)$, *then the following are true:*

(1) $F = E$;
(2)

$$F(z) = \sum_{n=0}^{\infty} \frac{z^n}{n!}$$

   converges for all $z \in \mathbb{C}$;
(3) $F(x + y) = F(x) * F(y)$;
(4) *there exists* $\pi \in \mathbb{R}, \pi > 0$ *such that* $F\left(\dfrac{\pi i}{2}\right) = i$ *and* $F(z) = 1$ *if and only if* $\dfrac{z}{2\pi i} \in \mathbb{Z}$;
(5) $F(z)$ *is periodic with period* $2\pi i$;
(6) $Dom\, F(it) = \mathbb{R}$ *and* $Ran\, F(it) = S^1$;
(7) *if* $w \in \mathbb{C}, w \neq 0$ *then there exists* $z \in \mathbb{C}$ *such that* $w = F(z)$.

Aside A.0.1. Another take on this material with further information is covered in [Cooper (1966)](no relation).

Appendix B

Recursive Definition of Stochastic Timing Machinery

Stochastic timing machinery relates to differential equations in at least two ways. This Appendix briefly reviews the idea of recursion starting with initial conditions for defining approximations to ordinary and stochastic differential equations. Then there is an analogous recursion for realizing behavior of stochastic timing machines. The main point is that whereas the recursions for approximating differential equations proceed serially by adding variable increments to one or more variables as time progresses, the algorithm for interpreting timing machines is intrinsically parallel.

B.1 Ordinary Differential Equation: Initial Value Problem

Algorithm B.8. Given $f : \mathbb{R}^N \to \mathbb{R}^N$ and $X_{\text{init}} : 1 \to \mathbb{R}^N$ to find an approximation to the solution $X : \mathbb{R} \to \mathbb{R}^N$ of the initial value problem

$$\frac{dX}{dt} = f(X)$$

$$X(0) = X_{init}$$

choose $\Delta t > 0$ and define (t_n, X_n) for $n \geq 0$ by

$$t_0 = 0$$

$$X_0 = X_{\text{init}}$$

$$t_{n+1} = t_n + \Delta t$$

$$X_{n+1} = X_n + f(X_n)\Delta t.$$

Remark B.9. This is the Euler Method of Forward Integration and is the simplest possible Method of solution, and is based directly on the definition of differentiation. Since each successive value X_{n+1}, which is an approximation to $X(t_{n+1})$, is obviously based on the previous value, this recursion cannot be implemented by executing the iterations of a computer programming loop in parallel.

B.2   Stochastic Differential Equation: A Langevin Equation without Inertia

Algorithm B.10. Given $\phi : \mathbb{T} \times \mathbb{R} \to \mathbb{R}$ where $\phi(t, X)$ is the potential of a (conservative) force at time t and location X, together with $X_{\text{init}} : 1 \to \mathbb{R}$ and positive constants $\zeta, D \in \mathbb{R}$, to find an approximation to the solution X of the Langevin Equation without Inertia ("overdamped" Langevin Equation [Papoulis (1965)] [Purcell (1977)] [Astumian and Bier (1996)] [Keller and Bustamente (2000)] [Bustamente *et al.* (2001)] [Reimann (2001)] [C.P. Fall and Tyson (2002)] [Wang *et al.* (2003)])

$$\zeta \frac{dX}{dt} = -\frac{d\phi}{dx} + \widetilde{g},$$

(where $\widetilde{g}$ is a fluctuation process (with covariance $2k_BT\zeta$ ([C.P. Fall and Tyson (2002)], p.346)), choose $\Delta t > 0$ and define (t_n, X_n) for $n \geq 0$ by

$$t_0 = 0$$
$$X_0 = X_{\text{init}}$$
$$t_{n+1} = t_n + \Delta t$$
$$X_{n+1} = X_n - \zeta \frac{d\phi}{dx}(t_n, X_n)\Delta t + \sqrt{2D\Delta t}\widetilde{G},$$

where $\widetilde{G}$ is a Gaussian random variable of mean 0 and variance 1.

Remark B.11. Essentially the same can be said for this stochastic problem as was said for the deterministic problem (Remark B.9).

Remark B.12. See Sections 5.8, 5.10 below for further discussion of the Langevin Equation.

B.3 Gillespie Exact Stochastic Simulation: Chemical Master Equation

Algorithm B.13. Let $a : \mathbb{R}^N \to \mathbb{R}^M$ and $v : [1 \;\cdots\; M] \to \mathbb{R}^N$. The "propensity function" a determines $a_j(X)dt$ which is "the probability, given [concentrations of chemical species at time t] x, that [chemical reaction] j will occur somewhere ... in the next infinitesimal time interval $[t, t + dt)$," and N represents the number of chemical species undergoing M chemical reactions which alter the concentrations $X \in \mathbb{R}^N$ according to the rule $X \mapsto X + v(j)$ for reaction $j = 1, \ldots, M$, and given $X_{\text{init}} : 1 \to \mathbb{R}^N$, to find an approximation to the solution of the Chemical Master Equation [Gillespie (1977)] [Gillespie (2007)], let (Ω, P) be probability space of "runs," let $X(t) : \Omega \to \mathbb{R}^N$ be the random variable of chemical concentrations, and let $N_j(t) : \Omega \to \mathbb{N}$ be the total number of occurrences of reaction j before time t. Then by definition there is an equality of events

$$[X_j(t + dt) - X_j(t) = v_j] = [N_j(t + dt) - N_j(t) = 1],$$

and the Chemical Master Equation may be derived:

$$\frac{d}{dt} P[X(t) = X \mid X(0) = X_{\text{init}}]$$

$$= \sum_{j=1}^{M} (P[X(t) = X - v_j \mid X(0) = X_{\text{init}}] a_j(X - v_j)$$

$$- P[X(t) = X \mid X(0) = X_{\text{init}}] a_j(X)).$$

An approximation to a run satisfying this equation is obtained by

$$t_0 = 0$$

$$X_0 = X_{\text{init}}$$

$$t_{n+1} = t_n + \frac{1}{\sum a_j(X_n)} \ln \frac{1}{\widetilde{U}}$$

$$X_{n+1} = X_n + v_{\widetilde{J}(X_n)}$$

where $\widetilde{U}$ is a uniformly distributed random variable on $[0, 1]$ and for any $X \in \mathbb{R}^N$ the random variable $\widetilde{J}(X) : \Omega \to [1, \ldots, M]$ satisfies

$$P[\widetilde{J} = j] = \frac{a_j(X)}{\sum a_j(X)}.$$

B.4   Stochastic Timing Machine: Abstract Theory

Let $P := (0, \infty) \subset \mathbb{R}$ denote the positive real numbers, and define the set of possible **remainders** $R := \{ \perp, 0 \} \cup P$. Let X be a non-void finite set of **states** and let V denote the set of real-valued variables. A map $\rho : X \to R$ called a **remainder map** assigns to a state $x \in X$ either $\rho(x) = \perp$ in which case x is called **inactive** in ρ, or $\rho(x) = 0$ so x is **timed out**, or $\rho(x) > 0$ and x is called **active with remainder** $\rho(x)$. A map $v : V \to \mathbb{R}$ assigns values to variables. Assume given a single element of structure called the **signal map**

$$2^X \times \mathbb{R}^V \xrightarrow{\sigma} R^X \times \mathbb{R}^V,$$

so that given a set $A \subset X$ of states and values for variables $v : V \to \mathbb{R}$,

$$\sigma(A, v) = (\sigma_X(A, v), \sigma_V(A, v)))$$

where $\sigma_X(A, v)$ is the signalled remainder as a function of A and v, and $\sigma_V(A, v)$ is the signalled new values for the variables as a function of A and v. Define the **minimizer** functional $m : R^X \to P$ by

$$m(\rho) := \min\{ \rho(x) \mid x \in X \ \& \ \rho(x) > 0 \},$$

and define the **decrement** function $z : P \times R^X \to R^X$ by

$$z(m, \rho)(x) := \begin{cases} \perp \text{ if } \rho(\mathrm{x}) = \perp \text{ or } \rho(\mathrm{x}) = 0 \text{ or } \rho(\mathrm{x}) < \mathrm{m} \ ; \\ \rho(x) - m \ \text{ otherwise } . \end{cases}$$

Algorithm B.14. Given $\rho_{\text{init}} : X \to R$ and $v_{\text{init}} : V \to \mathbb{R}$ as initial remainder and values, define (t_n, ρ_n, v_n) for $n \geq 0$ by

$$t_0 = 0$$

$$\rho_0 = \rho_{\text{init}}$$

$$v_0 = v_{\text{init}}$$

$$t_{n+1} = t_n + m(\rho_n) \tag{B.1}$$

$$\rho_{n+1} = z(m(\rho_n), \sigma_X(\rho_n^{-1}(0), v_n)) \tag{B.2}$$

$$v_{n+1} = \sigma_V(\rho_N^{-1}(0), v_n). \tag{B.3}$$

Remark B.15. Equation (B.1) defines the advance of time as the least positive remainder of all active states. Equation (B.2) applies the decrement function to the minimum time and the signals of the timed out states. Finally, Eq. (B.3) assigns new values to the variables depending on the timed out states and the previous values of the variables.

Appendix C

MATLAB Code

C.1 Stochastic Timing Machine Interpreter

The MATLAB function file **tmint.m** is the stochastic timing machine interpreter. It invokes the **minimm.m** function and the **zerdec.m** function. There are four input parameters for **tmint.m**:

dotdur Each dot is assigned a default duration, but the value may be altered by signals from dots that time out.

dotrem At any time during a simulation each dot is either inactive or active. If active it has a positive amount of time remaining before it times out. When first activated the dotrem of a dot is set equal to its dotdur. As the simulation progresses the dotrem of a dot is diminished by the minimum of all the dotrems of all active dots. That minimum is calculated by **minimm.m** (a parallelizable function). When the dotrem of a dot is 0 that is the moment it times out, and so its signals must be transmitted. A dot is de-activated by setting its dotrem equal to −1. The **zerdec.m** function performs decision-making and advancement of time.

itrcnt The total number of timeout events is declared with itrcnt, the iteration count. A timeout event is a set of dots all timing out exactly at the same time, that is, all of which have dotrem equal to 0 at the same time during the simulation. In other words, a simulation progresses by the smallest amount of time up to the next time any dot times out.

sghndl The interpreter is general purpose because it receives the MATLAB handle of the **XxxxSignal.m** function, which is specific to the model, **Xxxx**.

MATLAB C.1.1. tmint.m

```
function  tmint(dotdur,dotrem,sghndl)
% dotdur  vector of positive integers of duration per dot
% dotrem  vector of non-negative integer remaining time per dot
% sghndl  handle of signal function
% CALLS: minimm.m, zerdec.m
% USAGE: Let Xxxx denote the name of the model, e.g., "Game".
%        Define a XxxxSignal.m file for the model.
%        Define a XxxxModel.m file of the form
%     dotdur=[   ...   ];
%     dotrem=[   ...   ];
%     sghndl=@(dotdur,dotrem)(ModelSignal(dotdur,dotrem));
%     tmint(dotdur,dotrem,iterat,sghndl)
% CALL TREE:
% XxxxTry(itrcnt)           itrcnt is the number of timeout events
%     XxxxVariables         initialize global variables
%         XxxxGlobals       declare global variables
%     XxxxModel(itrcnt)     invoke model
%         XxxxSignal        define signals
%         tmint             interpret stochastic timing machine
%             minimm
%             zerdec
%     <graphics>            generate graphs from data
%----GLOBAL VARIABLES-------------------------------------------------
global ITRIDX
global ITRCNT
global TOTTIM
ITRIDX=0;
TOTTIM=0;
%----LOCAL VARIABLES--------------------------------------------------
minrem=inf;
%----PROCESS ALL DOTS------------------------------------------------
while not(max(dotrem) == -1)&&ITRIDX<ITRCNT
%While some dot is active and iterations remain,
%disp(dotrem);
        dotrem=sghndl(dotdur,dotrem);   %Send signals from timed out dots.
        minrem=minimm(dotrem);          %Calculate minimum remaining time.
        f=@(x)(zerdec(x,minrem));       %Function handle for "zerdec.m".
        dotrem=arrayfun(f,dotrem);      %Recalculate remaining times.
        ITRIDX=ITRIDX+1;                %Decrement iteration count.
        TOTTIM=TOTTIM+minrem;           %Accumulate advancement of time.
        %disp(sprintf('Total time = %d.',tottim));
end
if  ITRIDX==ITRCNT
    disp(sprintf('Iterations = %d.',ITRCNT));
else
    disp('There are no positive remaining times.');
end
end
```

MATLAB C.1.2. minimm.m

```
function m=minimm(dotrem)
% Given a vector dotrem of dot remaining times, find
% the minimum positive remaining time. This for loop
% shall become a parfor loop for parallelization
% provided an algorithm is implemented to subdivide
% the search into concurrent chunks of dotrem, and
% finishing off by taking the minimum of the results
% among the minima of the chunks.
    m=Inf;
    L=length(dotrem);
    if  L==0
        return;
    else
        for    i=1:L
                if (0<dotrem(i)&&dotrem(i)<m)
                    m=dotrem(i);
                end
        end
    end
end
```

MATLAB C.1.3. zerdec.m

```
function dotrem=zerdec(dotrem,minrem)
% Given a remaining time dotrem and a minimum remaining time
% minrem, if the remaining time is 0 this means the dot has
% just timed out and its signals have been transmitted. Hence,
% that dot should be deactivated by setting dotrem for it down
% to -1. Otherwise, if decrementing the remaining time by the
% minimum remaining time gets to 0 or below 0, then set that
% dotrem to 0 so that on the next cycle that dot times out.
% If decrementing the remaining time by the minimum remaining
% is above 0, then the new remaining time is the remaining time
% less the minimum remaining time.
if  dotrem==0||dotrem==-1
    dotrem = -1;
else
    dotrem=dotrem-minrem;
end end
```

Aside C.1.1. If there are many thousands of states then timeouts may be very frequent, which could slow down the interpreter. Instead of declaring two timeout events as simultaneous if and only if they occur at exactly the same time (within the tolerance implied by the default arithmetic precision of the computer system), we could implement a "simultaneity window" with width that adapts to the number of timeout events. Thus, the window could be widened if there are few but frequent timeout events, and conversely, made more narrow if there are many but infrequent events.

A different problem occurs if several states that timeout all signal the same target (state or variable) simultaneously. Which signal should get to the target? In MATLAB, for example, the par-for loop that implements parallelism cannot guarantee uniform probability across the "iterations" of the loop. One solution is that all states that signal the same state should all send the same set of signals, but select one with uniform probability. To prevent all but one of those signals from reaching the target, the first one that happens to send the signal according to the par-for loop algorithm would set a flag that all of those states check before sending their randomly chosen signal. This technique would guarantee uniformly random choice of a signal from among simultaneous timeouts.

C.2   MATLAB for Stochastic Timing Machinery Simulations

The basic pattern of files for creating a stochastic timing machine simulation model is as follows:

(1) Choose a good name for the model, call it "Xxxx".
(2) Create a function file **XxxxSignal.m** with a MATLAB **switch** statement that declares for each dot its behavior upon timeout. This behavior includes the possibilities of de-activating dots, activating dots, and calculating changes in variables.
(3) Create a function file **XxxxModel.m** that initializes `dotdur` and `dotrem`. These are vectors equal in length to the number of dots in the simulation. An infinite `dotrem` or `dotdur` is represented by 1e6; inactive is represented by -1. The `sghndl` parameter for the interpreter is given the value of a handle to the **XxxxSignal.m** function, with parameters `dotdur` and `dotrem`.
(4) Create a script file **XxxxGlobals.m** in which to *declare* any global variables, and a separate script file **XxxxVariables.m** in which to *initialize* the global variables in **XxxxGlobals.m**. Thus, any function of the simulation need only invoke **XxxxGlobals.m** to gain access to the global variables.
(5) Create a function file **XxxxTry.m** to invoke **XxxxVariables.m**, maybe do some bookkeeping on global variables, then invoke

XxxxModel.m followed by graphing routines that present data in global variables.

MATLAB code for the Bouncing Particle in a Force Field are provided below.

C.3 Brownian Particle in Force Field

MATLAB C.3.1. BoltGlobals.m

```
%----GLOBAL VARIABLES------------------------------------------------------
global PRBFLD; %Ensure access to "force" defining probabilities,
global VSTCNT; % and to vector of visit counts,
global LOCIDX; % and to index of particle location,
global LOCCNT; % and to number of particle locations,
global ITRCNT; % and to iteration count,
global ITRIDX; % and to iteration index in tmint.m ,
global HSTORY; % and to history of distribution,
global POTENT; % and to scalar potential field,
global MAXWEL; % and to theoretical equilibrium distribution.
%--------------------------------------------------------------------------
```

MATLAB C.3.2. BoltVariables.m

```
%----INITIALIZE GLOBAL
%VARIABLES-----------------------------------------------------------------
 VSTCNT=zeros(1,LOCCNT);
 LOCIDX=int32(LOCCNT/2);     % Start the particle in the middle
 VSTCNT(LOCIDX)=1;           % LOCIDX must be the location index
                             % of the 1 occurring in VSTCNT
%--------------------------------------------------------------------------
% constant.eps
PRBFLD=(0.49)*ones(1,LOCCNT);  % Small constant force field.
%--------------------------------------------------------------------------
% spike.eps
%PRBFLD(4+LOCIDX)=0.75;          % Same, with off-center spike to the right.
%--------------------------------------------------------------------------
%rampjump.eps
%[POTENT,PRBFLD]=rampit(LOCCNT,1,int32(LOCCNT/2),8); % Ramp with jump.
%PRBFLD=nmlize(PRBFLD,.2,0.49);
%--------------------------------------------------------------------------
%saw3.eps
POTENT=arrayfun(@(x)(pwlpot(x)),(0:LOCCNT)/LOCCNT); % Elston-Doering
PRBFLD=arrayfun(@(x)(eforce(x)),(0:LOCCNT)/LOCCNT); % piecewise-linear
PRBFLD=nmlize(PRBFLD,.2,0.465); % 3-term Fourier series approximation.
%--------------------------------------------------------------------------
```

MATLAB C.3.3. BoltTry.m

```
function BoltTry(itrcnt,loccnt)
BoltGlobals;
LOCCNT=loccnt;
ITRCNT=itrcnt;
ITRIDX=1;
BoltVariables;                              %Initialize global variables.
HSTORY=zeros(ITRCNT,LOCCNT);                %Array to hold history.
BoltModel();                                %Run the model.
MAXWEL=maxwell(PRBFLD);
VisitBar(VSTCNT/sum(VSTCNT),MAXWEL,LOCCNT);%Create bar graph of visits.
BoltStory(POTENT,PRBFLD,VSTCNT,MAXWEL,LOCCNT);
end
```

MATLAB C.3.4. BoltModel.m

```
function BoltModel()
dotdur=[ 0.1  1e6  0.1  0.1  ]; %Timeout durations.
dotrem=[ 0.1  -1   -1   -1   ]; %Initial remaining times.
%Create handle to model-specific signal function.
sghndl=@(dotdur,dotrem)(BoltSignal(dotdur,dotrem));
%Invoke stochastic timing machine interpreter function.
tmint(dotdur,dotrem,sghndl);
end
```

MATLAB C.3.5. BoltSignal.m

```
function dotrem=BoltSignal(dotdur,dotrem)
%----GLOBAL VARIABLES-----------------------------------------------------
BoltGlobals;
%-------------------------------------------------------------------------
L=length(dotdur); %Number of dots in model.
if   L==0
    disp('No signals to try.');
    return
else
    for   i=1:L
        if dotrem(i)==0
            switch i
                case 1
                    % State (a) has timed out
                    % to stochastic state (b),
                    % so de-activate (a) and
                    % ranomly choose (c) or (d)
                    % to activate, and increment
                    % the distribution
                    dotrem(1)=-1;
                    if rand(1)<PRBFLD(LOCIDX)
                        dotrem(3)=dotdur(3);
```

```
                        %disp('(c)');
                    else
                        dotrem(4)=dotdur(4);
                        %disp('(d)');
                    end
                    VSTCNT(LOCIDX)=VSTCNT(LOCIDX)+1;
                    %fprintf('%d',VSTCNT);
                    HSTORY(ITRIDX,1:LOCCNT)=VSTCNT;
                case 2
                    % State (b) is a stochastic
                    % state so does not really
                    % need to be activated.
                case 3
                    % State (c) has timed out
                    % to state (a) so de-activate
                    % (c) and activate (a) and
                    % increment LOCIDX
                    dotrem(3)=-1;
                    dotrem(1)=dotdur(1);
                    LOCIDX=min(LOCIDX+1,LOCCNT);
                case 4
                    % State (d) has timed out
                    % to state (a) so de-activate
                    % (d) and activate (a) and
                    % decrement LOCIDX
                    dotrem(4)=-1;
                    dotrem(1)=dotdur(1);
                    LOCIDX=max(1,LOCIDX-1);
            end
        end
    end
end
end
```

MATLAB C.3.6. maxwell.m

```
function e=maxwell(prbfld)
frcfld=prbfld-0.5;
loccnt=length(frcfld);
v=zeros(1,loccnt);
e=zeros(1,loccnt);
v(1)=-frcfld(1);
for i=2:loccnt
    v(i)=v(i-1)-frcfld(i);
    e(i)=exp(-4*v(i));
end
e=e/sum(e);
end
```

MATLAB C.3.7. VisitBar.m

```
function VisitBar(vstcnt,maxwel,loccnt)
%  Partly auto-generated by MATLAB
BoltGlobals;
figure1 = figure;
axes1 = axes(   'Parent',figure1,...
                'XTickLabel',cellstr(char(32*ones(1,loccnt)')')),...
                'XTick',(1:loccnt));
xlim(axes1,[0 loccnt]);
box(axes1,'on');
hold(axes1,'all');
xlabel({'Brownian Particle Location'},'FontSize',6,'FontName','Arial');
ylabel('Normalized Number of Visits','FontSize',6,'FontName','Arial');
%------------------------------------------------------------------
hold on;
bar(vstcnt,'FaceColor',[1 1 0],'DisplayName','Visits');
plot(maxwel,'Color',[0 0 1]);
```

MATLAB C.3.8. BoltStory.m

```
function BoltStory(potent, prbfld, vstcnt, maxwel,loccnt)
%  Partly auto-generated by MATLAB on 16-Jan-2010 07:27:20
figure1 = figure;
%-----------------------------------------------------------------------
subplot1 = subplot(3,1,1,...
             'Parent',figure1,...
             'XTickLabel',cellstr(char(32*ones(1,loccnt)')')),...
             'XTick',(1:loccnt));
box(subplot1,'on');
hold(subplot1,'all');
plot(potent,'Parent',subplot1,'Color',[1 0 0]);
title('Scalar Potential');
%-----------------------------------------------------------------------
subplot2 = subplot(3,1,2,'Parent',figure1,...
             'Parent',figure1,...
             'XTickLabel',cellstr(char(32*ones(1,loccnt)')')),...
             'XTick',(1:loccnt));
box(subplot2,'on');
hold(subplot2,'all');
plot(prbfld,'Parent',subplot2,'Color',[0 1 0]);
title('Probability Field');
%-----------------------------------------------------------------------
subplot3 = subplot(3,1,3,'Parent',figure1,...
             'Parent',figure1,...
             'XTickLabel',cellstr(char(32*ones(1,loccnt)')')),...
             'XTick',(1:loccnt));
hold(subplot3,'all');
bar(vstcnt/sum(vstcnt),'FaceColor',[1 1 0],...
    'Parent',subplot3,'DisplayName','Visits');
```

```
plot(maxwel,'Parent',subplot3);
title({'Yellow:normalized visit count',...
          'Blue:theoretical probability density function'});
```

MATLAB C.3.9. eforce.m

```
function force=eforce(x)
% Negative gradient of pwlpot(x) =
% Elston-Doering Equation (37): 3 terms of Fourier series expansion
% of a piecewise linear potential given by
% y:=5/(4*pi)*(sin(2*pi*x) - sin(4*pi*x)/2 + sin(6*pi*x)/3).
force=-(5/2)*(cos(2*pi*x) - cos(4*pi*x) + cos(6*pi*x));
end
```

MATLAB C.3.10. rampit.m

```
function [ramp,deriv]= rampit(loccnt,slope,jmploc,jmpsiz)
ramp=cumsum(ones(1,loccnt+1));
ramp1=ramp(1:jmploc);
ramp2=ramp(jmploc+1:loccnt+1)+jmpsiz;
ramp=slope*[ramp1,ramp2];
deriv=ramp(2:loccnt+1)-ramp(1:loccnt);
ramp=ramp(1:loccnt);
end
```

MATLAB C.3.11. nmlize.m

```
function normalizedarray = nmlize(array,amplit,offset)
m=min(array);
f=array-m;
normalizedarray=offset+(amplit*(f/max(f)));
end
```

MATLAB C.3.12. pwlpot.m

```
function potential = pwlpot( x )
% Elston-Doering Equation (37): 3 terms of Fourier series expansion
% of a piecewise linear potential.
potential=5/(4*pi)*(sin(2*pi*x) - sin(4*pi*x)/2 + sin(6*pi*x)/3);
end
```

C.4   Figures. Simulating Brownian Particle in Force Field

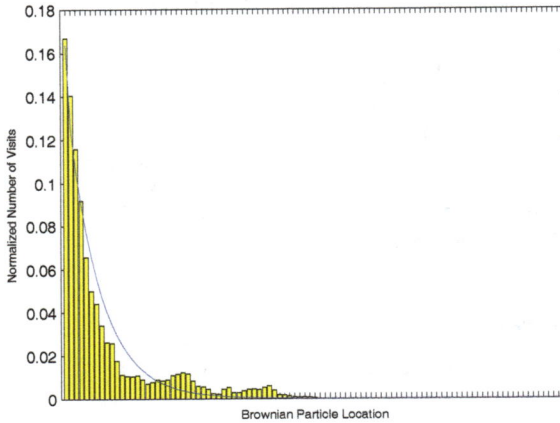

Fig. C.1 (`BoltTry(10000,100)`) Bar graph (yellow) of (normalized) number of visits among 100 locations of a bouncing particle after 5000 moves in a uniform force field. The blue curve is the theoretical equilibrium distribution computed by code `maxwell.m` in Appendix B based on Equation (7.5).

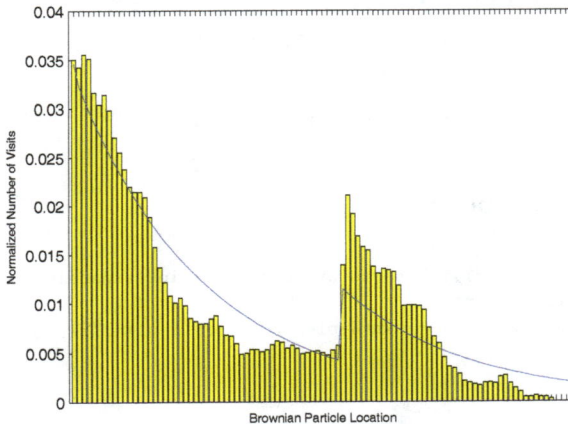

Fig. C.2 (`BoltTry(50000,100)`) Bar graph (yellow) of (normalized) number of visits among 100 locations of a bouncing particle after 25000 moves in a uniform force field with a central spike of force to the right. The blue curve is the theoretical equilibrium distribution.

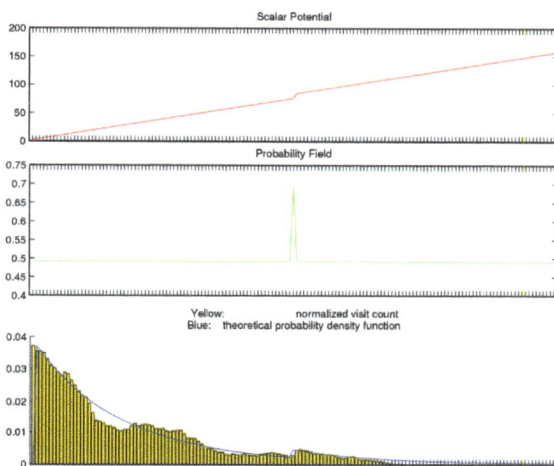

Fig. C.3 (`BoltTry(100000,150)`)The upper graph (red) is the scalar potential which ramps, jumps, and ramps again. The center graph (green) is the probability field corresponding to the scalar potential. The bottom bar graph (yellow) is the (normalized) number of visits among 150 locations of a bouncing particle after 50000 moves in the resulting force field. The blue curve is the theoretical equilibrium distribution.

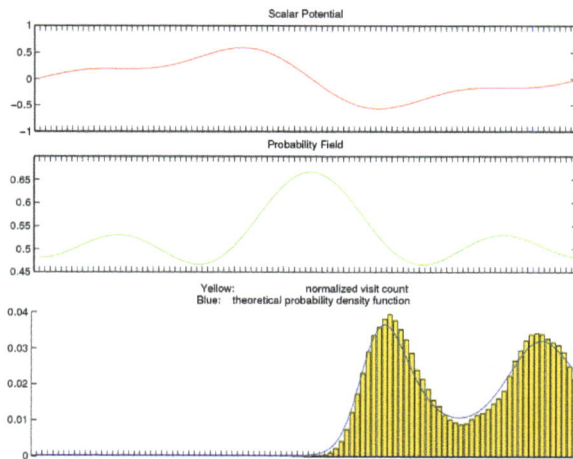

Fig. C.4 (`BoltTry(100000,100)`) The upper graph (red) is the scalar potential equal to the first 3 terms of the Fourier series approximation to a piecewise linear (sawtooth) curve. The center graph (green) is the probability field corresponding to the scalar potential. The bottom bar graph (yellow) is the (normalized) number of visits among 100 locations of a bouncing particle after 50000 moves in the resulting force field. The blue curve is the theoretical equilibrium distribution.

Appendix D

Fundamental Theorem of Elastic Bodies

Say B is a body in the universe with spatial region $B(u)$ in state $u \in \Sigma$ and recall that $\Sigma \xrightarrow{\vec{p}_B} \mathbb{R}^3$ denotes the total amount of linear momentum vector substance in B. The seemingly innocuous Balance Equation

$$\dot{\vec{p}}_B(u) = \vec{p}_B^{\mathcal{E}}(u) \quad [\mathbf{LMM}][\mathbf{TME}]^{-1} \qquad (\text{D.1})$$

actually has momentous consequences. For starters, this equation implies Newton's Three Laws of Motion. To be blunt, *force is momentum current*:

Newton's	Translated from Force	to Momentum Current[1]
First Law	If there are no forces acting on a body, the body will stay at rest or move uniformly in a straight line.	If no momentum currents are flowing into or out of a body, the momentum of the body will not change.
Second Law	The time rate of change of the momentum of a body $\frac{d\vec{p}}{dt} = \frac{dm\vec{v}}{dt} = m\frac{d\vec{v}}{dt} = m\vec{a}$ equals the force F acting upon the body: $m\vec{a} = \vec{F}$.	The time rate of change of the momentum of a body $\dot{\vec{p}}_B$ equals the momentum current $\vec{p}_B^{\mathcal{E}}$ flowing into the body: $\dot{\vec{p}}_B = \vec{p}_B^{\mathcal{E}}$.
Third Law	If body A exerts a force F upon body B, then body B exerts an equal but opposite force $-F$ upon A.	If a momentum current flows out of a body A and into a body B, the intensity of the current leaving A is the same as that entering B.

---

[1]http://www.physikdidaktik.uni-karlsruhe.de/publication/pub_fremdsprachen/englisch.html

Greater depth of the Momentum Balance Equation (D.1) surfaces in the

Theorem D.16. *(Cauchy Infinitesimal Tetrahedron) If $B(u) \xrightarrow{K} \mathbb{R}$ and $B(u) \times S^2 \xrightarrow{j} \mathbb{R}$ are continuous maps, and*

$$\int_R K \, dR = \int_{\partial R} j(\overrightarrow{n}) \, d\partial R \tag{D.2}$$

for all subregions $R \subseteq B(u)$, then j_O extends to a linear functional

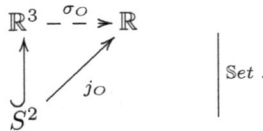

$$
\begin{array}{c}
\mathbb{R}^3 \xdashrightarrow{\;\sigma_O\;} \mathbb{R} \\[4pt]
\Big\uparrow \quad \nearrow \\[2pt]
\Big\downarrow \quad j_O \\[2pt]
S^2
\end{array}
\qquad \Big| \quad \text{Set .}
$$

Proof. Inside the spatial region $B(u)$ of the body choose a point $O = (0,0,0)$, and choose a unit vector $\overrightarrow{n} = (n_1, n_2, n_3)$. Let $\varepsilon > 0$ be a positive infinitesimal and let

$$A := (a, 0, 0) \tag{D.3}$$
$$B := (0, b, 0)$$
$$C := (0, 0, c)$$

be the intersections with the coordinate axes x, y, z of the plane orthogonal to $\overrightarrow{n}$ through the point $\varepsilon \overrightarrow{n}$, as in Fig.(D.1). Typically,

$$
\begin{aligned}
0 &= \langle (a, 0, 0) - \varepsilon \overrightarrow{n} \, | \, \overrightarrow{n} \rangle \tag{D.4} \\
&= (a - \varepsilon n_1) n_1 - \varepsilon n_2^2 - \varepsilon n_3^2 \\
&= a n_1 - \varepsilon (n_1^2 + n_2^2 + n_3^2) \\
&= a n_1 - \varepsilon
\end{aligned}
$$

and therefore

$$a = \frac{\varepsilon}{n_1} \qquad b = \frac{\varepsilon}{n_2} \qquad c = \frac{\varepsilon}{n_3} \,. \tag{D.5}$$

Thus, the equation of that plane is

$$\frac{x}{a} + \frac{y}{b} + \frac{z}{c} = 1 \tag{D.6}$$

and the points O, A, B, C are the vertices of an **infinitesimal Cauchy tetrahedron**. By the formula for volume of a cone learned by high school

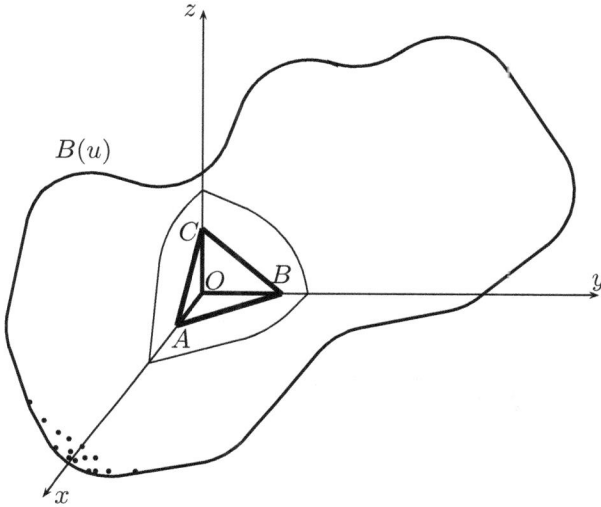

Fig. D.1 Cutaway view inside a body relative to coordinate system xyz with an infinitesimal Cauchy tetrahedron $OABC$ located at point O.

students – one third base area times height – there are four ways to compute the volume of the infinitesimal Cauchy tetrahedron:

$$V = \frac{1}{3}|ABC||OP| = \frac{1}{3}|ABC|\varepsilon \tag{D.7}$$

$$V = \frac{1}{3}|OAB||OP| = \frac{1}{3}|OAB|\frac{\varepsilon}{n_3}$$

$$V = \frac{1}{3}|OAC||OP| = \frac{1}{3}|OAC|\frac{\varepsilon}{n_2}$$

$$V = \frac{1}{3}|OBC||OP| = \frac{1}{3}|OBC|\frac{\varepsilon}{n_1}$$

where $|\ldots|$ stands in general for "measure of", which could be volume, or area. Solving these equations for the areas of the faces, the ratio of the area of the top face $|ABC|$ to the total surface area $|\partial OABC|$ is

$$\frac{|ABC|}{|ABC| + |OAB| + |OAC| + |OBC|} = \frac{3V/\varepsilon}{3V/\varepsilon + 3n_3V/\varepsilon + 3n_2V/\varepsilon + 3n_1V/\varepsilon} \tag{D.8}$$

$$= \frac{1}{1+N} =: \alpha$$

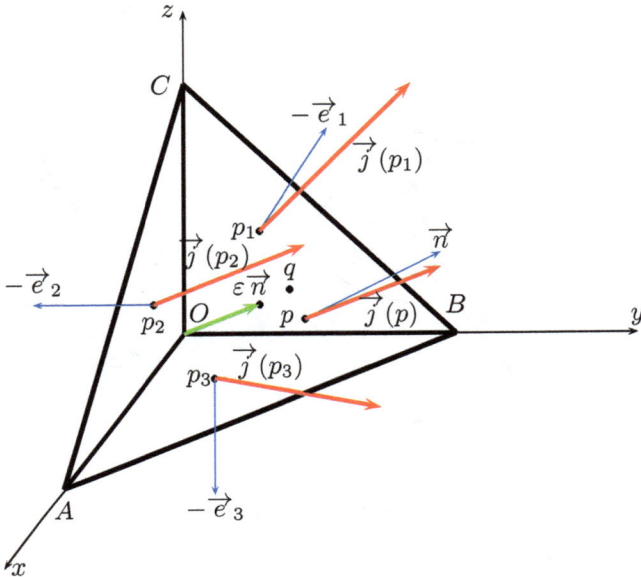

Fig. D.2   The infinitesimal Cauchy tetrahedron. The linear momentum density vectors are shown in red. Unit normal vectors at the four faces are shown in color blue. The green vector is of infinitesimal length ε and is perpendicular to the plane through coordinate points A, B, C. Points p_1, p_2, p_3 and p are in faces of the tetrahedron, point q is within its interior.

where $N := n_1 + n_2 + n_3$, and the ratios of the right-triangular faces to the total surface are

$$\frac{|OAB|}{|ABC| + |OAB| + |OAC| + |OBC|} = n_3\alpha \qquad (D.9)$$

$$\frac{|OAC|}{|ABC| + |OAB| + |OAC| + |OBC|} = n_2\alpha$$

$$\frac{|OBC|}{|ABC| + |OAB| + |OAC| + |OBC|} = n_1\alpha$$

By hypothesis for $R = OABC$, and by appeal to the Integral Mean Value Rule based on the continuity of K, there exists within $OABC$ a point q such that

$$\frac{\int\limits_{\partial OABC} j(\vec{n})\,d\partial OABC}{|\partial OABC|} = \frac{\int\limits_{OABC} K\,dOABC}{|\partial OABC|} \tag{D.10}$$

$$= \frac{J(q)|OABC|}{|\partial OABC|}$$

$$= J(q)\frac{\varepsilon}{3(1+N)} \approx 0\,.$$

Likewise, there exist within the faces points p, p_1, p_2, p_3 such that

$$0 \approx \frac{1}{|\partial OABC|}\int\limits_{\partial OABC} j(\vec{n})\,d\partial OABC \tag{D.11}$$

$$= \frac{1}{|\partial OABC|}\left(\int\limits_{ABC} j(\vec{n})\,dABC + \int\limits_{OAB} j(\vec{n})\,dOAB\right.$$

$$\left. + \int\limits_{OAC} j(\vec{n})\,dOAC + \int\limits_{OBC} j(\vec{n})\,dOBC\right)$$

$$= \frac{1}{|\partial OABC|}(j_p(\vec{n})|ABC| + j_{p_3}(-\vec{e_3})|OAB|$$

$$+ j_{p_2}(-\vec{e_2})|OAC| + j_{p_1}(-\vec{e_1})|OBC|)$$

which implies

$$j_p(\vec{n})\frac{|ABC|}{|\partial OABC|} = -j_{p_3}(-\vec{e_3})\frac{|OAB|}{|\partial OABC|} - j_{p_2}(-\vec{e_2})\frac{|OAC|}{|\partial OABC|}$$

$$- j_{p_1}(-\vec{e_1})\frac{|OBC|}{|\partial OABC|} \tag{D.12}$$

or

$$j_p(\vec{n})\alpha = -j_{p_3}(-\vec{e_3})\alpha n_3 - j_{p_2}(-\vec{e_2})\alpha n_2 - j_{p_1}(-\vec{e_1})\alpha n_1\,. \tag{D.13}$$

After cancelling α from both sides, observe that since the tetrahedron is infinitely small, the points p, p_1, p_2, p_3 are infinitely near to O. Therefore, after taking standard parts,

$$j_O(\vec{n}) = -j_O(-\vec{e_3})n_3 - j_O(-\vec{e_2})n_2 - j_O(-\vec{e_1})n_1\,. \tag{D.14}$$

In particular $j_O(\vec{e_i}) = -j_O(-\vec{e_i})$ by the substitutions $\vec{n} = \vec{e_i}$. Therefore,

$$j_O(\vec{n}) = j_O(\vec{e_1})n_1 + j_O(\vec{e_2})n_2 + j_O(\vec{e_3})n_3 . \tag{D.15}$$

Finally, for an arbitrary vector $\vec{x} = (x_1, x_2, x_3)$ the formula

$$j_O(\vec{x}) = j_O(\vec{e_1})x_1 + j_O(\vec{e_2})x_2 + j_O(\vec{e_3})x_3 . \tag{D.16}$$

delivers the promised extension σ_O. $\qquad\square$

Aside D.0.1. In my opinion this result by Baron Augustin-Louis Cauchy is the Fundamental Theorem of Elastic Bodies. I unpacked the above proof from [Marsden and Hughes (1983)]. More opaque expositions of this result appear in [Love (1944)], [Aris (1962)], [Butkov (1968)], [Jeffreys and Jeffreys (1972)].

> **To see the sorry difference between the standards of thermodynamics and its contemporary mathematical sciences we need only look at continuum mechanics, where the counterpart of FOURIER's unstated and half-implied flux principle is CAUCHY's theorem of the existence of the stress tensor, published in 1823. CAUCHY, who knew full well the difference between a balance principle and a constitutive relation, stated the result clearly and proudly; he gave a splendid proof of it, which has been reproduced in every book on continuum mechanics from that day to this; and he recognized the theorem as being the foundation stone it still is [Truesdell (1971)].**

Bibliography

Abraham, R. H. (1967). *Foundations of Mechanics* (W. A. Benjamin, Inc.), a mathematical exposition of classical mechanics with an introduction to the qualitative theory of dynamical systems and applications to the three-body problem, with the assistance of Jerrold E. Marsden, Princeton University.

Abraham, R. H. and Robbin, J. (1967). *Transversal Mappings and Flows* (W. A. Benjamin, Inc.).

Abraham, R. H. and Shaw, C. D. (1983). *Dynamics — The Geometry of Behavior* (Aerial Press, Inc.), Part One: Periodic Behavior, Part Two: Chaotic Behavior, and Part Three: Global Behavior.

Ahlfors, L. V. (1966). *Complex Analysis, an introduction to the theory of analytic functions of one complex variable, Second Edition* (McGraw-Hill Book Company, New York).

Aigner, M. and Heegaard, J. (1999). One-dimensional quasi-static continuum model of muscle contraction as a distributed control system, Annual research brief, Center for Turbulence Research.

Alberty, R. A. (2001). Use of Legendre transforms in chemical thermodynamics, *Pure and Applied Chemistry* **73**, 8, pp. 1349–1380, UPAC Technical Report.

Aris, R. (1962). *Vectors, Tensors, and the Basic Equations of Fluid Mechanics* (Dover Publications, Inc., New York).

Aris, R. (1989). *Elementary Chemical Reactor Analysis* (Dover Publications, Inc.).

Arnold, V. I. (1978). *Ordinary Differential Equations* (MIT Press), translated and edited by Richard A. Silverman. Second printing, August, 1980.

Arnolt, L. (1974). *Stochastic Differential Equations: Theory and Applications* (John Wiley & Sons).

Arons, A. B. (1965). *Development of Concepts of Physics, From the Rationalization of Mechanics to the First Theory of Atomic Structure* (Addison-Wesley Publishing Company, Inc.).

Astumian, R. D. and Bier, M. (1996). Mechanochemical coupling of the motion of molecular motors to ATP hydrolysis, *Biophysical Journal* **70**, pp. 637–653.

Astumian, R. D. and Hänggi, P. (2002). Brownian motors, *Physics Today*.

Baez, J. C. and Lauda, A. (2009). A prehistory of n-categorical physics, arXiv:0908.2469v1 [hep-th].

Baez, J. C. and Stay, M. (2009). Physics, topology, logic and computation: A Rosetta Stone, arXiv:0903.0340v3 [quant-ph] 6 Jun 2009.

Baierlein, R. (2001). The elusive chemical potential, *American Journal of Physics* **69**, 4.

Baker, J. E. (2003). Muscle force emerges from dynamic actin-myosin networks, not from independent force generators, *American Journal of Physiology, Cell Physiology* **284**, pp. 1678–1679.

Baker, J. E. and Thomas, D. D. (2000). A thermodynamic muscle model and a chemical basis for A. V. Hill's muscle equation, *Journal of Muscle Research and Cell Motility* **21**, pp. 335–344.

Beck, J. (1969). Distributive laws, *Lecture Notes in Mathematics* **80**, pp. 119–140.

Bent, H. A. (1965). *The Second Law, An Introduction to Classical and Statistical Mechanics* (Oxford University Press).

Berkeley, E. C. (1949). *Giant Brains: or, Machines that Think* (John Wiley & Sons), The original typed manuscript of this book sold for $19,200 at Christie's in a sale called "Origins of Cyberspace," 2005.

Billingsley, P. (1986). *Probability and Measure, Second Edition* (John Wiley & Sons, New York), p. 317.

Blackmore, S. (2004). *Consciousness, An Introduction* (Oxford University Press).

Bothamley, J. (1923). *Dictionary of Theories* (Gale Research International Ltd).

Brown, R. (2006). *Topology and Groupoids* (Ronald Brown, www.booksurge.com).

Buck, R. C. (1978). *Advanced Calculus, Third Edition* (McGraw-Hill, Inc.).

Bustamente, C., Keller, D. and Oster, G. (2001). The physics of molecular motors, *Accounts of Chemical Research* **34**, 6, pp. 412–420.

Butkov, E. (1968). *Mathematical Physics* (Addison-Wesley Publishing Company, Reading, MA).

Callen, H. B. (1985). *Thermodynamics and an Introduction to Thermostatistics*, 2nd edn. (John Wiley & Sons, New York).

Cantor, G. (1915,1955). *Contributions to the Founding of the Theory of Transfinite Numbers* (Dover Publications, Inc.), Translated and provided with an introduction and notes, by Philip E. G. Jourdain.

Cápek, V. and Sheehan, D. P. (2005). *Challenges to the Second Law of Thermodynamics, Theory and Experiment*, Fundamental Theories of Physics, Vol. 146 (Springer, New York).

Cariani, P. A. (2001). Neural timing nets, *Neural Networks* **14**, pp. 737–753.

Cellier, F. (1992). Hierarchical non-linear bond graphs: A unified methodology for modeling complex physical systems, *SIMULATION* **58**, 4, pp. 230–248.

Chen, E., Goldbeck, R. A. and Kliger, D. S. (1997). Nanosecond time-resolved spectroscopy of biomolecular processes, *Annu. Rev. Biophys. Biomol.Struct.* **26**, pp. 327–355.

Chin, L., Yue, P., Feng, J. J. and Seow, C. Y. (2006). Mathematical simulation of muscle cross-bridge cycle and force-velocity relationship, *Biophysical Journal* **91**, pp. 3653–3663.

Coecke, B. and Éric Oliver Paquette (2009). Categories for the practicing physicist, arXiv:0905.3010v2 [quant-ph].

Cooper, E. D. (1996). Three equations for a theory of consciousness, (Toward a Science of Consciousness, Tucson, Arizona).

Cooper, E. D. and Lengyel, F. (2009). Generalized Gillespie stochastic simulation algorithm and chemical master equation using timing machinery, Technical Report TR-2009013, The City University of New York, Ph.D. Program in Computer Science.

Cooper, R. (1966). *Functions of Real Variables, A Course of Advanced Calculus* (D. Van Nostrand Company, Princeton).

C.P. Fall, J., E.S. Marland and Tyson, J. (2002). *Computational Cell Biology* (Springer, New York).

Craig, N. C. (1992). *Entropy Analysis, An Introduction to Chemical Thermodynamics* (VCH Publishers, Inc.).

Curtis, C. W. (1974). *Linear Algebra, An Introductory Approach* (Springer-Verlag, New York).

Davis, M. and Etheridge, A. (2006). *Louis Bachelier's "Theory of Speculation," The Origins of Modern Finance* (Princeton University Press), translation of and commentary on Louis Bachelier's Doctoral Thesis, 1900.

De Groot, S. R. and Mazur, P. (1984). *Non-Equilibrium Thermodynamics* (Dover Publications, Inc.).

de Heer, J. (1986). *Phenomenological Thermodynamics, with Applications to Chemistry* (Prentice-Hall, Inc.).

Diemente, D. (1998). A closer look at the addition of equations and reactions, *Journal of Chemical Education* **75**, 3, pp. 319–321.

Dill, D. L. and Alur, R. (1990). A theory of timed automata, in *Proceedings of the 17th International Colloquium on Automata, Languages, and Programming*.

Döring, A. and Isham, C. (2008). 'What is a Thing?': Topos theory in the foundations of physics, arXiv:0803.0417v1 [quant-ph] 4 March 2008.

Duke, T. (1999). Molecular model of muscle contraction, *Proceedings of the National Academy of Sciences USA* **96**, pp. 2770–2775.

Duke, T. (2000). Cooperativity of myosin molecules through strain dependent chemistry, *Philosophical Transactions of the Royal Society of London B* **355**, pp. 529–538.

Eckmann, B. and Hilton, P. (1962). Group-like structures in categories, *Mathematische Annalen* **145**, p. 227–255.

Édouard Goursat (1959). *A Course in Mathematical Analysis, Volume I* (Dover Publications), translated by Earle Raymond Hedrick. This Dover edition is an unabridged and unaltered republication of the Hedrick translation Entered at Stationer's Hall, Copyright, 1904.

Eilenberg, S. and Mac Lane, S. (1986). General theory of natural equivalences, in *Eilenberg-Mac Lane Collected Works* (Academic Press).

Eilenberg, S. and MacLane, S. (1945). General theory of natural equivalences, *Transactions of the American Mathematical Society* **58**, 2, pp. 231–294.

Einstein, A. (1989). On the movement of small particles suspended in stationary liquids required by the molecular kinetic theory of heat, in *The Collected Papers of Albert Einstein, Volume 2, The Swiss Years: Writings, 1900–1909*, chap. Annalen der Physik 17, (1905): 549–560 (Princeton University

Press), Anna Beck, Translator; Peter Havas, Consultant.

Eisberg, R. and Resnick, R. (1985). *Quantum Physics of Atoms, Molecules, Solids, Nuclei, and Particles* (John Wiley & Sons, New York).

Érdi, P. and Tóth, J. (1989). *Mathematical Models of Chemical Reactions* (Princeton University Press, Princeton).

Falk, G., Herrmann, F. and Schmid, G. B. (1983). Energy forms or energy carriers? *American Journal of Physics* **51**, 12, pp. 1074–1077.

Fauconnier, G. and Turner, M. (2002). *The Way We Think, Conceptual Blending and the Mind's Hidden Complexities* (Basic Books).

Fermi, E. (1956). *Thermodynamics* (Dover Publications, New York), this book originated in a course of lectures held at Columbia University, New York, during the summer session of 1936.

Feynman, R. P., Leighton, R. B. and Sands, M. (1964). *The Feynman Lectures on Physics, Volume II, Mainly Electromagnetism and Matter* (Addison-Wesley Publishing Company, Reading).

Fuchs, H. U. (1996). *The Dynamics of Heat* (Springer-Verlag, New York).

Fuchs, H. U. (1997). The continuum physics paradigm in physics instruction, In Three Parts, Department of Physics, Technikum Winterthur, 8401 Winterthur, Switzerland.

Georgescu-Roegen, N. (1971). *The Entropy Law and the Economic Process* (Harvard University Press).

Gibbs, J. (1957). *The Collected Works of J. Willard Gibbs, In Two Volumes* (Yale University Press, New Haven).

Gillespie, D. T. (1977). Exact stochastic simulation of coupled chemical reactions, *The Journal of Physical Chemistry* **81**, 25, pp. 2340–2361.

Gillespie, D. T. (2007). Stochastic simulation of chemical kinetics, *Annual Review of Physical Chemistry*, 58, pp. 35–55.

Goldblatt, R. (1998). *Lectures on the Hyperreals, An Introduction to Nonstandard Analysis* (Springer).

Goldstein, H. (1980). *Classical Mechanics*, 2nd edn. (Addison-Wesley Publishing Company).

Goodsell, D. (1998). *The Machinery of Life* (Copernicus).

Gunawardena, J. (2003). Chemical reaction network theory for *in-silico* biologists, Lecture notes.

Haddad, W., Chellaboina, V. and Nersesov, S. (2005). *Thermodynamics, A Dynamical Systems Approach* (Princeton University Press, Princeton).

Halmos, P. R. (1960). *Naive Set Theory* (Springer-Verlag, New York).

Henle, J. M. and Kleinberg, E. M. (1979). *Infinitesimal Calculus* (The MIT Press, Cambridge).

Herrmann, F. and Job, G. (2006). Karlsruhe physics course, volume 1, energy, momentum, entropy, URL http://www.physikdidaktik.uni-karlsruhe. de/kpk/english/KPK_Volume_1.pdf.

Herrmann, F. and Schmid, G. B. (1984). Statics in the momentum current picture, *American Journal of Physics* **52**, 2, pp. 146–152.

Herrmann, F. and Schmid, G. B. (Year not available). Rotational dynamics and the flow of angular momentum, Institut für Didaktik der Physik, Universität

Karlsruhe, 76128 Karlsruhe, Germany.

Herstein, I. N. (1964). *Topics In Algebra* (Blaisdell Publishing Company, New York).

Hill, A. V. (1938). The heat of shortening and the dynamic constants of muscle, *Proceedings of the Royal Society of London, B Biological Science* **126**, pp. 136–195.

Hill, T. L. (1974). Theoretical formalism for the sliding filament model of contraction of striated muscle, Part I, *Progress in Biophysics and Molecular Biology* **28**, pp. 267–340.

Hirsch, M. W. and Smale, S. (1974). *Differential Equations, Dynamical Systems, and Linear Algebra* (Academic Press).

Holland, J. H. (1998). *Emergence, from Chaos to Order* (Addison-Wesley Publishing Co., Inc.).

Holmes, J. W. (2006). Teaching from classic papers: Hill's model of muscle contraction, *Advances in Physiology Education* **30**, pp. 67–72.

Hooper, S. L., Hobbs, K. H. and Thuma, J. B. (2008). Invertebrate muscles: thin and thick filament structure; molecular basis of contraction and its regulation, catch and asynchronous muscle, *Progress in Neurobiology* **86**, 2, pp. 72–127.

Horne, M., Farago, P. and Oliver, J. (1973). An experiment to measure Boltzmann's constant, *American Journal of Physics* **41**, 6, pp. 344–348.

Huxley, A. F. and Niedergerke, R. (1954). Structural chanages in muscle during contraction, *Nature* **173**, 4412, pp. 971–973.

Huxley, H. and Hanson, J. (1954). Changes in the cross-striations of muscle during contraction and stretch and their structural interpretation, *Nature* **173**, 4412, pp. 973–976.

Huxley, H. E. (1984). Fifty years of muscle and the sliding filament hypothesis, *American Journal of Physics* **52**, 9, pp. 794–799.

Huxley, H. E. (2004). Fifty years of muscle and the sliding filament hypothesis, *European Journal of Biochemistry* **271**, pp. 1403–1415.

Jarzynski, C. (1997). Nonequilibrium equality for free energy differences, *Physical Review Letters* **78**, 14.

Jeffreys, S. H. and Jeffreys, B. S. L. (1972). *Methods of Mathematical Physics*, 3rd edn. (Cambridge At the University Press).

Job, G. and Herrmann, F. (2006). Chemical potential — a quantity in search of recognition, *European Journal of Physics* **27**, pp. 353–371.

Jong, H. D. (2002). Modeling and simulation of genetic regulatory systems: a literature review, *Journal of Computational Biology* **9**, 1, pp. 67–103.

Jülicher, F., Ajdari, A. and Prost, J. (1997). Modeling molecular motors, *Reviews of Modern Physics* **69**, 4, pp. 1269–1281.

Kac, M. (1954). *Random Walk and the Theory of Brownian Motion* (Dover Publications, Inc.), pp. 295–317.

Keisler, H. J. (2002). Elementary calculus: An approach using infinitesimals, Free on-line edition vailable at http://www.math.wisc.edu/~keisler/calc.html.

Keller, D. and Bustamente, C. (2000). The mechanochemistry of molecular motors, *Biophysical Journal* **78**, pp. 541–556.

Kelley, J. L. (1955). *General Topology* (D. Van Nostrand Company, Inc.).

Kestin, J. (1979). *A Course in Thermodynamics* (Taylor & Francis).

Kuhn, T. S. (1978). *Black-Body Theory and the Quantum Discontinuity, 1894–1912* (The University of Chicago Press).

Lambek, J. (1994). Are the traditional philosophies of mathematics incompatible? *The Mathematical Intelligencer* **16**, 1, pp. 56–62.

Lan, G. and Sun, S. X. (2005). Dynamics of myosin-driven skeletal muscle contraction: I. steady-state force generation, *Biophysical Journal* **88**, pp. 4107–4117.

Landau, L. and Lifshitz, E. (1987). *Fluid Mechanics, Landau and Lifshitz, Course of Theoretical Physics*, Vol. 6 (Pergamon Press).

Lang, S. (1987). *Linear Algebra, Third Edition* (springer-Verlag).

Lang, S. (2002). *Short Calculus* (Springer-Verlag), the Original Edition of "A First Course in Calculus".

Lawvere, F. W. (1969). Adjointness in foundations, *Dialectica* **23**, pp. 281–296.

Lawvere, F. W. (1976). *Algebra, Topology, and Category Theory, A Collection of Papers in Honor of Samuel Eilenberg*, chap. Variable Quantities and Variable Structures in Topoi (Academic Press), pp. 101–131.

Lawvere, F. W. and Rosebrugh, R. (2003). *Sets for Mathematics* (Cambridge University Press).

Lawvere, F. W. and Schanuel, S. (1997). *Conceptual Mathematics* (Cambridge University Press).

Lemons, D. S. and Gytheil, A. (1997). Paul Langevin's 1908 paper "On the Theory of Brownian Motion", *American Journal of Physics* **65**, 11, p. 1081.

Lemons, D. S. and Gythiel, A. (1997). Paul Langevin's 1908 paper "On the Theory of Brownian Motion", *American Journal of Physics* **65**, 11, pp. 1079–1081.

Lengyel, S. (1989). Chemical kinetics and thermodynamics, *Computers & Mathematics with Applications* **17**, pp. 443–455.

Lewis, G. N. and Randall, M. (1923). *Thermodynamics and The Free Energy of Chemical Substances* (McGraw-Hill Book Company, New York).

Lin, C. C. and Segel, L. A. (1988). *Mathematics Applied to Deterministic Problems in the Natural Sciences* (Society for Industrial and Applied Mathematics, Philadelphia).

Lindén, M. (2008). Stochastic modeling of motor proteins, Doctoral Thesis, KTH School of Engineering Sciences, Stockholm, Sweden 2008.

Llinas, R. R. (1988). The intrinsic electrophysiological properties of mammalian neurons: insights into central nervous system function, *Science* **242**, pp. 1654–1664.

Love, A. E. H. (1944). *A Treatise on the Mathematical Theory of Elasticity*, 4th edn. (Dover Publications).

Lymn, R. W. and Taylor, E. W. (1971). Mechanism of adenosine triphosphate hydrolysis by actomyosin, *Biochemistry* **10**, 25, pp. 4617–4624.

Mac Lane, S. (1967). *Homology* (Springer-Verlag).

Mac Lane, S. (1971). *Categories for the Working Mathematician* (Springer-Verlag, New York).

Mac Lane, S. (1986). *Mathematics Form and Function* (Springer-Verlag, New York).

Manin, Y. I. (2007). *Mathematics as Metaphor, Selected Essays of Yuri I. Manin* (American Mathematical Society), with Foreword by Freeman J. Dyson.

Marsden, J. E. and Hughes, T. J. R. (1983). *Mathematical Foundations of Elasticity* (Dover Publications, Inc.).

Mateus, P., Morais, M., Nunes, C., A., A. P. and Sernadas, S. C. (2003). Categorical foundations for randomly timed automata, *Theoretical Computer Science* **308**, pp. 393–427.

Mayer, B. J., Blinov, M. L. and Loew, L. M. (2009). Molecular machines or pleiomorphic ensembles: signaling complexes revisited, *Journal of Biology* **8**, 81, pp. 1–8.

McAdams, H. H. and Arkin, A. (1997). Stochastic mechanisms in gene expression, *Proc. Natl. Acad. Sci. USA* **94**, pp. 814–819.

McAdams, H. H. and Arkin, A. (1998). Simulation of prokaryotic genetic circuits, *Annu. Rev. Biphys. Biomol. Struct.* **27**, pp. 199–224.

McCulloch, A. D. and Bae, W. C. (2006). BE112A: Biomechanics, A Course at the University of California San Diego.

McQuarrie, D. A. and Simon, J. D. (1997). *Physical Chemistry, A Molecular Approach* (University Science Books, Sausalito).

Metzinger, T. (2009). *The Ego Tunnel: The Science of Mind and the Myth of Self* (Basic Books).

Misra, V., Gong, W.-B. and Townsley, D. (1999). Stochastic differential equation modeling and analysis of TCP-windowsize behavior, in *Proceedings of IFIP WG 7.3 Performance*.

Moraczewska, J. (2002). Structural determinants of cooperativity in acto-myosin interactions, *Acta Biochimica Polonica* **49**, 4, pp. 805–812.

Muldoon, C. A. (2006). *Shall I Compare Thee to a Pressure Wave? Visualisation, Analogy, Insight and Communication in Physics*, Doctor of Philosophy, University of Bath.

Nadásdy, Z. (1998). *Spatio-Temporal Patterns in the Extracellular Recording of Hippocampal Pyramidal Cells, From Single Spikes to Spike Sequences*, Ph.D. thesis, Rutgers, The State University of New Jersey.

Nelson, D. L. and Cox, M. M. (2005). *Lehninger Principles of Biochemistry*, 4th edn. (W. H. Freeman and Company).

Nelson, E. (1967). *Dynamical Theories of Brownian Motion* (Princeton University Press).

Newburgh, R., Peidle, J. and Rueckner, W. (2006). Einstein, Perrin, and the reality of atoms: 1905 revisited, *American Journal of Physics* **74**, 6, pp. 478–481.

Newton, N. (1996). *Foundations of Understanding* (John Benjamins Publishing Company).

Ohanian, H. C. (2008). *Einstein's Mistakes, The Human Failings of Genius* (W. W. Norton & Company).

Oster, G. F., Perelson, A. S. and Katchalsky, A. (1973). Network thermodynamics: dynamic modelling of biophysical systems, *Quarterly Reviews of*

Biophysics **6**, pp. 1–134.

Pais, A. (1982). *'Subtle is the Lord..' The Science and the Life of Albert Einstein* (Clarendon Press).

Papoulis, A. (1965). *Probability, Random Variables, and Stochastic Processes* (McGraw-Hill Book Company).

Partridge, E. (1966). *Origins, A Short Etymological Dictionary of Modern English* (The MacMillan Company, New York).

Perrin, J. (2005). *Brownian Movement and Molecular Reality* (Dover Publications, Inc., Mineola), originally published by Taylor and Francis, London, 1910.

Planck, D. M. (1926). *Treatise on Thermodynamics*, 3rd edn. (Dover Publications, New York), translated from the Seventh German edition.

Pollack, G. H., Blyakhman, F. A., Liu, X. and Nagornyak, E. (2005). Sarcomere dynamics, stepwise shortening and the nature of contraction, in H. Sugi (ed.), *Sliding Filament Mechanism in Muscle Contraction, Fifty Years of Research*, Vol. 565 (Springer), pp. 113–126.

Present, R. (1958). *Kinetic Theory of Gases* (McGraw-Hill Book Company, Inc., New York).

Prigogine, I. (1980). *From Being to Becoming, Time and Complexity in the Physical Sciences* (W. H. Freeman and Company).

Prigogine, I., Outer, P. and Herbo, C. (1948). Affinity and reaction rate close to equilibrium, *Journal of Physical Chemistry* **52**, 2, pp. 321–331.

Purcell, E. M. (1977). Life at low Reynolds number, *American Journal of Physics* **45**, pp. 3–11.

R. L. Finney, M. D. W. and Giordano, F. R. (2003). *Thomas' Calculus, Updated Tenth Edition, Revised* (McGraw-Hill Book Company, Boston).

Rall, J. A. (1982). Sense and nonsense about the Fenn effect, *American Journal of Physiology* **242**, 1, pp. H1–6.

Reconditi, M., Piazzesi, G., Linari, M., Lucii, L., Sun, Y.-B., Boesecke, P., Narayanan, T., Irving, M. and Lombardi, V. (2002). X-Ray interference measures of structural changes of the myosin motor in muscle with Å resolution, *Notiziario Neutrone E Luce Di Sincrotrone* **7**, 2, pp. 19–29.

Reedy, M. C. (2000). Visualizing myosin's power stroke in muscle contraction, *Journal of Cell Science* **113**, pp. 3551–3562.

Reimann, P. (2001). Brownian motors: noisy transport far from equilibrium, ArXiv:cond-mat0010237v2 [cond-mat.stat-mech] 4 Sep 2001.

Romano, A., Lancellotta, R. and Marasco, A. (2006). *Continuum Mathematics using Mathematica, Fundamentals, Applications and Scientific Computing* (Birkhauser, Boston).

Rosch, E. (1994). Is causality circular? event structure in folk psychology, cognitive science and buddhist logic, *Journal of Consciousness Studies* **1**, 1, pp. 50–65.

Rudin, W. (1966). *Real and Complex Analysis* (McGraw-Hill Book Company, New York).

Schmid, G. B. (1984). An up-to-date approach to physics, *American Journal of Physics* **52**, 9, pp. 794–799.

Schutt, C. E. and Lindberg, U. (1992). Actin as the generator of tension during muscle contraction, *Proceedings of the National Academy of Sciences USA* **89**.

Schutt, C. E. and Lindberg, U. (1993). A new perspective on muscle contraction, *FEBS LETTERS* **325**, 1,2, pp. 59–62.

Schutt, C. E. and Lindberg, U. (1998). Muscle contraction as a Markov process I: energetics of the process, *Acta Physiologica Scandinavica* **163**, pp. 307–323.

Schutt, C. E. and Lindberg, U. (2000). The new architectonics: An invitation to structural biology, *The Anatomical Record (New Anat.)* **261**, pp. 198–216.

Shamos, M. H. (1959). *Great Experiments in Physics* (Holt, Rinehart and Winston).

Sherrington, C. S. (2009). *Man on his Nature* (Cambridge University Press), Based on the Gifford Lectures of 1937-8 in Edinburgh.

Simon, C. P. and Blume, L. (1994). *Mathematics for Economists* (W. W. Norton & Company, Inc.).

Slawnych, M., Seow, C., Huxley, A. and Ford, L. (1994). A program for developing a comprehensive mathematical description of the crossbridge cycle of muscle, *Biophysical Journal* **67**, pp. 1669–1677.

Spencer, J. N. (1974). δg and $\frac{\partial g}{\partial \xi}$, *Journal of Chemical Education* **51**, 9, pp. 577–579.

Steenrod, N. E., Halmos, P. R., Schiffer, M. M. and Dieudonné, J. R. (1973). *How to write mathematics* (American Mathematical Society).

Steinfeld, J. I., Francisco, J. S. and Hase, W. L. (1989). *Chemical Kinetics and Dynamics* (Prentice Hall, Englewood Cliffs).

Strong, J. (1936). *Procedures in Experimental Physics* (Lindsay Publications, Inc.), In collaboration with H. Victor Neher, Albert E. Whitford, C. Hawley Cartright, Roger Hayward, and illustrated by Roger Hayward. Original copyright 1938 by Prentice-Hall, Inc.

Sugi, H., Minoda, H., Inayoshi, Y., Yumoto, F., Miyakawa, T., Miyauchi, Y., Tanokura, M., Akimoto, T., Kobayashi, T., Chaen, S. and Sugiura, S. (2008). Direct demonstration of the cross-bridge recovery stroke in muscle thick filaments in aqueous solution by using the hydration chamber, *Proceedings of the National Academy of Sciences* **105**, 45, pp. 17396–17401.

Sun, S. X., Lan, G. and Atilgan, E. (2008). Stochastic modeling methods in cell biology, *Methods in Cell Biology* **89**, pp. 601–621.

Suppes, P. (1960). *Axiomatic Set Theory* (D. Van Nostrand Company, Inc., Princeton).

Szent-Györgyi, A. (1974). The mechanism of muscle contraction, *Proceedings of the National Academy of Sciences USA* **71**, 9, pp. 3343–3344.

Szent-Györgyi, A. G. (2004). The early history of the biochemistry of muscle contraction, *Journal of General Physiology* **123**, 9, pp. 631–641.

Tanenbaum, A. S. (1994). *Structured Computer Organization, 5th Edition* (Prentice-Hall, Inc., New York).

Tao, T. (2008). *Structure and Randomness, pages from year one of a mathematical blog* (American Mathematical Society).

Taylor, E. F. and Wheeler, J. A. (1966). *Spacetime Physics* (W. H. Freeman and Company, San Francisco), Fig. 9, p. 18.

Tester, J. W. and Modell, M. (2004). *Thermodynamics and Its Applications, 3rd Edition* (Pearson Education, Taiwan).

Thoma, J. U. (1976). Simulation, entropy flux and chemical potential, *BioSystems* **8**, pp. 1–9.

Tisza, L. (1961). The thermodynamics of phase equilibrium, *Annals of Physics* **13**, pp. 1–92.

Tolman, R. C. (1979). *The Principles of Statistical Mechanics* (Dover Publications, New York).

Truesdell, C. (1971). *The Tragicomedy of Classical Thermodynmaics* (Springer-Verlag), course held at the Department of Mechanics of Solids, July, 1971, International Centre for Mechanical Sciences *Motus est Vita*, Courses and Lectures, No. 70.

Van Kampen, N. (2007). *Stochastic Processes in Physics and Chemistry*, 3rd edn. (Elsevier, Amsterdam).

Voet, D. and Voet, J. G. (1979). *Biochemistry, Second Edition* (John Wiley & Sons, New York).

Waerden, B. L. V. D. (1931, 1937, 1940, 1949, 1953). *Modern Algebra, Volume I* (Frederick Ungar Publishing Co.), Translated from the second revised German edition by Fred Blum, with revisions and additions by the author.

Wang, H. (2008a). An improved WPE method for solving discontinuous Fokker-Planck equations, *International Journal of Numerical Analysis and Modeling* **5**, 1, pp. 1–23.

Wang, H. (2008b). Several issues in modeling molecular motors, *Journal of Computational and Theoretical Nanoscience* **5**, 12, pp. 2311–2345.

Wang, H., Peskin, C. and Elston, T. C. (2003). A robust numerical algorithm for studying biomolecular transport processes, *Journal of Theoretical Biology* **221**, 4, pp. 491–511.

Warren, J. W. (1983). Energy and its carriers: a critical analysis, *Physics Education* **18**, pp. 209–212.

Weinhold, F. (2009). *Classical and Geometrical Theory of Chemical and Phase Thermodynamics* (John Wiley & Sons, Inc.).

Woo, H. J. and Moss, C. L. (2005). Analytical theory of the stochastic dynamics of the power stroke in nonprocessive motor proteins, *Physical Review E* **72**, 051924, pp. 1539–3755.

Xing, J., Wang, H. and Oster, G. (2005). From continuum Fokker-Planck models to discrete kinetic models, *Biophysical Journal* **89**, pp. 1551–1563.

Zou, G. and George N. Phillips, J. (1994). A cellular automaton model for the regulatory behavior of muscle thin filaments, *Biophysical Journal* **67**, pp. 11–28.

Zwanzig, R. (2001). *Nonequilibrium Statistical Mechanics* (Oxford University Press).

Index